AVIAN DISEASE MANUAL

SEVENTH EDITION

American Association of Avian Pathologists

Edited by **M. Boulianne**

with

M. L. Brash

B. R. Charlton

S. H. Fitz-Coy

R. M. Fulton

R. J. Julian

M.W. Jackwood

D. Ojkic

L. J. Newman

J. E. Sander

H. L. Shivaprasad

E. Wallner-Pendleton

P. R. Woolcock

Seventh Edition Published by the
American Association of Avian Pathologists, Inc.

Printed by OmniPress

Library of Congress Control Number (LCCN)
2012940076

International Standard Book Number (ISBN)
9780978916343

Cover photo by Marco Langlois

Copies available from:
American Association of Avian Pathologists, Inc.
Additional information available at:
www.aaap.info/avian-disease-manual

AAAP, Inc.
12627 San Jose Blvd., Suite 202
Jacksonville, Florida 32223-8638
Email: AAAP@aaap.info
Website: www.aaap.info

Preface to the 7th Edition

The Avian Disease Manual has become the best selling publication of the AAAP. Its success is likely due to its ability to deliver at a reasonable cost, concise yet complete information on commonly encountered diseases affecting poultry. Not surprisingly, it has become an educational staple to North American veterinary and poultry science students, to those interested in avian diseases, but also a most useful reference in developing countries.

The world of commercial poultry production is a rapidly evolving one, new pathogens regularly emerge, microorganisms are reclassified and renamed, discoveries are being made, hence the need for regular re-edition of this manual. Putting together a new edition, presented this new editor with the challenges of keeping the great teaching qualities of past editions while updating the information and improving the format. This was made possible through a great team effort. The current editorial committee is made up of newcomers and experienced members. They all have extremely busy professional life, but all generously accepted to answer my call and share their knowledge and expertise. I would like to thank them for their timely diligence in reviewing and updating their chapters. Naturally, we are also indebted to a number of esteemed colleagues who, since the first edition in 1980, initially written by C.E. Whiteman and A.A. Bickford, provided us with a solid heritage on which we keep building.

The manual is divided in various chapters grouping diseases by agent (viral, bacterial, fungal, etc...). Within each chapter, diseases are listed alphabetically and the addition of an index will further help the reader to quickly locate the required information. The Appendix contains tables, each of which lists the most common diseases of a single body system. Our students have always appreciated the various tables and positively commented on the fact that you can quickly compare diseases at a glance. To these tables, we have added two new ones: diseases of the ducks and diseases of the upland game birds, to cover a wider spectrum of avian species. A poultry drug use list is also provided as a general guide, but medication recommendations should always be carefully verified with the manufacturer's label prior to use. Nowadays, most of our students come from an urban background and have never seen a live chicken or turkey, let alone been on a poultry farm. To fill this gap, as well as to put into perspective the work of a poultry veterinarian, a new chapter "How do we investigate a sick flock? " was added to the manual. The necropsy chapter underwent major revision to include the differential diagnosis procedure which goes on when a post-mortem examination is being performed.

Under the editorial guidance of Dr. Bruce Charlton, the previous edition incorporated the addition of electronic photos available on a CD. The 7th edition has now included them in the text while enhancing its library content. After all, an image is worth a 1000 words! Photos referenced in the text can be found after each disease section. The editorial committee is deeply indebted to all the authors of the photographs for the exceptional quality and historic significance of the photos in their collection. Special recognition needs to be offered to Dr. HL Shivaprasad (CAHFS, UC Davis) and Dr. HJ Barnes (North Carolina State University) for their passion and amazing photo collection, and also to the numerous colleagues who spontaneously accepted to go through their slides and photos to provide the readers with the highest quality images. We also used select photos from the AAAP Slide Sets and want to extend our gratitude to their authors. Readers of this manual are encouraged to investigate these sets and the book Diseases of Poultry for further excellent photos and information. Although every attempt has been made to correctly credit authors and institutions of the photographs, we apologize for any mistake that might inadvertently occur.

On a final note, I would like to thank the AAAP Board for their continuous support and willingness to endorse the editorial committee suggestions, as well as recognize the hard work of Mr. Bob Bevans-Kerr, AAAP Executive Director. His patience, availability and expertise in Photoshop and the editing process has made the whole process an enjoyable experience.

Martine Boulianne, Editor

Table of Contents

How do we investigate a sick flock?

Written by Martine Boulianne

The approach to investigating a sick poultry flock is one of population medicine. Not only do you need to look at the flock as the unit of interest, but you must also closely examine its immediate environment. Indeed, because poultry is most often kept in barns, it closely depends on housing conditions to be healthy. The Koch's postulate 'one infectious agent, one disease' thus can be remodelled into this schematization:

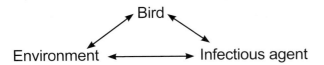

Optimal environmental conditions, access to high quality feed and water must therefore be provided for the bird comfort, which in turn will translate in maximum efficiency production and growth.

In the coming chapter, emphasis will be put on the various parameters one needs to evaluate when visiting a sick poultry flock.

The toolbox

Every veterinarian will tell you that essential instruments are the thermometer and stethoscope, all, but the poultry veterinarian. In this case, a knife will be more useful. But an acute sense of observation, the ability to collect good information when talking to the owner/animal caretaker and logical reasoning are as important, whatever the field of practice, to determine the necessary course of action.

In the poultry veterinarian's toolbox, you will find a necropsy kit (e.g. necropsy knife, necropsy shears to cut bones, enterotome to incise the gut, scalpel, forceps…), materials for sampling (e.g. needles and syringes, blood tubes, sterile plastic bags and swabs, specimen containers with and without 10% phosphate buffered formalin), possibly a microscope to look at *Eimeria* from gut scrapings, various instruments to measure air and water quality, and the appropriate gears (clean clothes or disposable coveralls, disposable plastic boots, head covering such as a disposable bouffant, disposable gloves, hand sanitizer) to comply with biosecurity measures.

Flock visit

Reasons for visit

A flock visit can be done either because there is a problem and the owner has requested your presence, or as part of a routine health check. Compared to the other fields of veterinary practice, there is no real emergency in poultry medicine i.e., rare are the after hour calls. However, in the case of a marked increase in mortality, or in case of a suspicious reportable disease, the veterinarian should visit the affected flock as soon as possible.

The most common reasons for a call to the poultry veterinarian are the following: 'increased mortality', respiratory clinical signs, presence of diarrhea, lameness or loss in performances. Due to the structure of the poultry industry, a technician might be called first to later notify the veterinarian and maybe request his/her expertise.

'Normal mortality'

In the case of an increased mortality, a greater number of birds than the usual expected daily rate die. This expected daily mortality does vary according to the age of the flock, the type of birds, the type of production and housing. For example, a certain number of chicks/poults/ducklings unable to find feed and water in the barn where placed will die of starvation/dehydration once their yolk sac, hence nutrient reserve, is exhausted.

During the growth period some meat birds might also die of a heart–related condition (sudden death syndrome in chickens and turkeys, ascites in chickens and bilateral cardiomyopathy of turkeys) or develop lameness. Inability to get to the feeder or the drinker will eventually lead to dehydration/inanition. These animals should be killed humanely to end their suffering.

Mortality is also expected in breeding/laying flocks, from reproductive-related conditions (often associated with obesity), or again from lameness. Cannibalism is another cause of death observed in many flocks.

These are part of the so-called expected mortality in a flock and this daily mortality is recorded on a chart which you must examine to determine the magnitude, onset and duration of the problem.

The following mortality rates are provided as general guidelines and only apply to meat type birds kept in a closed barn and laying hens kept in cages.

Table 1. Expected mortality rates according to the bird type

Type of bird	Brooding period	Growth	Total end-flock mortality
Broiler chicken	≤ 1% (from 0 to 2 weeks of age)	0.5 bird/1000 birds/day (between 2 and 4 weeks of age) < 1 bird/1000 birds/day (from 4 weeks and up)	≤ 2.5%
Broiler turkey	≤ 2% (from 0 to 2 weeks of age)	0.5 to 1/1000/day (between 2 and 10 weeks of age) 1 to 2/1000/day (from 10 weeks and up)	4 to 6% (broiler turkey) 6 to 12% (tom turkey)
Broiler breeder	1 to 2% (from 0 to 2 weeks of age)	≤0.25/1000/day during growth 0.1% to 0.25%/ month during lay (for a total of approx 6% during that period)	10%
Laying hen	1 to 2% (from 0 to 2 weeks of age)	≤0.25/1000/day during growth < 0.5% / month during lay	2 to 5%
Meat duck	<2%	0.5 to 1/1000/day (between 2 and 10 weeks of age) 1 to 2/1000/day (from 4 weeks and up)	3 to 7% when slaughtered at 7 weeks of age
Quail		≤0.25/1000/day	
Guinea fowl		≤0.25/1000/day	

Normal birds

During the flock visit it is very important to closely observed the birds and look for unhealthy or sick birds. General physical and behavioural observations provide a good indication. Healthy birds will be alert, active with bright round and open eyes. Mature turkeys and laying chickens should have a bright red comb. Birds should be clean with smooth feathers and bright neat scaly legs. Feces should be well-formed, brown or grey with a white 'cap' (the urates). During the visit you might also see on the litter, pale brown frothy feces which are from the ceca and can regularly be observed in the barn.

Depending on the housing system, birds should be able to stand, walk, even run, scratch and sit only for short periods. Turkeys and ducks do not scratch the litter and birds in cage cannot unless provided with a sand box. Chickens will quickly walk away from the unusual visitors while turkeys will follow them. Young broiler chickens will often be found fighting i.e jumping at each other with spread wings. Turkeys can show a belligerent behaviour to their mates but will be hissing, walking slowly in the crowd with fluffed up feathers, a blue colored head with an elongated snood. Breeder birds should be also found mating during a visit. Many birds but not all will be drinking and eating at the time of your visit once the stress caused by your presence has decreased.

Evaluating the body condition

Numerous growth charts are available and vary according to the genetic, management system and feeding company. Such charts should be consulted to verify if the bird's body weight and growth rate are within normal goals. One can also tell the body condition of a bird by palpating the breast muscle. With the bird held by the legs in one hand in an upside down position, use the palm of the other hand to palpate the protuberance of the keel, the development of the breast muscles alongside that keel, and the convexity or concavity of the breast muscle contour. A nice growing bird will show a convex (rounded) contour of the breast with plump breast muscle and no protuberance of the keel, while an emaciated bird will show a marked concavity of the breast contour caused by a prominent keel with barely no pectoral muscles being felt.

Clinical signs

Clinical signs will vary according to the disease and affected system(s) and will vary in severity. Not all birds in a flock will exhibit clinical signs. Early in the course of the disease only a few individuals might be affected and care should be taken to find them.

In general, sick birds are listless, will sit for long periods, their head held close to the body, tail and possibly wings dropping. Comb and wattles may be paler and shrunken. Eyes will be dull and sometimes closed. They might not drink nor eat, hence slowing/stopping growth and eventually losing weight. Anorexic chickens will often have green colored feces (due to bile stain) which might stain the feathers of the pericloacal area. Dehydrated birds will show darker and thinner looking legs, they will feel lighter and the skin will not move freely over the keel. If cold or pyrexic, feathers will be fluffed, and birds will huddle in corners or with others to keep warm. Uncomfortable chicks will initially be chirping loudly before becoming depressed if source of discomfort is not corrected.

If a respiratory disease is suspected, early on the course of the disease, chickens will shake their head and scratch it with their feet. As the disease progresses watery eyes and/or nasal exudate might be observed and will make the birds look dirty with the dust and dirt adhering to the wet feathers and beak. Swollen infra-orbital sinuses will affect the shape of the eye and might even force its closure. Respiratory sounds can also be heard from a light 'snick' to loud rales. Birds do not have a diaphragm and will not cough. Severe respiratory difficulties might even cause the bird to extend its neck and abdominal wall movements can be observed. If you want to hear the light snick sick chickens make during a respiratory disease episode, you can gently whistle and the chickens will stop cackling and will raise their head intrigued by this new sound. This trick does not work in turkeys since they will respond to you with loud gobbles.

If an enteritis is suspected, some birds might have dirty feathers around the vent that might be even soiled with blood or sulphur colored feces depending on the infectious agent. These blood or sulphured colored feces will also be found on the litter.

Lame birds will spend more time sitting, and will walk with difficulties, spreading their wings. Traumatic lesions will be observed on the ventral aspect of the carpo-metacarpal joint as well as a sternal bursitis in chronically recumbent birds. Depending on the cause, joints might or might not be swollen and hot. The plantar surface of the feet might be dirty, crusted, cracked and/or reddened.

When investigating a loss in performances, a reduced body weight, a higher feed conversion, a drop in egg production, a decreased hatchability, flock results should be carefully examined and compared to expected result in order to define the problem perceived by the flock manager and answer the basic questions: who, what, when, where, how?

Barn environment

As stated earlier, the quality of housing will greatly impact on the birds' health. Poultry barns should provide clean feed and water, fresh air, protection against predators, shelter from cold, rain, wind, sun and excessive heat; as well as a source of heat when birds are young. During the visit, you can verify the most important elements using the acronym F-L-A-W-S. F is for feed, L for light/litter, A for Air, W for water, S for Sanitation/safety/space/staff.

Feed. Feed and water are usually available *at libitum* in meat birds, but quantities are controlled in breeders and layers. Feeders and drinkers must be located at the right height to optimize access. Variations in feed consumption can be indicative of a disease, but also associated with hot and cold weather, the feed itself (energy, fiber, particle size) or with the birds' needs (e.g. point of lay).

Light. The lighting schedule and light intensity are very important parameters in laying birds, since light stimulates egg production. In many meat type birds, daylength will be shortened early in life to control the growth rate.

Litter. The litter is a mixture of feces and bedding material. The latter should be made of absorbent material and in enough quantity for comfort. If too dry, respiratory problems will arise while a too humid litter might trigger intestinal and skeletal pathologies. A litter is too humid if it keeps its shape once you have squeezed a handful in your fist.

Air. One of the most important elements of managing the environment inside a chicken barn is air quality and, in particular, airflow. Ventilation in the majority of commercial barns is mechanical and of the outmost importance since any power shortage will rapidly cause death due to hyperthermia. Not only good ventilation will bring fresh air into the barn, but it will take out noxious gases (CO_2, ammonia…), dust and humidity. Poor air quality will increase in respiratory problems. Furthermore, if the ventilation is poor, the litter will be more humid, creating an ideal milieu for certain bacteria and parasites. You might then end up with a coccidial challenge or lame birds. Barn temperatures are electronically controlled and monitored with probes. There is a comfort zone at which growth is optimal. Since newly hatched birds are poikilotherm, a heat source must be provided. Since birds do not have sweat glands and use evaporative cooling via their breath, temperature superior to 40°C are very uncomfortable and might be lethal when more than 46°C.

Water. Drinking water should be of quality and present in adequate quantity. Birds generally drink approximately twice as much water as the amount of feed consumed on a weight basis. Any water restriction will impair feed intake. Water consumption will often decrease a day or two before the onset of clinical signs. Consumption is also closely associated with environmental temperatures. For example, during periods of extreme heat stress, water requirements may easily quadruple.

Sanitation. Information regarding cleaning, disinfection, pest control, downtime, as well as previous history of disease, routine and current medication, and vaccination program should be collected during the visit. Biosecurity measures should also be in place in order to minimize the risk of disease introduction and spread.

Space. Birds also need adequate space for movement and exercise, access to feeders and drinkers. Space requirements vary with the species, type or breed of birds that are raised, as well as the type of production system used.

Staff. An attentive and skilled farm manager and employee are of the outmost importance in the rearing and keeping of a healthy poultry flock. Any changes to the management can adversely affect the birds.

Many books and extension services factsheets will provide you with the appropriate information regarding the housing criteria that must be respected in order to make the birds comfortable.

Conducting a necropsy

Once you have closely observed the birds and their housing conditions and look at the mortality chart and other performance data, you have probably listed all the possible differential diagnosis given the clinical facts. In order to verify your hypothesis, you have the possibility of opening carcasses on the farm to verify for the presence of lesions. The necropsy is essential to quickly observe the internal lesions, establish a differential diagnosis and decide on the course of action. Ideally, necropsy should be performed on animals representative of the condition. Indeed, the challenge of a good poultry diagnosis is to identify the most significant flock problem(s) rather than focusing on individual bird pathologies. For large poultry flocks, approximately five dead birds as well as five individuals showing clinical signs should be selected for necropsy. Euthanasia of the sick birds should be performed rapidly and humanely in accordance to ethical standards. Necropsy procedures are described in another chapter of the current manual. For further analysis and confirmation of your tentative/preliminary diagnosis, birds or samples should be sent to an animal diagnostic laboratory.

Taking samples

Some samples can be taken when birds are alive (e.g. blood samples, tracheal swabs...) or at post-mortem. Blood samples are usually collected for serology. Paired sera taken two weeks apart will be desired if seroconversion to some disease is expected. In adult birds, collecting eggs will also serve this purpose since antibody titers can be measured from the yolk.

Blood can easily be sampled from the brachial vein in most birds, such as young and mature chickens, while the tibiotarsal vein is a good option for turkeys and ducks. Since the avian skin is very thin, it is easy to visualize the vein once a few feathers have been plucked and pressure applied proximally to the puncture site. Dampening the skin with 70% alcohol will help to better define the vein.

In most birds, a ½ to 1 inch, 21 to 22 ga needle (depending on the size of the bird), with a 5 ml syringe, will suffice. Do not use a vacutainer for avian blood collection as the vein will simply collapse, but apply a gentle steady negative pressure on the syringe plunger to withdraw blood. Avian blood will easily coagulate during sampling.

For most serological analysis, a 2 ml blood sample will be adequate. Blood should be collected aseptically in a vial and laid horizontally until it clots. Placing vials in warm water right after collection will hasten the clotting, while refrigeration will hinder the coagulation process. Sera can then be transferred in vials, put on ice and shipped to the lab. Never freeze sera if agglutination tests are planned since this might cause false positive reactions.

For some biochemical or other analysis requiring unclotted blood samples, please enquire to the diagnostic lab as to the preferred anticoagulant (e.g. heparin, sodium citrate...). Samples should be sent as soon as possible on ice to the diagnostic laboratory.

Upon performing the necropsy, numerous tissues and organs can be sampled depending on observed lesions. If bacterial cultures or viral samples are needed, they should be collected as aseptically as possible using sterile scalpel blade (e.g. to collect joint/sinus exudates). Shipping whole or parts of organs to the laboratory is also an option.

Tissues for histopathologic examination can be immersed into 10% formalin (or other fixative), immediately after death. Specimens should be small for a quick fixative penetration and preserved in ten times their own volume. Ensure the container is tight and leak free for shipping purposes. (Alternative shipping: remove specimens from the jar and ship in a Ziploc bag with a paper towel dampened in formalin to keep the tissue moist, but that will not crack or leak or spill in transit due to the sensitivity of shippers. Label the bags, of course).

Feed samples should be collected from the feeders if a problem with an ingredient, drug level, etc... is suspected, when there is feed refusal, or whenever mortality, drop in egg production, poor growth performances, are unexplained.

Water from the wells and from the end of the water line should be analyzed at least once a year to determine microbiological and biochemical characteristics. Water pH and chlorine levels can be estimated with specific color strips.

If ventilation is suboptimal during the visit, ammonia, CO_2, relative humidity can easily be measured with the appropriate instruments. Barn temperature can be measured and comfort zones can be established with an infra-red thermometer. Many barns are also equipped with computer controlled ventilation systems monitoring and logging minimal, maximal and mean barn temperatures as well as relative humidity.

The visit report

The visit report should include farm/barn identification, description of the problem (who, what, when, where, how...), clinical observations, necropsy findings, conclusions and recommendations based on available facts. A tentative diagnosis can be offered pending further laboratory test results. Confirmation should be given as soon as possible via a phone call to the flock owner/manager or technician.

VIRAL DISEASES

New parts and revision by Davor Ojkic, Marina L. Brash, Mark W. Jackwood and H.L. Shivaprasad

ARBOVIRUS INFECTIONS

DEFINITION
Arbovirus is an abbreviation of "arthropod-borne virus" which describes viruses that replicate in arthropods and are then transmitted by blood-sucking to their hosts.

OCCURRENCE
Four arboviruses have been described in poultry and farmed birds in North America: Eastern equine encephalitis virus (EEEV) Western equine encephalitis virus (WEEV) Highlands J virus (HJV) and West Nile virus (WNV). This chapter is limited to arbovirus infections in North America.

HISTORICAL INFORMATION
EEEV – first identified in pheasants and pigeons in 1938.

WEEV – first identified in turkeys in 1957.

HJV – first recognized in blue jays 1960 in Florida.

WNV – first identified the northeastern US in 1999.

ETIOLOGY
EEEV, WEEV and HJV – genus *Alphavirus*, family Togaviridae. Viral particles are spherical, enveloped, 70 nm in diameter. Genetic material is positive sense single stranded RNA.

WNV – genus *Flavivirus*, family Flaviviridae. Viral particles are spherical, enveloped, 40-60 nm in diameter. Genetic material is positive sense single stranded RNA.

EPIZOOTIOLOGY
1. The spread is seasonal when infected mosquitoes transmit the infection among susceptible birds (and horses or people for EEEV, WEEV and WNV) while feeding on them. Birds are important source of virus for mosquitoes because they carry a higher titer of virus than most mammals.

2. Cannibalism of viremic, sick, or dead birds by other susceptible birds may be an important method of transmission of virus within infected flocks. Also, certain biting insects (gnats, deerflies, horseflies, etc.) may transmit the virus mechanically.

3. Susceptible wild birds and poultry flocks can have transient infections and show no clinical signs. Antibodies can be demonstrated in their sera.

CLINICAL SIGNS
EEEV – neurologic disease and increased mortality have been described in turkeys, pheasants, chukar partridges, ducks and chickens. EEEV infection can also cause drop in egg production in breeder turkeys.

WEEV – associated with neurologic disease in turkeys in the past. WEEV was isolated from turkeys with egg production drops in California in 1999.

HJV – Highlands J virus infection has been associated with neurologic disease in chukar partridges and egg production drops in turkey breeders.

WNV – Outbreaks of naturally occurring West Nile virus infections have been reported in geese and domestic ducks (Fig. 1b). Affected ducks display general weakness and inability to stand and increased flock mortality. Day-old-chickens develop are susceptible and develop a neurologic disease following experimental inoculation. Turkeys appear to be resistant.

LESIONS
Eastern Equine Encephalitis
1. Pheasants:
   Gross lesions are not observed. Nervous lesions histologically include vascular endothelial hypertrophy (Fig. 1a), vasculitis, multifocal necrosis, perivascular lymphoid cuffing, gliosis, degeneration of neurons and meningitis or meningoencephalitis (Fig. 2a and 3a). Splenic fibrinous necrosis, myocardial necrosis (Fig. 4a) and hepatic necrosis have also been reported.

   Partridge: At necropsy, lesions include multifocal myocardial necrosis and mottled enlarged spleens. Histologically, nervous lesions include perivascular lymphoid cuffing, gliosis, satellitosis and non-suppurative myocarditis.

2. Turkeys:
   In young turkeys, brain lesions include lymphoid perivascular cuffing, neuronal degeneration and endothelial cell hypertrophy. With experimental infections in young turkeys, at necropsy, lesions included dehydration, lack of feed in crops and thymic and bursal atrophy. Histologically, multifocal myocardial, renal and pancreatic necrosis and thymic, splenic and bursal lymphoid depletion were reported. In breeding hens, there is decreased egg production with white, thin-shelled and shell-less eggs.

3. Chickens:
   With experimental infection, non-suppurative myocarditis is the predominant lesion. Histological lesions of the brain were variable and included necrosis and mild lymphoid perivascular cuffing.

Multifocal hepatic necrosis and splenic, thymic and bursal lymphoid depletion also may be observed.

Western Equine Encephalitis
No significant lesions have been described.

Highlands J Virus
1. Chukar Partridge
Splenomegaly is a common finding at necropsy whereas multifocal myocardial necrosis is occasionally reported. Histologically, commonly reported lesions included multifocal myocardial necrosis with mineralization and fibrinous splenic necrosis with lymphoid depletion and rare brain lesions which include mild lymphoid perivascular cuffing, endothelial cell hypertrophy and lymphocytic meningitis.

2. Turkeys
In experimental infections of young turkeys the lesions identified are similar to those seen with EEE infection and include dehydration, lack of feed in crops and thymic and bursal atrophy. Microscopic lesions include bursal, thymic and splenic lymphoid depletion, with occasional splenic fibrinous necrosis. Multifocal myocardial necrosis and mineralization and pancreatic and renal necrosis is also described.

West Nile Virus
Enlarged, flaccid heart with mild pale streaking of the myocardium is described at necropsy (Fig. 2b). Histologically, there is multifocal nonsuppurative myocarditis (Fig. 3b), splenic necrosis with lymphoid depletion, pancreatic necrosis and occasionally mild multifocal hepatic necrosis. Brain lesions include nonsuppurative meningoencephalitis, perivascular lymphoid cuffing, focal gliosis, neuronal degeneration and satellitosis. Cerebellar lesions were those of multifocal malacia of the grey matter with necrosis of Purkinje cells and edema of the Purkinje cell layer.

DIAGNOSIS
1. Virus isolation is not recommended for routine diagnosis for EEEV, WEEV and WNV because work with these live agents requires level 3 biocontainment facilities.
2. Antigen detection ELISA has been used for detection of EEEV, WEEV and WNV, but more sensitive and specific real-time RT-PCR methods are now available.
3. Immunohistochemistry (IHC) for EEEV (Fig. 5a) and WNV (Fig. 4b and 5b) will reveal positive staining for viral antigens in various tissues, such as myocardium, intestines and brain, when used. Infections with EEEV, WEEV and WNV are reportable in many jurisdictions.

CONTROL
1. Protect birds against insects by raising them where mosquitoes do not thrive, or by the use of screens, sprays, or other mosquito control methods.
2. Avoid overcrowding and keep the houses or pens at a comfortable temperature.
3. Keep the houses rather dark and use only red light bulbs.

TREATMENT
No treatment is available.

ZOONOTIC POTENTIAL
EEEV, WEEV and WNV are zoonotic agents.

Eastern Equine Encephalitis

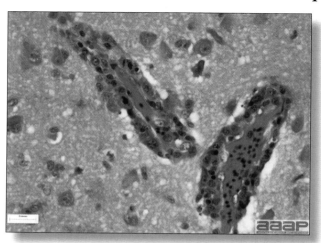

Fig. 1a
Narrow predominantly lymphoid perivascular cuffs surrounding vessels lined by hypertrophied endothelial cells in the brain of a pheasant.

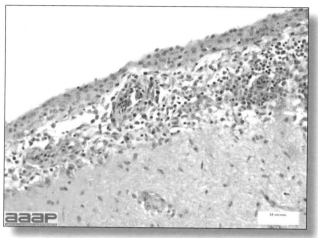

Fig. 2a
Meningoencephalitis in a EEEV affected pheasant. One can observe low to moderate numbers of plasma cells, lymphocytes, heterophils within meninges.

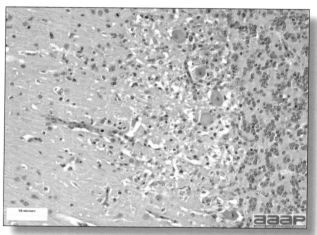

Fig. 3a
Multifocal gliosis, mild loss of Purkinje cells, neuronal necrosis in a pheasant that died of EEEV.

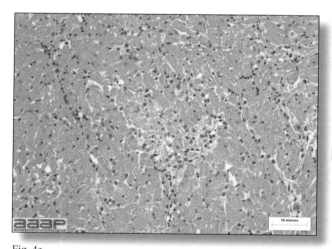

Fig. 4a
Myocardial necrosis in a ring-neck pheasant affected with EEEV.

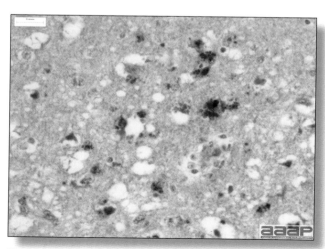

Fig. 5a
EEEV antigen positive neurons and glial cells at immunohistochemistry.

West Nile Virus

Fig. 1b
General weakness and inability to stand in a WNV affected duck.

Fig. 2b
Enlarged, flaccid heart with mild pale streaking of the myocardium.

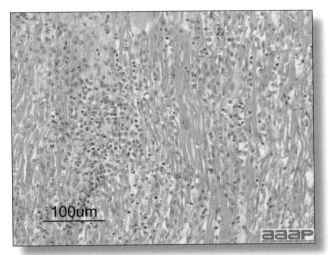

Fig. 3b
Myocarditis: histologic examination reveals degenerative cardiac myocytes with fibrosis and an infiltrate of predominantly mononuclear inflammatory cells (H&E, X100um).

Fig. 4b
Abundant staining for West Nile viral antigen present in the myocardium of affected ducks (IHC, X100um).

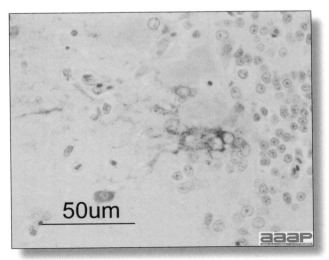

Fig. 5b
Positive staining for West Nile viral antigen present in the brain of affected ducks (IHC, X50um).

[handwritten margin notes: Always -include- Ham + Egg? / Mainthing is liver Gross gout homocard Inclusions in liver / Histo degen + necrosis / turkeys / Adenoviruses ① Inclusion body hepatitis-kidneys + gout ② Hemorrhagic enteritis ③ Egg drop syndrome]

AVIAN ADENOVIRUS INFECTIONS

DEFINITION

Adenovirus infections are common in poultry and some can be defined in terms of clinical and pathologic characteristics. However, many adenovirus infections are either subclinical or associated with nondescript clinical syndromes.

OCCURRENCE

Serologic surveys indicated that most poultry flocks have been exposed to infection with one or more adenoviral serotypes. Adenoviruses play a primary or secondary role in a variety of syndromes including inclusion body hepatitis and hepatitis/hydropericardium syndrome in chickens; hemorrhagic enteritis of turkeys; egg production declines in laying chickens (egg drop syndrome—1976); bronchitis in quail and other respiratory, arthritic, encephalitic, and enteric syndromes including gizzard erosions, pancreatitis and proventriculitis. However, the frequent presence of these viruses even in healthy birds means that their role in disease must be critically examined.

HISTORICAL INFORMATION

The first recognized adenovirus infection of birds was quail bronchitis described in 1951.

ETIOLOGY

1. Adenoviruses are DNA viruses that replicate and produce inclusion bodies in the nuclei of infected cells. The viruses are non-enveloped and range in size from 70 to 90 nm.

2. Adenovirus classification is not intuitive. The family Adenoviridae is divided into 5 genera and adenoviruses infecting birds are in 3 genera;

    a. Aviadenovirus (previously known as Group I Avian adenoviruses),

    b. Siadenovirus (Group II Avian adenoviruses) and

    c. Atadenovirus (Group III Avian adenoviruses).

3. Fowl adenoviruses (FAdV), goose adenoviruses, falcon adenovirus 1, duck adenovirus-2, pigeon adenovirus-1 and turkey adenovirus-1 and -2 are in the genus Aviadenovirus. Turkey adenovirus-3 (Hemorrhagic enteritis) and raptor adenovirus-1 are in the genus Siadenovirus. Duck adenovirus-1 (Egg drop syndrome virus) is in the genus Atadenovirus.

EPIZOOTIOLOGY

Adenoviruses can be transmitted both vertically and horizontally.

CLINICAL SIGNS

Diseases with well-established adenoviral etiologies, namely inclusion body hepatitis, hemorrhagic enteritis, egg drop syndrome - 1976 and quail bronchitis, are presented in detail later in this section. Reports of other diseases attributed to adenoviral causation should be scrutinized closely for solid evidence of a definitive etiologic role.

LESIONS

Lesions vary depending on the virus/syndrome involved and are presented later in this section.

DIAGNOSIS

Routine diagnosis of infection is typically carried out by a combination of virus isolation and post-mortem examination/histopathology, sometimes augmented with electron microscopy or polymerase chain reaction.

CONTROL

Licensed and autogenous vaccines are available in some countries.

TREATMENT

Not available.

ZOONOTIC POTENTIAL

Human infection with avian adenoviruses has never been documented. However, one controversial report has suggested, based on a serological survey, a possible role of an avian adenovirus in human obesity.

I. INCLUSION BODY HEPATITIS

[handwritten: This is the one w/the hemo pericardia]

DEFINITION

Inclusion body hepatitis (IBH) is a disease of young chickens characterized by sudden onset, increased mortality and hepatitis accompanied with intranuclear inclusion bodies.

OCCURRENCE

FAdV-caused hepatitis has a worldwide distribution and has been described as IBH in North America, Europe, Australia and New Zealand and as hepatitis/hydropericardium syndrome (HHS) in South America and Asia.

Hepatitis associated with adenovirus infection has also been reported in turkeys, quail, pigeons, falcons and psittacines. Many other animal species such as snakes, dogs, chimpanzees and humans have their "own" hepatitis-associated adenoviruses.

HISTORICAL INFORMATION

1. In 1963 hepatitis with inclusion bodies was described in chickens but the causative agent was not identified. That outbreak probably was the disease we now call IBH. In the early 1970s a similar disease occurred in many flocks in Canada and the United States.

Adenovirus was isolated from an Indiana outbreak and, eventually, from flocks in many other locations.

2. Historically, IBH occurred in immunologically deficient flocks as a consequence of earlier infection with IBD or CIA virus. However, IBH is now recognized as a primary disease and often does not require a preceding immunosuppressive event.

ETIOLOGY

1. Most commonly IBH cases involve FAdV8 and FAdV11, but sporadic cases associated with FAdV2 have been documented.

2. HHS has been associated with FAdV4.

3. Outbreaks of IBH are sometimes associated with immunosuppression or exacerbated if affected flocks are immunosuppressed.

EPIZOOTIOLOGY

1. IBH can be transmitted both vertically and horizontally.

2. Egg-transmitted adenoviruses may remain inactive in infected chickens or poults until maternal antibody wanes. Outbreaks in young birds (1-2 weeks) are typically associated with vertical transmission while outbreaks in older birds are most often due to a horizontal transmission. When broilers from multiple sources are mixed it is sometimes impossible to determine the source of infection.

3. In exposed birds the virus enters via the alimentary tract (and, in some cases, by the conjunctiva and nasal passages) and primary replication occurs in the nasopharynx and intestine. There is frequently a viremic stage in the infection with widespread dissemination of virus to secondary sites of replication. As antibody is produced, viral activity wanes but the virus may persist in a latent state in some organs.

4. There may be periods of virus reactivation throughout life especially during episodes of immunosuppression or stress.

5. Exposure to one serotype does not confer immunity to other serotypes within the group or other groups. Thus, birds can (and do) suffer repeated infections with antigenically unrelated adenoviruses.

6. Adenoviruses are relatively resistant to physical and chemical factors and can remain infective in a contaminated environment.

CLINICAL SIGNS

1. A sudden marked increase in mortality is often the first indication of the disease. Mortality increases for 3-5 days, levels off for 3-5 days, and then decreases to normal levels over another 3-5 days. Total mortality may be as high as 30% but is typically considerably lower.

2. There are few specific signs. There may be pallor of the comb, wattles, and facial skin. The affected birds are depressed and listless. In some outbreaks the clinical signs are masked by other diseases in the flock.

LESIONS

1. The skin is pale and may be discoloured yellow (Fig. 1). Petechial and ecchymotic hemorrhages may be present in the skeletal muscles of the legs.

2. The liver is swollen, enlarged, yellow to tan, and there may be mottling with focal soft areas with petechial and ecchymotic hemorrhages under the capsule and in the parenchyma (Fig. 2).

3. The kidneys frequently are swollen and pale or mottled (Fig. 3).

4. The bursa of Fabricius can be reduced in size.

5. Microscopically, there is multifocal to locally extensive degeneration and necrosis of hepatocytes (Fig. 4) often with the characteristic large basophilic intranuclear adenoviral inclusions in the hepatocytes (Fig. 5) within the foci of necrosis and elsewhere. The renal lesions include membranous and membranoproliferative glomerulitis and less frequently cortical tubular degeneration and necrosis with intraluminal inflammatory cells. In the bursa, there may be reduced follicular size with mild to moderate follicular lymphoid depletion.

DIAGNOSIS

1. In young, growing flocks a sudden increase in mortality is suggestive of IBH. Typical gross lesions and a history of prior outbreaks from the same parental flock(s) or on the premises are helpful.

2. Histopathology – Demonstration of typical microscopic lesions in the liver, including the characteristic intranuclear inclusions, is required for a diagnosis of IBH.

3. Virus isolation – Isolation of FAdV from the liver of affected chickens

4. PCR – detection of FAdV DNA in the liver of affected chickens

5. Serology – Seroconversion to IBH-associated serotypes (FAdV2, FAdV8, FAdV11) can be detected by a micro-neutralization assay. Group antigens can be detected by agar gel immunodiffusion or ELISA, but these tests are of limited value because adenovirus infection is widespread and they do not differentiate among non-pathogenic and pathogenic serotypes/ strains.

6. Genotyping - Analysis of the nucleotide sequences encoding adenovirus hexon protein, the most abundant viral surface protein that contains major antigenic determinants, has been used for genotyping of fowl adenoviruses.

CONTROL

Live-licensed vaccine against FAdV8 has been used in Australia. Killed autogenous vaccines have been used with various degrees of success. In North America autogenous vaccines are typically bivalent and contain FAdV8 and FAdV11.

TREATMENT

Not available.

ZOONOTIC POTENTIAL

Not recorded.

II. HEMORRHAGIC ENTERITIS OF TURKEYS

DEFINITION

Hemorrhagic enteritis (HE) is a viral disease of young turkeys characterized by sudden onset, depression, bloody droppings, and variable but often high mortality. A subclinical form characterized by an enlarged, mottled spleen occurs and is more common than the acute form.

OCCURRENCE

HE has a worldwide distribution and typically occurs in 6-12-week-old turkeys, but has been seen in poults as young as 2 weeks. It is rare in turkeys less than 4 weeks of age, presumably because of maternal antibody.

HISTORICAL INFORMATION

HE in turkeys was first reported in 1937 but the cause was unknown. Only a few reports of the disease were published during the next 30 years. In 1972 the disease was demonstrated to be caused by a viral infection. Since 1970 there have been numerous reports on research and field aspects of the disease and HE is recognized as a common and important disease of turkeys.

ETIOLOGY

HE is caused by a turkey adenovirus, hemorrhagic enteritis virus.

EPIZOOTIOLOGY

1. The virus is very resistant to environmental factors and is shed in feces, hence the transmission route is fecal-oral. Infection frequently reoccurs on the same farm in successive flocks.

2. There is no evidence of egg transmission.

3. Infection of turkeys with HE virus results in a transient immunosuppression, often involving secondary colibacillosis.

CLINICAL SIGNS

1. Sudden deaths are often the first sign of HE in a flock. A concurrent drop in feed and water consumption may be noted. Droppings containing fresh blood or melena can be seen, especially around waterers.

2. A few birds exhibit signs of depression and have bloody feces. Blood may be seen oozing from the vent of dead or moribund birds or may be adhered to feathers around the vent. Blood may be expelled from the vent if the abdomen is squeezed. Most birds with bloody feces die.

3. The disease usually runs its course in a flock in 10-14 days. Most mortality occurs over a 10-day period. Mortality averages 5-10% but may exceed 60%.

4. Outbreaks of colisepticemia often follow clinical and subclinical infections with hemorrhagic enteritis virus 12 to14 days later. Colisepticemia may be the only indication of prior HE subclinical infection.

LESIONS

1. Dead poults often appear pale due to intestinal blood loss but are well fleshed with feed in their crops. The skin and feathers around the vent can be stained with blood or blood stained feces.

2. The intestinal tract, especially the duodenal loop, is distended, dark purple, and filled with hemorrhagic content (Fig. 1). The intestinal mucosa, especially of the duodenum is congested, and may be covered with a yellow layer of fibrinonecrotic exudates.

3. Early in the course of the disease, the spleen is typically very enlarged and mottled (Fig. 2) and as the disease progresses, the spleen becomes smaller and pale. Experimentally infected birds have splenic enlargement only during the first 4 days of illness. Lungs may be congested.

4. Microscopically, early in the course of the disease, reticuloendothelial cells of the spleen contain numerous large intranuclear adenoviral inclusions and the condensed nuclear chromatin around the inclusions often resembles a signet ring (Fig. 3). In later stages, the white pulp undergoes widespread necrosis and involution. Lymphoid depletion also occurs in the thymus and the bursa of Fabricius. Intestinal lesions are most prominent in the duodenum with marked mucosal congestion, degeneration and exfoliation of the epithelium lining the villus tips and hemorrhage from the tips of the villi into the lumen with increased number of mixed mononuclear cells, mast cells and heterophils in the lamina propria. Rarely, intranuclear adenoviral inclusions are seen in the epithelial cells lining the villi. In addition, intranuclear adenoviral inclusions can be seen in the liver, bone marrow, circulating white blood cells, lung, pancreas, brain and kidney.

DIAGNOSIS

1. Typical history and gross lesions strongly suggest the diagnosis. Demonstration of intranuclear inclusions in reticuloendothelial cells in the spleen or intestine confirms the diagnosis unless the turkeys have received HE vaccine.

2. The disease can be reproduced in 6-week-old or older, susceptible poults by giving minced splenic tissue or its supernate intravenously, orally, or intracloacally. Typical intestinal content also will reproduce the disease when given orally or cloacally.

3. If known-positive antiserum and known-infectious splenic tissue are available, it is possible to use the agar-gel diffusion test to demonstrate antigen in the spleen of an infected turkey or to demonstrate antibody in the convalescent sera of recovered birds.

4. HE must be differentiated from acute bacterial septicemia including colisepticemia, salmonellosis, fowl cholera and erysipelas. Gastrointestinal hemorrhage/mucosal congestion may be associated with acute septicemic/viremic/bacteremic conditions. Intestinal coccidiosis should also be considered. HEV infection in growing turkeys results in immunosuppression predisposing birds to secondary infections such as *Escherichia coli* septicemia.

CONTROL

1. Avirulent strains of HEV and related marble spleen disease (of pheasants) virus are used as vaccines.

2. Vaccines are prepared as crude splenic homogenates or are cell culture derived.

TREATMENT

No treatment is available. Good care and management will reduce mortality and economic loss. Radical changes in feed or management should be avoided.

ZOONOTIC POTENTIAL

Not recorded.

III. EGG DROP SYNDROME

DEFINITION

Egg drop syndrome (EDS) is an infectious disease of laying hens caused by a hemagglutinating adenovirus characterized by loss of color in pigmented eggs and failure to achieve production targets, or by production of thin-shelled or shell-less eggs in otherwise healthy-looking birds.

OCCURRENCE

EDS in chickens has not been described in North America, but is present in Europe, Africa and Australasia. However, the causative agent of EDS appears to be widespread in its natural host, waterfowl.

HISTORICAL INFORMATION

This syndrome was first described as a unique problem in laying hens in Holland in 1976, hence the initial name egg drop syndrome-1976. It appears that the EDS virus was first introduced to chickens through a contaminated vaccine.

ETIOLOGY

EDS is caused by duck adenovirus-1 (DAdV-1) or egg drop syndrome virus.

EPIZOOTIOLOGY

The virus is spread both vertically and horizontally. Wild birds represent a potential source of infection, but this mode of transmission appears to be less common. The primary site of virus replication is the pouch shell gland. In infected embryos or young birds the virus is latent until they start laying eggs.

CLINICAL SIGNS

Drop in egg production, loss of color in pigmented eggs and production of eggs with tin or no shells are early symptoms. Once established in a flock the egg shell-related problems are less common, but birds typically fail to reach expected production peaks. Infections of waterfowl are mostly asymptomatic. However, outbreaks of an acute respiratory disease in goslings in Hungary and ducklings in Canada have been described.

LESIONS

Gross lesions other than inactive ovaries and atrophied oviducts are not seen in natural infections. Edema and swelling of the uterine mucosal folds and exudate in the shell gland lumen have been described in experimentally infected hens. With experimental infections, histologically, oviduct changes include proprial edema, infiltration of mixed mononuclear leukocytes (lymphofollicular aggregates in some cases) and heterophils changing to predominantly mixed mononuclear cells in the later stages of infection, atrophy of tubular glands, and degeneration/desquamation and attenuation of uterine epithelium. Intranuclear adenoviral inclusions may be seen in epithelial cells of the uterus, isthmus, and vagina.

Most descriptions of the pathology from naturally occurring disease outbreaks do not include the acute oviduct inflammation or necrosis or the identification of viral inclusions as the lesions are transient.

DIAGNOSIS

Reduction in production with the occurrence of depigmented, soft-shelled eggs in the absence of other clinical signs should trigger consideration of EDS. Isolation and identification of the virus is best achieved using EDS76-

free embryonated duck or goose eggs or cell culture of duck or goose origin.

Harvested allantoic fluid or cell culture supernatant can be checked for hemagglutinating activity, which is inhibited by specific EDS antiserum or viral DNA is detected by PCR.

The hemagglutination inhibition test in suspect flocks is most helpful immediately after egg changes are observed because many infected flocks do not have demonstrable antibody during the growing period.

CONTROL
An inactivated vaccine has been successfully used against clinical EDS. Eradication programs can be used to eradicate the disease.

TREATMENT
No treatment is available.

ZOONOTIC POTENTIAL
Not recorded.

IV. QUAIL BRONCHITIS

DEFINITION
Quail bronchitis (QB) is an acute, contagious and sometimes highly lethal respiratory disease of bobwhite quail (Colinus virginianus) characterized by catarrhal tracheitis and airsacculitis.

OCCURRENCE
QB has been documented sporadically in captive quail throughout the United States. There is evidence suggesting occurrence in wild quail as well.

HISTORICAL INFORMATION
A respiratory disease (bronchitis) of quail caused by a virus was first described in 1951.

ETIOLOGY
The causative agent of quail bronchitis is fowl adenovirus-1.

EPIZOOTIOLOGY
1. The sources of the causative adenovirus are infected breeders (via transovarial passage), carrier birds, contaminated feces, or fomites.

2. Once established in a flock the QB virus spreads rapidly primarily by the fecal-oral route. Morbidity usually reaches 100% in susceptible birds.

3. The disease frequently occurs in succeeding broods of quail reared on contaminated premises owing in great part to the environmental resistance and persistence of the causative adenovirus.

CLINICAL SIGNS
1. QB occurs with sudden onset of severe respiratory signs including tracheal rales, coughing, and sneezing. Lacrimation, conjunctivitis, and neurologic disorders may also be seen, but are less consistent signs.

2. The disease is most severe in young quail (under 4 weeks of age). Infections are milder or subclinical in birds over 8 weeks of age.

3. The incubation period of QB is 2-7 days, which explains the explosive spread of the disease in susceptible flocks. Morbidity and mortality can be substantial, ranging from 10 to 100% in young birds and the course of the disease in affected flocks varies from 1 to 3 weeks.

LESIONS
1. Excess mucus with thickening and roughening of the mucosa are the major lesions in the trachea (Fig. 1) and bronchi. Air sacs may be mildly thickened and cloudy. Ocular and nasal discharges are occasionally noted. Lungs are congested. Multiple small pale foci are randomly distributed over the liver. The spleen may be mottled and slightly enlarged.

2. Microscopically, tracheal and bronchial lesions include epithelial deciliation, necrosis, exfoliation and proliferation with epithelial cells containing large basophilic intranuclear adenoviral inclusions. A mild to moderate cellular infiltrate is in the lamina propria composed of primarily lymphocytic/plasmacytic. Luminal exudate is composed of exfoliating epithelial cells frequently containing intranuclear adenoviral inclusions, erythrocytes, a mixture of inflammatory cells and necrotic cellular debris. Pulmonary lesions include focally extensive pneumonia. Liver has multiple foci of necrosis with variable numbers of mononuclear cells and fewer heterophils. Large basophilic intranuclear adenoviral inclusions are often in the hepatocytes adjacent to the areas of necrosis or inflammation.

3. There is multifocal to locally extensive splenic lymphoid necrosis with fibrin exudation and very mild leukocyte infiltration and rare adenoviral inclusions. The severity of the bursal lesions is variable and range from single cell lymphocyte necrosis with lymphoid depletion and follicular atrophy to severe follicular lympholysis. Intranuclear adenoviral inclusions are frequently in the mucosal epithelium.

DIAGNOSIS
1. Acute respiratory disease with high mortality in young quail chicks is highly suggestive of QB and is confirmed histologically with a severe catarrhal tracheitis and bronchitis and respiratory epithelium containing the characteristic intranuclear adenoviral inclusions.

2. Isolation of the causative adenovirus confirms the diagnosis of QB. Isolation is accomplished in 9-11-day-

old specific-pathogen-free embryonating eggs or cell cultures.

3. Serologic tests are of limited value unless flock sampling is done on both an acute and convalescent basis to demonstrate definitive seroconversion

CONTROL
No licensed vaccines are available.

TREATMENT
No treatment is available but increasing brooding house temperature, elimination of drafts, and expanding floor space may be helpful as supportive measures in the face of an outbreak.

ZOONOTIC POTENTIAL
Not recorded.

Adenovirus

INCLUSION BODY HEPATITIS

Fig. 1
Jaundiced chicken dead with IBH.

Fig. 2
IBH in a chicken: the liver is swollen, enlarged, tan, with mottling with petechial hemorrhages under the capsule and in the parenchyma.

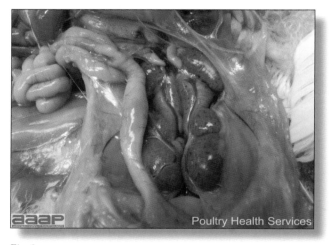

Fig. 3
The kidneys of this IBH chicken are markedly swollen, pale and mottled.

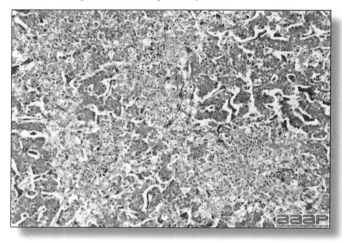

Fig. 4
At microscopy, presence of multifocal to locally extensive degeneration and necrosis of hepatocytes.

Add image of hemopericard

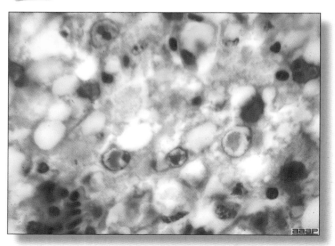

Fig. 5
Microscopically, characteristic large basophilic intranuclear adenoviral inclusions in the hepatocytes are observed.

HEMORRHAGIC ENTERITIS OF TURKEYS

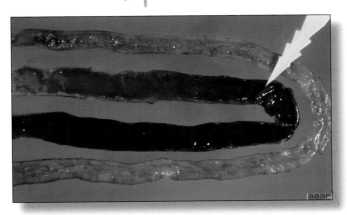

Fig. 1
The intestinal tract of a turkey affected with HEV, especially the duodenal loop, is distended, dark purple, and filled with hemorrhagic content.

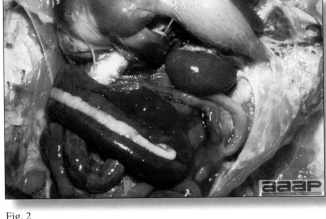

Fig. 2
Turkey affected with HEV: the spleen is typically very enlarged and mottled and the intestinal content is filled with hemorrhagic content.

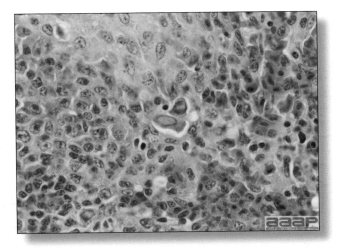

Fig. 3
Reticuloendothelial cells of the spleen containing large intranuclear adenoviral inclusions.

QUAIL BRONCHITIS

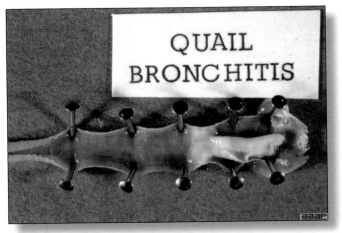

Fig. 1
Excess mucus and necrotic exudate in the trachea of a quail affected with QB.

AVIAN ENCEPHALOMYELITIS
(AE; Epidemic Tremor)

DEFINITION
Avian encephalomyelitis (AE) is a viral infection of chickens, turkeys, pheasants, and coturnix quail characterized in young birds by ataxia progressing to paralysis and, usually, by tremors of the head and neck. Infected adults usually show no signs.

OCCURRENCE
Clinical outbreaks are usually observed in chickens and most outbreaks are in 1-3-week-old chicks. Turkey poults, pheasants, and coturnix quail are also infected naturally. Experimental infection has been induced in ducklings, guinea fowl, and pigeon hatchlings. Infection can occur in older birds but usually is clinically inapparent. AE is worldwide in distribution.

HISTORICAL INFORMATION
1. In 1930 AE was first seen in 2-week-old Rhode Island Red commercial chicks. Within a few years the disease was present in most of the other New England states and was referred to as "New England disease". Between 1955 and 1970 the disease was described successively in coturnix quail, pheasants, and turkeys.

2. A nationwide testing program for AE antibody revealed that many chicken flocks in the United States have antibody to AE virus.

3. Hatcheries once replaced baby chicks that had AE or developed AE shortly after delivery. This practice caused considerable loss to the hatcheries. Vaccination of the breeders was first successfully implemented in the 1950s and AE largely became controlled in commercial flocks by the 1960s.

ETIOLOGY
1. AE is caused by a hepatovirus belonging to the Picornaviridae family. There appear to be no serologic differences among isolates although they vary in their tissue tropisms. All field strains are enterotropic but some strains are more neurotropic than others and pathogenicity varies.

2. The virus can be grown in the yolk sac of chick embryos free of maternal antibodies and in a variety of tissue culture systems. Embryo-adapted strains are not infectious by the oral route, are highly neurotropic, and cause muscular dystrophy in inoculated embryos.

3. Virus is present in the feces of infected birds and will survive there for at least 4 weeks.

4. The virus survives treatment with ether and chloroform and is fairly resistant to various environmental conditions.

EPIDEMIOLOGY
1. During the acute phase of infection in laying chickens, a period up to 1 month, some layers shed virus in some of the eggs they lay. Although vertically transmitted AE may affect hatchability, many of the chicks will hatch and can show clinical signs of the disease as early as the 1st day of age. The infected chicks will shed virus in their feces resulting in horizontal spread to other chicks. Younger chicks tend to shed virus for a longer period of time than older chicks.

2. The method of transmission of AE to susceptible adult flocks is unknown but is probably via fomites. Multiage farms are more likely to be infected than those with single age groups.

CLINICAL SIGNS
1. In chicks, signs may be present at the time of hatch but usually occur between the 1st and 2nd week of age. Age resistance is marked if exposure is after 2-3 weeks of age.

2. In chicks, signs include dull expression, ataxia progressing to paralysis and prostration (Fig. 1) and tremors of the head and neck. Tremor may be inapparent but often can be accentuated if the bird is frightened or held inverted in the hand. Prostrate birds are soon trampled and killed by the other birds.

3. The morbidity in chicks is quite variable but may go as high as 60%. If most chicks in the flock come from immune dams, morbidity is usually low. Mortality averages 25%. Few birds with signs recover completely. Those that survive often fail to grow or produce eggs normally. Many survivors later develop a bluish opacity to the lens of the eye and have impaired vision (Fig. 2 and 3).

4. Layers seldom show signs when infection is going through the flock. However, good production records often reveal a significant decline in egg production generally lasting no more than 2 weeks.

LESIONS
1. Generally, there are no gross lesions. In chicks, whitish areas in musculature of the gizzard can sometimes be observed (Fig. 4). No gross lesions are seen in adult birds.

2. Microscopic lesions, if typical, have special diagnostic value. There is a disseminated, nonpurulent encephalomyelitis with widespread and marked perivascular cuffing (Fig. 5). Two microscopic changes are especially helpful: swelling and chromatolysis of neurons (Fig. 6) in nuclei (nucleus rotundus and nucleus ovoidalis) in the midbrain and cerebellum, and dense lymphoid aggregates in the muscle of the proventriculus (Fig. 7) and/or gizzard as well as the myocardium and pancreas.

DIAGNOSIS

1. In chicks, the history, age of the birds, and typical signs of central nervous system (CNS) lesions permit a strong presumptive diagnosis. The diagnosis can often be strengthened by histopathologic examination. Alternatively, the direct fluorescent antibody technique can be used to demonstrate AE viral antigen in infected chicks.

2. Isolation and identification of the virus from the brains of infected chicks is possible but is time consuming and expensive. Also, there must be a source of susceptible chick embryos and this usually necessitates a layer flock that has never been exposed to AE.

3. Antibodies to AE can be detected as early as 4 days postinfection and persist for at least 28 months. Serologic assays include the ELISA, immunodiffusion test, virus neutralization test, passive hemagglutinin test and the indirect FA test. Rising titers in sequential samples are highly suggestive of active infection.

4. AE must be differentiated from other diseases that cause signs of CNS disease in young birds. These include:

Newcastle disease
Arboviral infection
Vitamin deficiencies
(E, A and Riboflavin)
Equine Encephalomyelitis
Virus (salt, some pesticides, etc.)

Mycotic encephalitis
Brain abscesses
Marek's disease
Toxicities

CONTROL

1. Chicks from immune hens are usually protected by parental immunity during the critical first few weeks after hatching. Breeding flocks can be vaccinated to provide maximum protection to their chicks. Although vaccination is usually conducted prior to the onset of lay, some killed vaccines can be used during production. The embryo susceptibility test, which involves inoculation of AE into the yolk sac of embryonated eggs from the breeders can be used to determine the immune status of the flock.

2. Both killed and live vaccines are used for vaccination and are effective. Live virus vaccines must not be embryo adapted as they lose their ability to infect orally and can cause clinical disease when administered parenterally. Live vaccine is given by the wing web stick method in combination with pox, via the drinking water, or by spray. Birds that will serve as breeders should not be vaccinated until they are at least 8 weeks old. One vaccination is usually adequate for the life of the bird. Live vaccines should be applied at least 4 weeks prior to production; vaccines used in chickens can be protective for turkeys.

3. Chicks from flocks that have been naturally infected will probably receive enough parental immunity so that they will not develop the disease.

TREATMENT
Treatment is of no value.

AVIAN ENCEPHALOMYELITIS

Fig. 1
Chicks with dull expression and paralysis due to AE.

Fig. 2
Surviving poult who has developed a bluish opacity to the lens of the eye.

AVIAN ENCEPHALOMYELITIS

survios (handwritten)

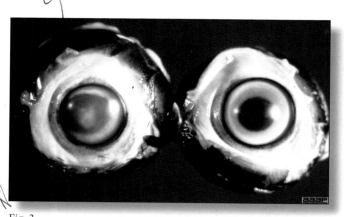

Fig. 3
Marked bluish opacity to the len of the eye of an AE affected chicken (left) vs a normal eye (right).

After they recover (handwritten)

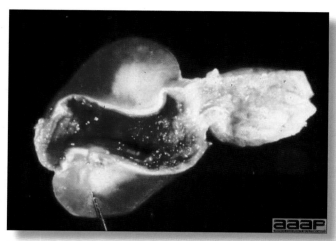

Fig. 4
Whitish areas in musculature of the gizzard of a chick.

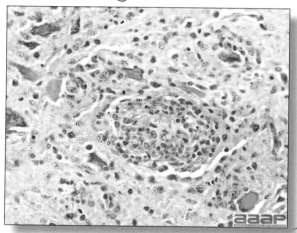

Fig. 5
Non-purulent encephalomyelitis characterized by perivascular cuffing. Also note the gliosis and the chromatolysis.

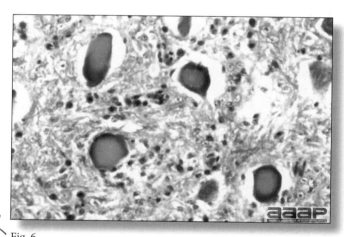

Fig. 6
Diffuse gliosis and central chromatolysis of neurons of one of the nuclei of the brain stem.

Fig. 7
Lymphoid aggregates in the muscle of the proventriculus.

Avian encephalomyelitis Heptavirus (handwritten)

AVIAN INFLUENZA

DEFINITION

Avian influenza (AI) is a viral disease characterized by respiratory signs, depression and reduced feed and water intake. In egg laying birds there is a decline in egg production.

There are many strains of AI viruses and generally they can be classified into two categories: low pathogenic (LPAI) that typically causes little or no clinical signs in birds and highly pathogenic (HPAI) that can cause severe clinical signs and/or high mortality in birds. Those virulent AI viruses are also classified as highly pathogenic notifiable avian influenza (HPNAI) viruses. Moreover, subtype H5 or H7 viruses with a hemagglutinin cleavage site similar to those in virulent viruses are also considered HPNAI viruses, regardless of their virulence *in vivo*.

The H5 and H7 isolates which are not highly pathogenic and do not have the hemagglutinin cleavage site amino acid sequence similar to HPNAI viruses are classified as low pathogenicity notifiable avian influenza (LPNAI) viruses.

Non-H5 or non-H7 AIVs which are not highly pathogenic are classified as low pathogenicity avian influenza (LPAI) viruses.

OCCURRENCE

AI viruses are spread worldwide in their hosts, wild waterfowl and shorebirds. AI outbreaks in commercial birds have also occurred throughout the world. In the past HPNAI was relatively infrequent, but the 'Asian' H5N1 has spread throughout 56 countries in Asia, Europe and Africa between 2004 and 2010. HPNAI outbreaks caused by other subtypes (H5N2, H7N3 and H7N3) were less common during the same period.

HISTORICAL INFORMATION

The most virulent form of AI was once called fowl plague and was first documented in Italy more than 100 years ago. In the United States highly pathogenic AI first occurred in 1924-1925. The current HPNAI/LPNAI/LPAI classification has been updated in 2009.

ETIOLOGY

Avian influenza is caused by a type A influenza virus belonging to the Orthomyxoviridae family. Influenza viruses have segmented RNA genome and two major surface antigens, hemagglutinin (H) and neuraminidase (N) that give rise to subtype names for specific viruses (eg. H4N6). There are 16 hemagglutinins and 9 neuraminidases making 144 possible virus subtypes. Influenza viruses are subtyped by hemagglutination inhibition and neuraminidase inhibition tests. Cross-protection does not occur between subtypes.

EPIDEMIOLOGY

1. Wild waterfowl and shorebirds are the major natural reservoir of AI viruses. Wild waterfowl are asymptomatic, may excrete virus in the feces for long periods, may be infected with more than one subtype, and often do not develop a detectable antibody response. AI virus has been recovered directly from lake and pond water used by infected wild ducks. Contact of these birds with range-reared commercial flocks is an important factor in some outbreaks. This source of infection often results in a seasonal incidence in some states.

2. Two man-made reservoirs are live bird markets and commercial swine facilities.

3. Live bird markets have existed in large cities, but they are an emerging phenomenon in some areas. They serve as a focal point for gathering and housing many species of birds that are then sold in or around large cities. These facilities are often neither cleaned nor depopulated. The continuous supply of susceptible poultry in such markets enhances opportunity for viral replication and mutation, and this in turn enhances the opportunity for viruses to be carried back to susceptible poultry flocks.

4. Swine have been known to be infected with swine flu (H1N1) since the 1930s, but recently another subtype (H3N2) has been spreading in swine populations. Transmission of influenza from swine to turkeys has been documented.

5. AIVs have been isolated from imported exotic birds. These infected birds are a potential threat to cage birds, wild birds, and poultry.

6. Although waterfowl shed virus in their droppings for long periods, most viral shedding from infected gallinaceous poultry stops after seroconversion. Influenza virus is released in respiratory secretions and excretions and droppings of infected birds where it is protected by organic material. The virus is labile in warm conditions, but can survive for months in a cold environment. Influenza virus has been isolated from turkey eggs and semen, but there is no evidence of vertical transmission. Improper disposal of infected eggs could potentially expose other susceptible birds, but such transmission has not been observed.

7. Once AIV is introduced into the poultry industry it is transmitted from farm to farm by direct and indirect contact. AI viruses can be transmitted on contaminated shoes, clothing, crates, and other equipment and by movement of birds and manure.

CLINICAL SIGNS

1. Most outbreaks are caused by LPAI viruses. The LPAI signs vary greatly and depend on many factors, including the age and species, the virulence of the virus, concurrent infections, and husbandry. In most outbreaks, signs are predominantly those of a respiratory disease with coughing, sneezing, rales, lacrimation, sinusitis (Fig. 1), and depression. In egg layers decreased egg production and quality are seen.

2. In young growing turkeys the disease may be subclinical or severe, particularly where secondary infection with live *Pasteurella* vaccine, *E. coli*, or *Bordetella* occurs. Outbreaks in egg laying turkeys often reduce production markedly and frequently are associated with abnormal eggshell pigmentation and quality.

3. Morbidity and mortality are highly variable, depending upon the same factors that determine clinical signs noted above.

4. HPAI is a severe form of influenza usually seen in chickens. Viruses of high pathogenicity may cause fatal infections preceded by few signs. Onset is sudden, the course is short, affected birds are quite ill, and mortality may approach 100%. Signs may relate to the respiratory, enteric, or nervous systems. There may be diarrhea, edema of the head and face (Fig. 1), or nervous disorders.

LESIONS

1. With LPAI outbreaks in poultry there is mild to moderate inflammation of the trachea, sinuses (Fig. 2), air sacs (Fig. 3) and conjunctiva. In laying birds there often is ovarian atresia (Fig. 4) and involution of the oviduct or egg yolk peritonitis (Fig. 5). Fibrinopurulent bronchopneumonia (Fig. 6) can occur with secondary infection. Various degrees of congestive, hemorrhagic, transudative, and necrotic lesions have been described.

2. In HPNAI infection, gross lesions in chickens are the most extensive and severe. Fibrinous exudates may be found on the air sacs, oviduct, pericardial sac, or on the peritoneum. Small foci of necrosis may be apparent in the skin, comb, and wattles or in the liver, kidney, spleen, or lungs. Indications of vascular damage often include congestion, edema, and hemorrhages at many sites.

3. Classical lesions of HPNAI in chickens include cyanosis and edema of the head (Fig. 7 and 8), vesicles and ulceration on the combs, edema of the feet, blotchy red discoloration of the shanks (Fig. 9), petechiae in the abdominal fat and various mucosal and serosal surfaces, and necrosis or hemorrhage in the mucosa of the gizzard and proventriculus (Fig. 10).

4. Lesions of HPNAI in turkeys are not well described, but encephalitis and pancreatitis have been reported.

DIAGNOSIS

1. History, signs, and lesions may be suggestive of LPAI, but are similar to other diseases.

2. Confirmation of suspect AI cases requires laboratory tests such as serology (AGID and/or ELISA) and virus detection (real-time RT-PCR and/or virus isolation).

3. Rapid testing by an influenza A cross-reactive real-time RT-PCR test is typically carried out first. Reactive samples are then subjected to H5 and H7 subtype-specific real-time RT-PCR tests.

4. Confirmation of HPNAI requires molecular characterization of the virus and/or inoculating susceptible chickens with the virus.

5. Influenza virus usually can be isolated in chick embryos from tissue or swab samples of trachea, lung, air sac, sinus exudate, or cloaca. Viruses from some species such as geese may not grow well in embryonated chicken eggs. The virus hemagglutinates chicken red blood cells.

6. Serological tests can be used to demonstrate seroconversion between acute and convalescent sera.

7. Influenza must be differentiated from other poultry diseases including Newcastle disease, other paramyxovirus infections, mycoplasmosis, chlamydial infections, and fowl cholera. HPNAI should be differentiated from velogenic viscerotropic Newcastle disease. Because AI viruses causing highly pathogenic AI are considered exotic, they are reportable and confirmation by virus isolation is essential.

CONTROL

1. Prevention of LPAI is largely through prevention of exposure to influenza viruses by direct or indirect contact with waterfowl and shorebirds, live bird markets and swine farms.

2. Once LPAI non-H5 or non-H7 virus is introduced into the poultry industry, control is largely dependent on voluntary, industry efforts since there are no official state eradication programs.

3. Routine serologic monitoring of blood or egg yolk antibody is used in areas where AI has been a problem. This effort provides early detection of an outbreak and permits other measures such as isolation and sanitation to be used early.

4. Reporting outbreaks to industry personnel who are in direct or indirect contact with poultry is necessary so that appropriate measures can be taken.

5. Voluntary isolation of infected flocks is the responsibility of the owner and is necessary to prevent transmission to other flocks. Rigorous measures to prevent the contamination of and control the movement of people and equipment are required in order to stop this disease.

6. Different states and industries take different approaches to the next step. Controlled marketing of flocks after they have recovered from infection is common in the turkey industry. In some broiler producing states, voluntary destruction of infected flocks is encouraged.

7. Rescheduling flocks is necessary to make sure there is no active AI virus on the farm before another flock is placed.

8. History has proven that prevention of HPNAI is based on successful control of H5 or H7 LPAI.

9. H5 or H7 LPNAI results in a response that varies according to the response plan developed by each state. Generally all outbreaks of H5 or H7 LPAI will result in a rapid aggressive response, although the means that are used to bring it under control will vary according to species involved, density and the state plan.

10. H5 or H7 HPNAI results in a uniform response plan under the direction of government agencies and there also will be input from public health, occupational health and pollution control agencies.

11. All outbreaks of influenza should be reported immediately to the state veterinarian or other appropriate health authorities.

12. Vaccines - Immunity is hemagglutinin subtype specific. Because birds are susceptible to all 16 hemagglutinins, preventive vaccination is not practical. Once an outbreak occurs and the subtype is identified, however, vaccination is a tool that may be used to help bring the infection under control. Because influenza viruses are unstable, little research has been done on live influenza virus vaccines for poultry. Killed, injectable and recombinant vaccines are available against H3, H5 and H7 subtypes.

TREATMENT
There is no effective treatment. However, good husbandry may reduce losses from secondary infections.

ZOONOTIC POTENTIAL
Although infection of humans with AIV is rare, human cases caused by avian influenza subtypes H5, H7 and H9 have been documented.

A pathogenic H5N1 virus that spread in poultry and wild birds over most of Asia and parts of Europe and Africa has caused 247 confirmed human deaths from 2003 - 2010.

Some H1N1 and H3N2 influenza virus strains that circulate in swine and humans can also infect birds and vice versa.

AVIAN INFLUENZA

Fig. 1
Turkey affected with LPAI showing sinusitis.

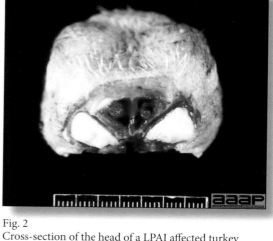

Fig. 2
Cross-section of the head of a LPAI affected turkey showing the fibrinopurulent exudate filled sinuses.

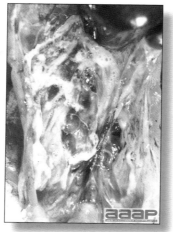

Fig. 3
Airsacculitis in a LPAI infected turkey.

Fig. 4
Atretic ovarian follicles.

Fig. 5
Egg yolk peritonitis in a LPAI positive laying hen.

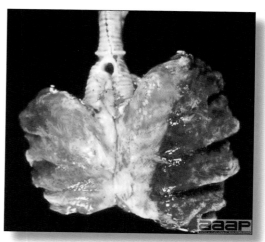

Fig. 6
Bronchopneumonia in a LPAI positive turkey.

AVIAN INFLUENZA

HPAI classic lesions

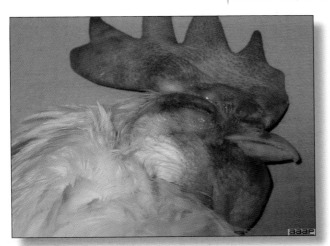

Fig. 7
Edema of the head in a HPAI infected layer.

Fig. 8
Subcutaneous edema of the head.

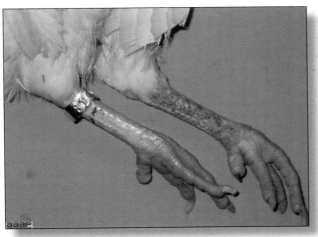

Fig. 9
Blotchy red discoloration of the shanks in a HPAI positive layer.

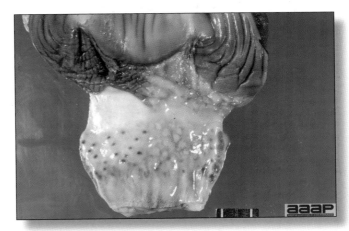

Fig. 10
Proventricular hemorrhages.

AVIAN METAPNEUMOVIRUS INFECTION

DEFINITION

Avian metapneumovirus (aMPV) infection is a highly contagious infectious respiratory disease of turkeys and chickens characterized by coughing, swollen sinuses, nasal discharge and lowered feed and water consumption. The disease caused by aMPV infection was originally referred to as avian pneumovirus infection or turkey rhinotracheitis in turkeys and as swollen head syndrome in chickens.

OCCURRENCE

1. aMPV infections have been described throughout most of the world. Only Canada and Australia appear to be free of aMPV infection. No chicken aMPV cases have been reported in the U.S.A.

2. There are four subtypes of aMPV, A, B, C and D. In the U.S.A. all isolates have been type C.

3. The virus has been isolated from turkeys in many parts of the world. Experimental studies have shown chickens, ducks, guinea fowl and pheasants to be susceptible. Serological studies have detected antibody in ostriches and herring gulls. Using RT-PCR, APV RNA has been detected in geese, coots, sparrows, swallows, starlings and an owl. A seasonal incidence with peaks in the spring and fall has been observed.

HISTORICAL INFORMATION

1. Turkey rhinotracheitis caused by aMPV was first identified in South Africa in the late 1970s.

2. In the U.S.A. aMPV was first identified in Colorado turkeys in 1997. Subsequently the disease was detected serologically, using an ELISA developed at the National Veterinary Services Laboratories, in Minnesota turkeys in the spring of 1997. It is likely that the infection was present in Colorado and Minnesota prior to 1997. It has remained in Minnesota turkeys infecting 40 to 50% of the flocks each year since 1997, and there has only been limited spread to neighbouring states.

ETIOLOGY

1. aMPV is an enveloped, single stranded RNA virus, member of the genus Metapneumovirus of the Paramyxoviridae family. aMPVs have a fusion (F) protein and a glycoprotein (G) as surface antigens.

2. The virus is present in respiratory secretions and excretions of infected birds where it is protected by organic material. The virus is susceptible to detergents and disinfectants. However, the virus can remain infective for long periods of time under cool and moist environmental conditions, e.g. in poultry litter for three days at 20 to 25°C and for 14 to 30 days at 8°C.

EPIDEMIOLOGY

1. Turkeys and chickens appear to be natural reservoirs of aMPV, but limited studies have found the evidence of aMPV infection in several species of wild and domestic birds. Whether the virus persists in recovered turkey flocks is also unknown. The virus has been detected in oviduct tissue, and neonatal infection has been observed, but no proof of egg borne transmission has been shown.

2. The virus is transmitted by direct contact between infected and susceptible birds and is believed to be transmitted by indirect contact including exposure to aerosol droplets or to virus-contaminated boots, clothing or equipment. Airborne transmission has been demonstrated in the laboratory, but such transmission from farm to farm is unproven.

CLINICAL SIGNS

1. The clinical signs are variable and depend on the age, gender, concurrent infections, and management factors. In young turkeys, clinical signs include nasal discharge (Fig. 1), foamy conjunctivitis, swollen infraorbital sinuses (Fig. 2) with submandibular edema (Fig. 3). As clinical signs subside birds may begin to die. Mortality rates in market turkeys range from nil to 80%, but death is usually due to secondary infections. Condemnation rates are usually elevated if turkeys are infected within two weeks of processing.

2. Turkey breeder hens may have a decline in egg production of 10 to 30% and lay increased numbers of cull eggs. Mortality in breeders is usually 0 to 2% but may be higher if live *Pasteurella* vaccine has been used in the flock.

3. In chickens aMPV infection may be subclinical and/or associated with other agents in swollen head syndrome (Fig. 4) and egg production problems.

LESIONS

1. In mature female turkeys, a marked decrease in egg production with increased production of eggs with poor shell quality and increased numbers of birds with peritonitis and uterine prolapsed are described. Experimental infections produced a watery mucoid nasal discharge and increased amounts of tracheal mucus in addition to egg peritonitis and production of poor quality and abnormal eggs. Secondary bacterial infections may increase the severity of the lesions with airsacculitis, pericarditis, perihepatitis and pneumonia.

2. Microscopically, with experimental infections, focal deciliation, epithelial necrosis, hyperemia and submucosal inflammation with a mixture of mononuclear cells of the nasal mucosa and transient deciliation of the tracheal mucosa have been described.

3. In chickens, the infection has been associated with secondary *Escherichia coli* infection with subsequent development of swollen head syndrome which

presents yellow gelatinous to heterophilic exudates in the subcutaneous tissues of the head and swollen periorbital and infraorbital sinuses. Transient localized inflammation of the upper respiratory tract has been observed.

DIAGNOSIS

1. The history and signs of coughing, nasal discharge and swollen sinuses may be suggestive of aMPV infection but are similar to other respiratory infections. Confirmation of the diagnosis requires laboratory tests.

2. The virus is difficult to isolate from swabs or tissues of affected birds so other laboratory tests used are: immunohistochemistry on formalin fixed turbinate tissues (Fig. 5), RT-PCR to detect viral RNA in tracheal swabs, choanal swabs or turbinates, and ELISA to detect pneumovirus-specific antibodies.

3. Avian pneumoviruses are difficult to isolate, but once recovered from affected birds aMPVs grow in embryos and tissue culture systems. Unlike other members of the paramyxovirus family, pneumoviruses do not hemagglutinate.

CONTROL

1. The reservoir of metapneumovirus in nature is not known but wild birds are suspected. Whether or not infected flocks remain a source of virus for their whole life is also not known, but they should be considered a potential source for life. Prevention of metapneumovirus infection requires preventing the introduction of the virus by direct or indirect contact from these possible reservoirs (wild birds and infected flocks).

2. Limited studies indicate that there might be partial cross protection between APV types.

3. Since the disease is spread by direct and indirect contact, strict biosecurity and a good sanitation program are imperative. A minimum biosecurity program for controlling aMPV would include:

4. Crews that handle birds (vaccination, moving, live haul, insemination) must be controlled. Crew members should wear disposable or freshly laundered clothing including footwear.

5. Equipment that moves from farm to farm and comes in contact with poultry or birds (rendering, moving, live haul trucks and dumpsters, loaders, vaccination equipment) should be washed with detergent and disinfectant.

6. Poultry facilities should be wild bird proofed.

7. A live attenuated vaccine is available in the Midwestern USA. Killed autogenous oil emulsion vaccine has been used after live vaccines in turkey breeders.

8. A routine monitoring program is suggested for areas where aMPV infection has been a problem. Serological screening of blood samples and PCR testing of choanal swabs can provide early detection so that other control measures can be instituted.

TREATMENT

No treatment is available. Reduced density, increased supplemental heat and good management conditions are associated with reduced financial loss due to the disease. Antibiotic treatment has been used to reduce the effects of concurrent bacterial infections.

AVIAN METAPNEUMOVIRUS INFECTION

Fig. 1
Turkey poult with nasal discharge.

Fig. 2
Sinusitis in a turkey affected with avian metapneumovirus.

Fig. 3
Submandibular edema in a young turkey. Note the dyspnea.

Fig. 4
Swollen head syndrome in a 29 day-old broiler chicken.

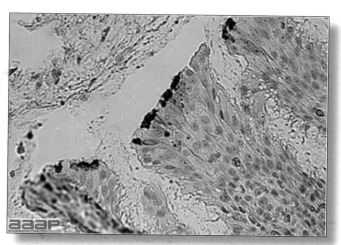

Fig. 5
Immunohistochemical findings for nasal turbinates, showing peroxidase staining of viral antigen in the epithelial surface of a turkey poult exposed to metapneumovirus.

AVIAN NEPHRITIS VIRUS INFECTION IN CHICKENS

DEFINITION
Avian nephritis virus (ANV) is an astrovirus that causes infection of the kidneys in young chickens. The infection is acute, highly contagious but usually subclinical in nature characterized by nephritis and urate deposits in the kidneys and abdominal viscera.

OCCURRENCE
ANV was first reported from Japan and the disease or the antibodies have been reported in Europe and USA.

ETIOLOGY
ANV has been classified as an astrovirus distinct from Duck Hepatitis type 2 and 3, turkey and chicken astroviruses. Astroviruses are non-enveloped single-stranded positive sense RNA icosahedral virus, 28-30 nm in diameter that may exhibit five or six pointed star-like surface when viewed by electron microscopy. ANV has been placed in a new genus *Aviastrovirus* in the family *Astroviridae*. There are genetic differences among various isolates of ANV.

EPIDEMIOLOGY
Transmission of ANV appears to be by direct contact with infected birds. Vertical transmission of the virus has also been suggested.

CLINICAL SIGNS
1. The only clinical sign reported due to ANV in 1-day-old chicks is transient diarrhea but not all birds will show this. But some of the characteristic lesions develop in chicks up to 4 weeks of age.
2. Runting stunting of chicks and decrease in body weights are observed.
3. Mortality can range from negligible to 6 %.

LESIONS
1. Grossly the chicks suffering from ANV may show enlarged and pale kidneys with increased urates (Fig. 1).
2. Visceral urate deposits (gout) in the pericardium, capsule of the liver, abdominal cavity, subcutaneous tissues, etc... can be observed.
3. Histologically, degeneration and necrosis of the epithelial cells of the proximal convoluted tubules and dilation accompanied by lymphocytic interstitial nephritis with tophi formation can be observed. Tophi and associated inflammation can also be observed in various other organs.

DIAGNOSIS
1. A presumptive diagnosis can be made based on clinical signs, mortality patterns combined with gross and microscopic lesions in young chicks.
2. Virus can be difficult to isolate. It can be isolated in 6-day-old chicken egg embryos by inoculation through yolk sac route and in chicken embryo kidney cells.
3. Other techniques such as RT-PCR, immunofluorescence, ELISA and electron microscopy have been used to diagnose ANV.

CONTROL AND TREATMENT
Currently there are no control or treatment measures that are available and biosecurity implementation may help limit the spread of the virus.

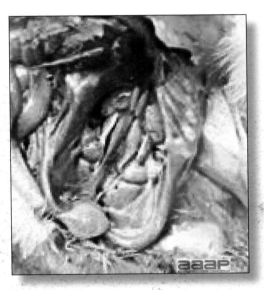

Fig. 1
Chick suffering from ANV showing enlarged and pale kidneys with increased urates.

AVIAN VIRAL TUMORS

DEFINITION

The viral neoplastic diseases of chickens and turkeys, although previously considered a "complex", are actually distinct disease entities. In some cases a single tumor virus strain can induce multiple disease syndromes, thus causing indecision whether these neoplasms should be classified by etiology or by lesion type. Furthermore, some of the lesion types are so rare as to be of little concern.

In an attempt to simplify this situation, we will consider here only the four neoplastic disease syndromes that have economic importance: Marek's disease, a common lymphoproliferative disease of chickens caused by an alpha herpesvirus; avian leukosis/sarcoma, common retroviral diseases characterized by lymphoid or other neoplasias and lowered egg production in adult chickens; reticuloendotheliosis, a nondefective retrovirus which causes a runting disease and a chronic lymphoma in turkeys, chickens, and a variety of other avian species; and lymphoproliferative disease, a retrovirus-induced disease of turkeys characterized by chronic lymphomas that although not yet reported in the United States, is found elsewhere and must be considered in a differential diagnosis.

I. MAREK'S DISEASE

DEFINITION

Marek's disease (MD) is a herpesvirus-induced neoplastic disease of chickens characterized by infiltration of various nerve trunks and/or organs with pleomorphic lymphoid cells.

OCCURRENCE

Marek's disease is important primarily in chickens, to a much lesser degree in quail, and has been rarely observed in turkeys, pheasants and jungle fowl. Turkeys and other species have limited susceptibility. The disease most commonly occurs in young, sexually immature chickens 2-7 months old, but can occur at virtually any age beyond 3 weeks. The disease occurs throughout the world and virtually all flocks are exposed to the causative virus.

HISTORICAL INFORMATION

A report in 1907 by a Hungarian veterinarian, Jozsef Marek, of paresis in roosters is the first description of the disease now called MD. The disease was first reported in the United States in 1914. Although forms of MD were an important cause of mortality in chickens prior to 1950, a sudden increase in mortality in the late 1950s and 1960s accelerated research. Reliable experimental transmission was achieved in 1962 and the causative herpesvirus was isolated and identified in 1967. Vaccines became available for use in the United States by 1970 and have been very effective in preventing the disease. However, sporadic losses and the fear of increased virulence of the virus have kept MD among the most important poultry diseases.

ETIOLOGY

1. Marek's disease virus is a cell-associated alpha herpesvirus of subgroup a3. The herpesviruses associated with MD are classified into three serotypes. Serotype 1 isolates are ubiquitous in chickens and pathotypes vary from very virulent plus (vv+) (oncogenic) to nearly avirulent (mild). Serotype 2 isolates are common in chickens and are nononcogenic. Serotype 3 isolates, also known as turkey herpesvirus, are ubiquitous in turkeys and are nononcogenic. The three serotypes have considerable antigenic cross-reactivity.

2. The serotype 1 virus can be grown in cultured chick kidney cells prepared from 1-3 week old chicks and in duck embryo fibroblasts. It produces a distinct cytopathic effect with intranuclear inclusions in those cells. Embryonal chick kidney cells and chick embryo fibroblasts are less effective for low-passage virus. Serotype 2 and 3 viruses can be isolated and propagated in chick embryo fibroblasts. The virus is usually tightly bound to living cells and in this form is very labile, but cell-free virus is released from the feather follicle epithelium and is relatively resistant to environmental factors. Both cell-associated and cell-free viruses are susceptible to a number of common disinfectants.

EPIDEMIOLOGY

Infected chickens shed virus-containing feather follicle dander, which is a source of infection for other chickens by the respiratory route. Infected carriers may or may not be clinically ill, and carrier birds can sporadically shed virus throughout their lifetimes. The disease is very contagious and infectious dander can be disseminated over long distances. Although excretions and secretions of infected chickens may contain virus, dander containing infectious enveloped virus particles is the most important means of transmission. Transmission of the virus through the egg does not occur. Hatchery transmission through shell contamination is also unlikely due to adverse environmental conditions for the virus.

CLINICAL SIGNS

Clinical signs occur in chickens affected with MD but are of little help in establishing a diagnosis. Birds with visceral tumors are depressed and often cachectic prior to death. Birds with lymphoid infiltration of peripheral nerves may demonstrate asymmetric partial paralysis (Fig. 1) and / or dilation of the crop due to vagus nerve paralysis. Blindness is associated with lymphoid infiltration of the iris (Fig. 2). Clinical signs usually do not appear prior to 3 weeks of age and peak between 2 and 7 months.

LESIONS

1. At least four different lesion patterns are recognized: gross enlargement (Fig. 3) and/or yellowing and loss of cross-striations of peripheral nerves (Fig. 4); discoloration of the iris (Fig. 5); enlargement of feather follicles (Fig. 6) with reddening (skin leukosis); and visceral tumors (Fig. 7) involving the liver (Fig. 7&8), heart (Fig. 7&9), spleen (Fig. 8), gonad, kidney (Fig. 10), proventriculus (Fig. 11), and other organs and tissues. Visceral tumors are the most frequent lesions, but combinations of lesion patterns are common.

2. Microscopically, the lymphomas are characterized by a mixture of pleomorphic lymphocytes (Fig. 12). Some of these probably are true tumor cells that carry T-cell surface antigens and a MD tumor-associated surface antigen (MATSA). Others are probably host cells reacting against viral or tumor antigens and represent both T- and B-cells.

DIAGNOSIS

1. A diagnosis can usually be made after careful consideration of the history, the ages of the birds affected, and the location of the neoplastic lesions in a generous sample of typically affected chickens. Few epornitic diseases resemble MD with the exception of lymphoid leukosis and reticuloendotheliosis.

2. Marek's disease often occurs in 2-5-month-old (sexually immature) chickens but can also occur after the onset of egg production. Outbreaks after the onset of egg production in vaccinated stock have been called "late Marek's" and are often associated with newer, more highly virulent vv+ pathotypes.

3. Characteristics of MD lesions of importance in differential diagnosis include nerve involvement (when present), the absence of bursal lesions or, rarely, diffusely thickened bursas, and pleomorphic lymphocytes comprising lesions, some of which exhibit MATSA and only few of which are positive for immunoglobulin M (IgM). The ubiquitous nature of MD virus renders virology and serology of little value in diagnosis.

CONTROL

1. Commercial flocks are usually immunized via injection at 18 days of embryonation or at hatching. Care must be taken to insure that an effective dose is administered to every embryo or chicken. Because immunity from vaccination is not fully developed for 7-10 days, it is crucial to minimize early exposure. This requires careful sanitation and disinfection, particularly because MD virus survives well for months in poultry houses. Revaccination is not necessary and immunity is usually life-long. Appearance of the disease at older ages has been attributed to immunodepression due to environmental stress or infection with the vv+ pathotype.

2. The most common vaccines consist of turkey herpesvirus (HVT), a serotype 3 virus as a cell-associated preparation, or a bivalent vaccine consisting of turkey herpesvirus and a serotype 2 virus (SB-1 or 301 B). Attenuated serotype 1 vaccines (CVI988, RM1 and 648A80) are also used. Care must be taken in handling cell-associated vaccines as they are highly susceptible to adverse environmental conditions.

3. Genetic differences associated with the major histocompatibility (B) complex can aid both in resistance to MD as well as the response to vaccination.

TREATMENT

There is no effective treatment for MD. Birds with tumors or multiple skin lesions are condemned at slaughter.

II. AVIAN LEUKOSIS/SARCOMA VIRUSES

DEFINITION

The avian leukosis (ALV)/sarcoma group are retrovirus-caused, neoplastic diseases of semimature or mature chickens. Strains of this group are classified by the pathological lesion they cause and by their subgroup. The most common, lymphoid leukosis (LL) is characterized by a gradual onset in a flock, persistent low mortality, and neoplasia of the bursa of Fabricius with metastasis to many other internal organs, especially the liver, spleen, and kidney. A relatively new strain of ALV, "J", probably resulting from the recombination of endogenous and exogenous viruses, primarily causes myeloid leukosis (myelocytomatosis).

OCCURRENCE

Lymphoid leukosis associated mortality is most common in chickens 16 weeks of age or older. The disease is worldwide in distribution and widespread in the United States. Virtually all flocks are considered to be exposed to the virus but infection rates within some flocks have decreased due to efforts at eradication by primary breeder companies. Overall, the incidence of LL is low (1 or 2%), although occasional heavy losses can occur. A higher incidence of bursal disease virus may be associated with a reduced incidence of LL. With ALV-J, meat-type chickens appear to be more susceptible than layers.

HISTORICAL INFORMATION

The first report of LL is attributed to Roloff in 1868. However, the disease was not well characterized until a basis for its separation from MD was established in 1962.

ETIOLOGY

Avian leukosis is caused by a family of retroviruses known as avian leukosis viruses (alpha retroviruses), which have been classified into 10 subgroups—A, B, C, D, E, F, G, H, I and J. In the United States, subgroup A viruses are most common and are most frequently associated with LL with ALV-J myelocytomatosis next in frequency. Subgroup B viruses are occasionally isolated, whereas subgroups C and D are rare. Subgroup E viruses are common and are considered «endogenous» because they are derived from proviral genes permanently integrated into the host cell DNA; they rarely are associated with neoplasms. Subgroup F, G, H and I viruses primarily cause leukosis in species other than chickens. The viruses produce a group-specific antigen that can be detected in albumen of eggs and body tissues or fluids. ALV-J viruses have extensive antigenic variation within the strain. The avian leukosis viruses can be cultured in chicken embryo fibroblasts but most produce no cytopathology and are detected by antigen tests. Simple tests for antigen detection are available and are used in eradication programs in breeders. Antibody tests are also available and are used to monitor the status of flocks from which the virus has been eradicated.

EPIDEMIOLOGY

Egg transmission is an important mechanism of spread of avian leukosis viruses. The frequency of infected eggs is usually low but chicks hatched from infected eggs are permanently viremic (immune tolerant), do not develop antibody, have an increased risk of death from LL, may lay fewer eggs, and will probably shed virus into their own eggs thus perpetuating the infection. Chickens also can become infected by contact exposure, particularly with ALV-J, which is efficient at horizontal transmission. In meat type chickens, ALV-J viremia negative/antibody positive birds can shed virus and post hatch infected birds become tolerant shedders. Some chickens, particularly those of greater susceptibility due to endogenous virus infection or absence of maternal antibody, may transmit virus to progeny as a result of contact infection soon after hatch.

CLINICAL SIGNS

Chickens with LL may present with nonspecific or no clinical signs of disease. Many birds with tumors are unthrifty or emaciated and have pale combs and wattles. Enlargement of the abdomen may result from massive enlargement of the liver. Some birds with tumors can be detected prior to death by palpation of an enlarged and lumpy bursa of Fabricius by insertion of a finger into the cloaca. Birds with skeletal myelocytomatosis may have observable masses on the shanks, head and thorax. Osteopetrosis of the long bones (Fig. 1) or "boot" shanks may occur. Flocks with high infection rates experience depressed egg production.

LESIONS

1. There are no unique external lesions. Lymphomas (Fig. 2) are seen in many organs in chickens 16 weeks of age or older, but are especially common in the liver, kidney, ovary, and bursa of Fabricius (Fig. 3). The white-to-gray neoplastic lesions can be diffuse or are sometimes focal. If the bursa of Fabricius is incised, small nodular lesions can often be detected that would

not otherwise be obvious. Myelocytomatosis (Fig. 4) is most common with ALV-J; however, other tumor types such as hemangiomas (Fig. 5) can also be seen.

2. Microscopically, the neoplastic cells in lymphoid tumors are uniformly lymphoblastic and the cells are pyroninophilic. Also, they are nearly all positive for surface immunoglobulin M (IgM). The tumors originate from bursal lymphocytes (B-cells) in which the proviral DNA of the virus integrates during the process of replication at a site in the host cell genome close to the c-myc gene, a normal host cell gene with homology to an oncogene present in avian retroviral strain MC29. Activation of this oncogene is believed to be the primary event in starting the neoplastic process.

DIAGNOSIS

1. Lymphoid Leukosis can usually be diagnosed after careful consideration of the age of the affected chickens, the course of the disease and the pattern of mortality in the flock, and the location of gross lesions in a moderate number of typically affected chickens. Involvement of the bursa of Fabricius is nearly always present, although the lesions may not be detected without incision of the organ and examination of the epithelial surface. In contrast to MD, bursal tumors are intrafollicular, generally causing a more nodular enlargement (Fig. 6). The characteristic tumor cell has B-cell and IgM surface markers. Molecular methods are available in research laboratories to detect in the DNA of tumor cells the proviral DNA of ALV located in close proximity to the c-myc gene.

2. Diagnosis is made more difficult because the lesions of LL often appear similar to those of MD, and can appear identical to those induced experimentally by reticuloendotheliosis virus. Because ALV is very widespread in chickens, virological and serological methods offer little help in confirming a diagnosis.

3. Diagnosis of ALV-J is achieved by gross and histopathologic examination of tumors and by virus isolation from cloacal or vaginal swabs or tumors. Although PCR tests have been developed, the virus mutates frequently requiring the production of new primers.

CONTROL

1. With LL, because egg transmission is so important and the disease is not very contagious, eradication is the preferred method of control. Most of the eradication efforts have been conducted by the primary breeding companies. Many breeders of egg-type chickens have reduced markedly the rate of congenital transmission in their primary breeders and grandparent stocks through a program of testing dams prior to egg production and removal of those considered likely to transmit virus to progeny. Some breeders have flocks from which the virus apparently has been eradicated. Commercial progeny from such breeders should have

lower infection rates and thus should experience less tumor mortality and greater egg production.

2. Although LL is not a disease of commercial broilers, ALV-J is a problem and breeders have made significant progress in their eradication programs. However, due to the efficient horizontal transmission of ALV-J, control by eradication is more difficult.

3. Genetic resistance to infection with subgroup A viruses is common in meat-type chickens, but quite rare in egg-type chickens. When present, this resistance offers an alternative approach to control.

4. There is no vaccine that can protect against tumors and mortality. Congenitally infected chicks are immunologically tolerant and cannot be immunized. Vaccines to immunize parent stock where ALV has been eradicated is being considered as a means to provide maternal immunity to progeny.

TREATMENT
There is no effective treatment for LL.

III. RETICULOENDOTHELIOSIS (RE)

DEFINITION
Reticuloendotheliosis (RE) is a term used for a variety of syndromes caused by retroviruses that may be either defective or nondefective for replication in cell culture. Only a runting syndrome and a chronic lymphoma, both caused by nondefective RE virus, are of economic importance.

OCCURRENCE
Nondefective RE virus is not ubiquitous, but infection is fairly widespread in chickens and turkeys, particularly in the southern region of the USA. The disease is uncommon. Runting disease has been associated with the use of RE virus-contaminated vaccines in chickens. Chronic lymphomas occur naturally in turkeys, including wild turkeys, ducks, quail, pheasants, geese, peafowl, prairie chickens and chickens but are rare. Exportation of seropositive birds to some countries is not permitted.

HISTORICAL INFORMATION
A virus was isolated from a field case of turkey lymphomas in 1958 that, after rapid serial passage in chickens and turkeys caused high neoplastic mortality within 3 weeks. Although this isolate, strain T, has been considered a prototype, it is not typical of field strains. Other isolates from ducks and chickens were recognized in 1974 to comprise a family of RE viruses.

ETIOLOGY
RE virus is a retrovirus with an unusually wide host range. It can be grown in cells from chickens, ducks, turkeys, quail, and other species, even some mammalian cells. It infects a variety of avian species. Non avian species are resistant to infection. All isolates are of a single serotype, but minor subtype differences have been noted.

EPIDEMIOLOGY
The virus is transmitted horizontally. Mosquitoes have been incriminated as passive carriers. Fowl pox viruses have also been found to harbor infectious REV. Transmission through the egg has also been identified, but usually occurs at a very low rate.

CLINICAL SIGNS
The runting syndrome, usually induced by inoculation of chicks at 1 week of age or less with RE virus-contaminated biologics, produces severe stunting and a feather abnormality characterized by compression of barbules to the shaft in its proximal portion. Signs associated with chronic lymphomas are few but birds may become depressed prior to death.

LESIONS
1. The runting syndrome is characterized by severe atrophy of the thymus and bursa of Fabricius. The birds are immunodepressed and may show lesions of concurrent infections. Generally no tumors are noted but some birds may have enlarged nerves, proventriculitis, enteritis, and anemia.

2. The chronic lymphoma syndrome as produced experimentally in chickens with nondefective RE virus is identical in all respects with lymphoid leukosis. A different lymphoma has been induced experimentally in certain chicken strains that closely resembles MD because of nerve lesions, tumors in the liver (Fig. 1), thymus, and heart, and its occurrence as early as 6 weeks of age. Chronic lymphomas in species other than chickens are characterized by tumors of the liver and spleen, but bursal tumors are not particularly common.

DIAGNOSIS
1. A diagnosis of RE is probably best made on the basis of typical lesions and demonstration of infection with the causative agent by virus or antibody tests. Currently, the PCR test, an immunoperoxidase plaque assay, and an enzyme immunoassay are available.

2. In chickens, the disease must be distinguished from both LL and MD. Thus far, however, naturally occurring chronic lymphomas in chickens have not been documented. The runting syndrome must be distinguished from other immunodepressive conditions, especially infectious bursal disease and chicken infectious anemia.

3. In turkeys, the disease must be distinguished from lymphoproliferative disease in countries of occurrence. This can usually be accomplished by noting the age of onset, the absence of greatly enlarged spleens, the uniform lymphoblastic morphology of the tumor cells

on histology and detection of viral nucleic acid using PCR assays.

CONTROL

No methods for control or treatment have been reported, most likely because of low rates of vertical transmission, and the sporadic incidence and the self-limiting nature of the disease. Strict sanitation and insect control may help prevent infection from environmental sources. Eradication programs patterned after those developed for LL may be useful in breaking the egg transmission cycle.

DIFFERENTIAL DIAGNOSIS OF AVIAN TUMORS

The differential diagnosis of tumors in chickens and turkeys is difficult and requires an adequate history and a careful postmortem examination of a representative sample of birds with typical lesions. In some cases, additional tests such as histology, immunofluorescent tests for surface antigens, in situ hybridization tests and molecular techniques (PCR) will be helpful. The characteristics in the following table may be helpful in arriving at the correct diagnosis.

Characteristic	Chickens			Turkeys	
	MD[A]	LL	RE[B]	RE	LPD
Gross lesions					
Liver	+++	+++	+++	+++	+++
Spleen	+++	+++	+++	+++	+++
Nerves	+++	-	++	+++	+
Skin	+++	+	+	+	+
Gonads	+++	+++	+++	+++	+
Heart	+++	+	+++	++	+
Bursa	+	+++	+++/--	+++	+
Intestine	+	++	+++	+++	+++
Lungs	+++	++	++	++	+++
Kidneys	+++	+++	+++	+++	+++
Histology					
Pleomorphic cells	+	-	-/+	-	+
Uniform blast cells	-	+	+/-	+	-
Antigens					
MATSA +	-	-	?	?	
IgM	+	+++	+/-	?	?
B-cell	+	+++	+/-	?	?
T-cell	+++	+	-/+	?	?
Age of occurrence (months)					
Peak time	2-7	4-10	2-6	4-6	2-4
Limits	> 1	> 3	> 1	> 4	> 2

[A]Abbreviations: MD = Marek's disease, LL = lymphoid leukosis, RE = reticuloendotheliosis, LPD = lymphoproliferative disease.

[B]Two experimental syndromes are recognized: a bursal lymphoma with characteristics similar to LL, and a nonbursal lymphoma with characteristics similar to MD.

MAREK'S DISEASE

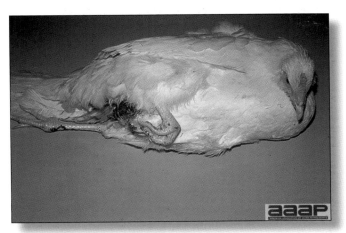

Fig. 1
Typical partial paralysis associated with peripheral nerve involvement in Marek's Disease.

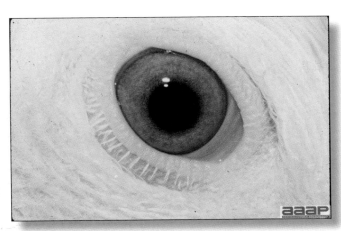

Fig. 2
Iris infiltration.

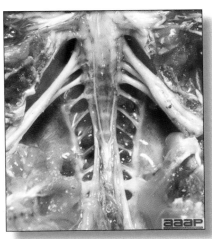

Fig. 3
Enlarged sciatic plexus on the right.

Fig. 4
Enlarged sciatic nerve at the bottom with swelling and loss of cross-striations compared to a normal sectioned sciatic nerve at the top.

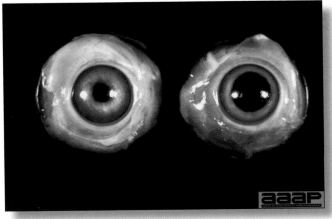

Fig. 5
Ocular lesions of MD. Note the normal eye (right) with a well-defined pupil and well-pigmented iris while the MD affected eye (left) has a discolored iris and irregular pupil as a result of mononuclear cell infiltration.

Fig. 6
Skin leukosis: enlarged feather follicles due to lymphoid infiltration.

MAREK'S DISEASE

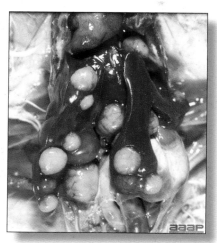

Fig. 7
Visceral lymphomas in the liver and the
heart occurring as focal, nodular growths
of varying sizes.

Fig. 8
Visceral lymphomas: diffuse hepatic and splenic
enlargements on the left vs normal liver and spleen
on the right.

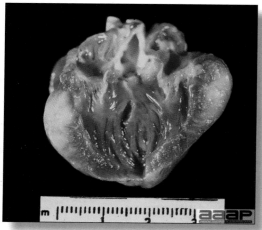

Fig. 9
Multiple lymphomas in heart.

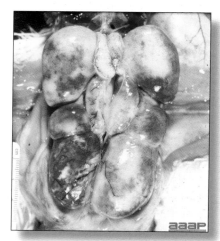

Fig. 10
Renal tumor.

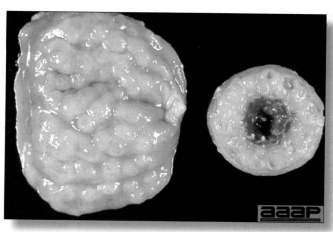

Fig. 11
Proventricular tumor.

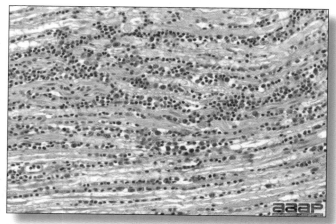

Fig. 12
Section of a nerve of a chicken with MD; infiltration of rapidly
proliferating pleomorphic lymphoid cells between the neurons.

AVIAN LEUKOSIS/SARCOMA VIRUSES

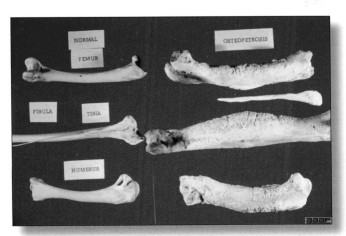

Fig. 1
Bones on the right are affected with osteopetrosis.

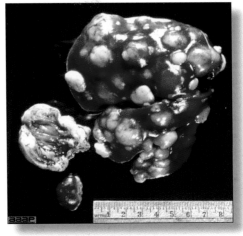

Fig. 2
Visceral tumors involving liver, heart and spleen.

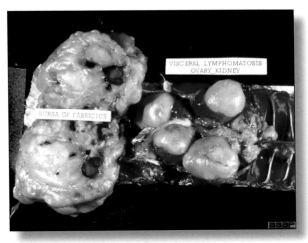

Fig. 3
Visceral tumors involving the bursa of Fabricius, kidneys and ovary.

Fig. 4
Myelocytomatosis.

Fig. 5
Hemangioma in a ALV-J positive broiler breeder.

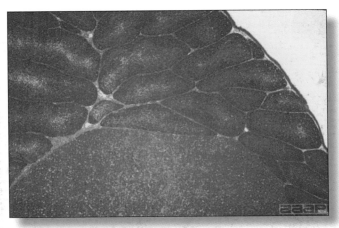

Fig. 6
Transformed LL bursal follicle at the bottom with normal appearing follicles in the remainder of the bursa of Fabricius.

RETICULOENDOTHELIOSIS (RE)

Fig. 1
Gross lymphoma in the liver.

CHICKEN INFECTIOUS ANEMIA
(CIA; Chicken Anemia Virus; Chicken Anemia Agent; Blue Wing Disease)

DEFINITION
Chicken infectious anemia (CIA) is a disease of young chickens characterized by aplastic anemia, generalized lymphoid atrophy, subcutaneous and intramuscular hemorrhage, and immunodepression.

OCCURRENCE
CIA is ubiquitous in all major chicken-producing countries in the world.

HISTORICAL INFORMATION
CIA virus (CIAV) was first isolated by Yuasa in Japan in 1979. It has also been called chicken anemia agent, chicken anemia virus, and parvovirus-like virus. The clinical signs and lesions previously described as blue wing disease, anemia-dermatitis syndrome, and hemorrhagic anemia may have been caused by CIAV.

ETIOLOGY
1. CIAV is classified into genus Gyrovirus of the family Circoviridae.
2. Viral particles are non-enveloped and are environmentally very resistant.
3. They have a diameter of approximately 25 nm and contain a single-stranded circular DNA genome.
4. Although viral isolates may differ at a molecular level antigenic or pathogenicity differences have not been reported.

EPIDEMIOLOGY
1. All ages are susceptible to infection but clinical disease is typically seen only during the first 2 to 4 weeks. However, age resistance may be delayed by simultaneous infection with infectious bursal disease virus.
2. The virus is spread both vertically and horizontally. The most important method of transmission is vertical from infected hens. Antibody-negative chicks are most susceptible to clinical disease. CIA virus also easily spreads via feces among birds in a population.

CLINICAL SIGNS
1. The only specific sign of CIAV infection is anemia characterized by hematocrit values ranging from 6 to 27% (normal hematocrit values are generally 29-35%) (Fig. 1).
2. Nonspecific clinical signs include depression, pale tissues, depressed weight gain, and secondary bacterial, mycotic, and viral infections.

3. Morbidity and mortality rates depend on various viral, host and environmental factors and concurrent infection with other agents. Uncomplicated CIA may only cause low mortality and poor performance. When complicated with other factors mortality can be 30% or even higher.
4. Early infections with CIAV can interfere with vaccination against Marek's disease or infectious bursal disease.

LESIONS
1. Marked thymic atrophy is the most consistent lesion (Fig. 2).
2. Fatty yellowish bone marrow, particularly in the femur, is characteristic (Fig. 3).
3. Bursal atrophy can also be seen in a small number of birds.
4. Hemorrhages in the mucosa of the proventriculus, subcutis, and muscles may also be observed (Fig. 4).
5. Secondary bacterial infections may occur and include gangrenous dermatitis or blue wing disease if the wings are affected (Fig. 5).
6. Histologically, there is marked thymic lymphoid depletion (Fig. 6) and marked atrophy of all cell lines in the bone marrow (Fig. 7). The bursal lymphoid follicles are mildly to severely depleted (Fig. 8) and spleen and other tissues with lymphoid aggregates are variably depleted. There may be histological evidence of secondary bacterial infections including gangrenous dermatitis.

DIAGNOSIS
1. A presumptive diagnosis is based upon clinical signs and gross lesions.
2. Isolation in cell cultures (MDCC-CU147 or MSB1) and identification of the virus from most tissues, buffy coat cells, and cloacal contents.
3. Serologic assays to detect antibodies such as the ELISA, virus neutralization test, and indirect immunofluorescence.
4. PCR is the test of choice for identification of CIA virus in cell cultures and chicken tissues.

CONTROL
Best prevention is by immunization of breeder flocks prior to the onset of egg production (between 13-15 weeks of age but no closer to egg production than 4 weeks).

TREATMENT
No treatment is available.

ZOONOTIC POTENTIAL
None reported.

CHICKEN INFECTIOUS ANEMIA

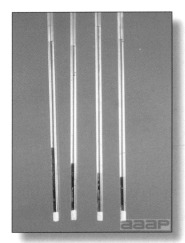

Fig. 1
Hematocrit tubes showing low Packed Cell Volume (15 to 22%) in 3 naturally infected chickens with CIA compared to a normal hematocrit (35%) on the left.

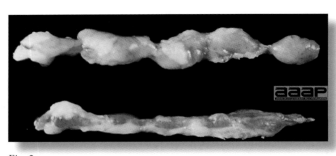

Fig. 2
Severely atrophied thymus in a chick affected with CIA compared to a normal thymus (top).

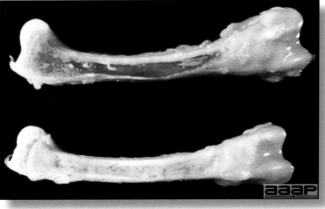

Fig. 3
Pale yellow, fatty bone marrow in the femur of a chick infected with CIA compared to a normal bone marrow (top).

Fig. 4
Petechiae and ecchymotic hemorrhages on the pectoral muscles in two chicks with hemorrhagic syndrome probably secondary to CIA.

Fig. 5
A 15-day-old chick with dermatitis, naturally infected with CIA.

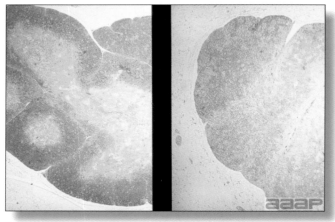

Fig. 6
Thymus: atrophy of lobules and loss of distinction between medulla and cortex with chick infected with CIA compared to normal on the left (H&E, 9,5X).

CHICKEN INFECTIOUS ANEMIA

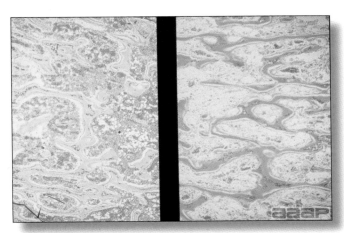

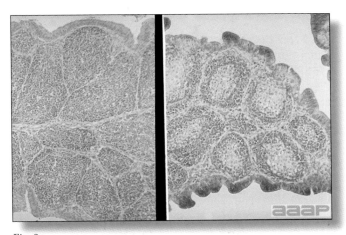

Fig. 7
Bone marrow: severe hypoplasia of both erythroid and myeloid cells and replacement by adipose tissue in a chick infected with CIA compared to normal on the left (H&E, 12,6X).

Fig. 8
Bursa of Fabricius: a plica showing lymphoid depletion and increased interstitium in a chick infected with CIA compared to normal on the left (H&E, 30,2X).

DUCK VIRUS ENTERITIS
(DVE; Duck Plague)

DEFINITION
Duck virus enteritis (DVE) is an acute herpesvirus disease of ducks, geese, and swans characterized by weakness, thirst, diarrhea, short duration, high mortality, and by lesions of the vascular, digestive, and lymphoid systems.

OCCURRENCE
1. Wild and domestic ducks, geese, and swans (order Anseriformes) are affected. All age groups and many varieties are susceptible; however, mostly adults are affected. The blue-winged teal is the most susceptible and the pintail duck is the least susceptible.

2. In the United States the disease has occurred in New York, Pennsylvania, Maryland, California, and South Dakota. The disease has been reported in Canada as well as the Netherlands, France, China, Belgium, and India. Because wild waterfowl are migratory, it seems likely that the disease may occur in other countries that have migratory waterfowl.

HISTORICAL INFORMATION
1. The disease was first observed in the Netherlands in 1923. Initially it was mistaken for avian influenza but in 1942 it was clearly differentiated from avian influenza and was termed duck plague. Subsequently the disease was identified in many other countries.

2. In 1967 DVE appeared in white Pekin ducks raised commercially on Long Island, New York. It also was identified in wild waterfowl. An effort to eradicate the disease in the domesticated white Pekin duck appeared to be successful.

3. Multiple outbreaks of DVE have been recognized in California and it is classified as a reportable disease in this state. In the spring of 1973 the disease appeared in congregated wild waterfowl in South Dakota and resulted in the death of approximately 48,000 waterfowl, mostly ducks.

4. DVE is now considered to be enzootic in North America. Prior to the 1973 outbreak, DVE was considered an exotic disease by the USDA. It is being watched with interest because of its ability to kill congregated, susceptible waterfowl.

ETIOLOGY
1. The etiologic agent is a herpesvirus. Although strains vary in pathogenicity, all appear to be identical immunologically.

2. The virus is nonhemagglutinating. This differs from the viruses causing Newcastle disease and avian influenza, which do hemagglutinate and which must be differentiated in diagnostic work.

3. The virus grows best on the chorioallantoic membrane of 9-14-day-old embryonated duck eggs or on duck embryo fibroblasts. Initially it does not grow in chicken eggs although it can be adapted to them. The virus also can be isolated in ducklings, with Muscovy ducklings being the most sensitive.

4. The virus produces intranuclear inclusion bodies in a variety of cells in the host.

EPIZOOTIOLOGY
1. The virus can be transmitted horizontally from infected to susceptible bird by direct contact or through contact with the contaminated environment (particularly water). Natural infection is limited to ducks, geese, and swans.

2. A carrier state for as long as 1 year has been demonstrated in wild ducks. And there is some evidence that carrier birds under stress shed virus intermittently, thus exposing susceptible birds.

3. Because viremia occurs in affected birds, arthropods feeding on those birds may transmit the disease. However, this method of transmission is unproven.

4. Vertical transmission has been reported experimentally.

CLINICAL SIGNS
1. In young commercial ducklings, signs appear 3-7 days after exposure. Ducklings have diarrhea, a blood-stained vent, dehydration, and a cyanotic bill. Death usually occurs in 1-5 days.

2. In domestic breeder ducks there is a marked drop in egg production (25-40%) a sudden, high persistent mortality. Sick birds show inappetence, weakness, ataxia, photophobia, adhered eyelids, nasal discharge, extreme thirst, prolapsed penis, and watery diarrhea. They soon become exhausted and unable to stand. They then maintain a position with their head down and drooping outstretched wings. Tremors may be apparent. Morbidity and mortality are usually high but vary from 5 to 100%. Most birds that develop clinical signs die. Wild waterfowl are said to have similar signs. They often conceal themselves and die in vegetation near the water.

LESIONS
1. Hemorrhages are present at many sites and there may be free blood in body cavities, gizzard, or intestine. Hemorrhages often occur on the liver, in the mucosa of the gastrointestinal tract (including the esophageal-proventricular junction), throughout the heart, in the pericardium and ovary. There may be edema in the cervical region.

2. There is severe enteritis. There may be elevated, crusty plaques in the esophagus, ceca, rectum, cloaca, or bursa of Fabricius. In young ducklings the esophageal mucosa may slough.

3. There is hemorrhage and/or necrosis in the annular bands or discs of lymphoid tissue along the intestine. The spleen is usually of normal or reduced size.

4. Initially the liver may be discolored and contain petechial hemorrhages. Later it may be bile-stained and contain scattered small, white foci as well as many hemorrhages.

5. Microscopically there may be intranuclear inclusion bodies in degenerating hepatocytes, epithelial cells of the digestive tract, and in reticuloendothelial cells.

DIAGNOSIS

1. Typical signs and lesions, along with epizootic losses, are highly suggestive of duck plague. The diagnosis can be strengthened if intranuclear inclusion bodies can be demonstrated or if the virus can be demonstrated in the tissues through fluorescent antibody tests.

2. The virus should be isolated and identified for confirmation. The virus will grow initially in embryonated duck eggs but not in chick embryos. Using known antisera to DVE, the virus can be identified by a neutralization test.

3. Retrospectively, it is possible to identify an outbreak of DVE if acute and convalescent sera are used to demonstrate an increasing antibody titer to duck plague virus. However the antibody response to natural infection can be low and transient.

4. DVE must be differentiated from duck viral hepatitis, pasteurellosis, Newcastle disease, avian influenza, coccidiosis, and other causes of enteritis.

CONTROL

1. Owners should prevent cohabitation or contact of their waterfowl with wild waterfowl. All appropriate quarantine and sanitary practices should be followed to prevent disease.

2. All suspected outbreaks should be reported immediately to state authorities. They, with federal authorities, will decide how an outbreak is to be handled. In commercially raised waterfowl, outbreaks were once controlled by combining slaughter with indemnification and by the application of quarantine measures. Presently, slaughter with indemnification has been discontinued and vaccination has been authorized on certain premises.

3. Both killed and live attenuated vaccines are available for prevention but approval by animal health authorities is required before they can be used. It has not been authorized for general use.

4. A monitoring system for detection of DVE has been established in the United States. Suspected outbreaks should be processed through official state or federal diagnostic laboratories.

TREATMENT
There is no effective treatment.

DUCK HEPATITIS
(DH, DHV, Duck viral hepatitis)

DEFINITION
Duck hepatitis (DH) is a peracute, rapidly spreading viral infection of young ducklings characterized by a short duration, high mortality, and by punctate or ecchymotic hemorrhages in the liver. Three different viruses are known to cause DH.

OCCURRENCE
Duck Hepatitis virus (DHV) type 1 occurs primarily in commercially raised Pekin ducklings and is seen almost exclusively in ducklings less than 5 weeks of age. Natural outbreaks have not been reported in other species. The disease is probably present in all major duck-raising areas of the world. DHV type 2 is seen exclusively in the United Kingdom and affects ducklings up to 6 weeks of age. The United States is the only country in which DHV type 3 has been observed. Ducklings up to 5 weeks of age are susceptible to DHV type 3.

HISTORICAL INFORMATION
A disease that probably was DHV type 1 first appeared in New York in 1945. A similar disease, called duck viral hepatitis, appeared on Long Island in 1949 and killed an estimated 750,000 ducklings. Subsequently, the disease was reported in many other states and from many countries throughout the world. In the United States the disease remains one of the major diseases of the duck-raising industry. DHV type 2 was first reported in DHV type 1-vaccinated ducklings in Great Britain in 1965. In 1969, DHV type 3 was reported to occur in DHV type 1-immune ducklings on Long Island.

ETIOLOGY
1. The etiologic agent of DHV type 1 is an enterovirus in the family Picornaviridae. It is chloroform resistant and does not hemagglutinate; features that help separate it from most other viral diseases of ducks. The virus is rather stable and difficult to eliminate from contaminated premises. Serologic variants of DHV type 1 have been reported.

2. DHV type 2 has been identified as an astrovirus. As with DHV type 1, the virus is fairly resistant. DHV type 3 is also caused by an astrovirus, but it is unrelated to DHV type 1.

3. DHV type 1 can be isolated from affected livers in embryonated chicken or duck eggs, 1-day-old ducklings, or duck embryo kidney or liver cell cultures. DHV type 2 is difficult to isolate, whereas DHV type 3 can be isolated on chorioallantoic membranes of 9-10-day-old duck embryos.

4. DHV stimulates a high degree of immunity in ducklings that survive infection and in inoculated adult ducks.

EPIZOOTIOLOGY
1. DHV type 1 is a highly contagious disease. The virus is excreted by recovered ducklings for up to 8 weeks after onset of infection. Susceptible ducklings can be infected by direct contact with infected ducklings or their contaminated environment. The virus can survive for 10 weeks in contaminated brooders and for 37 days in feces. DHV type 2 is transmitted via both the oral and cloacal routes. Survivors excrete virus for up to 1 week postinfection. DHV type 3 is similar to but less severe than DHV type 2.

2. Wild birds have been suspected of acting as mechanical carriers of virus over short distances. The viruses do not appear to be transmitted through the egg and there are no known vectors of the disease.

CLINICAL SIGNS
DHV type 1
1. The incubation period is very short, often around 24 hours in experimental infections, and morbidity is close to 100%. Onset of disease and spread within a flock are very rapid and most mortality occurs within 1 week of onset.

2. Affected ducklings at first are slow and lag behind the flock. Within a short time they squat with their eyes partially closed, fall on their side, kick spasmodically, and soon die. They often die in the opisthotonos position (Fig. 1). Death often occurs within 1 hour of the appearance of signs.

3. Mortality is age related and occurs as follows: ducklings less than 1 week old—up to 95% mortality; ducklings 1-3 weeks old—up to 50% mortality; ducklings over 4 weeks and older ducks—negligible mortality.

4. In older or partially immune ducklings, signs and losses may be so limited that the disease may go unrecognized.

DHV type 2 and 3
Affected ducklings die within 1-2 hours after the onset of clinical signs. Clinical signs usually appear within 1-4 days postinfection. Signs include convulsions and opisthotonos. Mortality ranges from 10 to 50% and nearly all birds with clinical signs die. DHV type 3 is similar to DHV type 1 but mortality is rarely over 30% and morbidity is higher.

LESIONS
1. The lesions observed with all three viruses are similar.

2. The cadaver may be in opisthotonos, the position in which many of the ducklings die.

3. The liver is swollen and contains punctate or diffuse hemorrhages (Fig. 2). The kidneys may be swollen and the spleen enlarged. Microscopically, there may be areas of hepatic necrosis, bile duct proliferation, and some degree of inflammatory response (Fig. 3).

DIAGNOSIS

1. The sudden onset, rapid spread, short course, and focal, hemorrhagic hepatitis in young ducklings suggest a diagnosis of DHV.

2. DHV type 1 can usually be isolated in embryonated chick or duck embryos or 1-day-old susceptible ducklings. Once the virus is isolated, it can be identified by serum neutralization using known hepatitis antiserum. Identification is also possible by inoculation of the virus into both susceptible and immune ducklings. DHV type 2 can be identified through electron microscopy on liver or blood. DHV type 3 cannot be isolated in chicken embryos and is difficult to reproduce in ducklings. The chorioallantoic membranes of duck embryos are the preferred route of inoculation. A direct fluorescent test on duckling liver has been reported.

3. The disease must be differentiated from duck viral enteritis, Newcastle disease, and avian influenza. In contrast to the viruses of DHV, the viruses causing those diseases are susceptible to chloroform; also the viruses of influenza and Newcastle disease hemagglutinate erythrocytes.

CONTROL

DHV type 1

1. In the initial stages of outbreak, all susceptible ducklings should be inoculated intramuscularly with duck hepatitis viral antiserum. One inoculation should be adequate if the antiserum is potent. A potent antiserum can be made from the blood of naturally or experimentally infected ducks. The blood can be collected at slaughter and the sera harvested. Antibodies for prophylactic use may also be obtained from the yolk of eggs produced by immune breeders, or from the eggs of chickens hyperimmunized with the virus.

2. Unexposed ducklings can be actively immunized using a chicken embryo-adapted apathogenic vaccine. However, young ducklings with parental immunity may not respond to vaccination.

3. Many duck breeders prefer to vaccinate their breeding stock at 3-4-month intervals to maintain a high antibody titer. Those birds then will transmit antibody through their eggs to the progeny. The progeny will usually be protected through the critical early weeks of life. Breeder birds should be vaccinated at least 2 weeks before their eggs are to be saved for hatching. Both live and inactivated vaccines are available.

DHV type 2 and 3

Vaccines for breeders are in the experimental stage only. Strict biosecurity procedures must be employed. Information concerning vaccines or antiserum can be obtained by contacting the Duck Research Laboratory, Eastport, Long Island, NY 11941.

TREATMENT

There is no treatment for the disease.

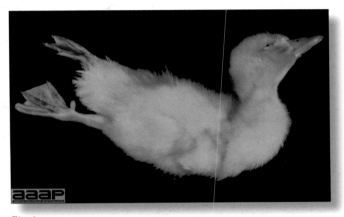

Fig. 1
Duckling affected with DVH, dead in the opisthotonos position.

Fig. 2
Enlarged liver with petechial to diffuse hemorrhages.

Fig. 3
Microscopic lesions in the liver of a duckling affected with DVH. Note the massive liver cell necrosis and hemorrhage.

FOWL POX
(Pox; Avian Pox)

DEFINITION

Fowl pox is a slow-spreading viral disease of chickens, turkeys, and many other birds characterized by cutaneous lesions on unfeathered skin of the head, neck, legs, and feet and/or by diphtheritic lesions in the mouth, upper digestive and respiratory tracts.

OCCURRENCE

Among poultry, pox occurs frequently in chickens and turkeys. Among other birds, pigeons, canaries, and psittacines are frequently infected and the disease is seen occasionally in many wild birds. Perhaps all birds are susceptible. The disease occurs in all age groups except the recently hatched and is worldwide in distribution.

HISTORICAL INFORMATION

Fowl pox is an ancient disease and in the distant past was mistakenly thought to be related to small pox and chicken pox of man. The characteristic pox inclusion bodies (Bollinger bodies) and the smaller elementary bodies within them (Borrel bodies) were studied by Drs. Bollinger and Borrel, respectively, in 1873 and 1904. In the United States, pox has been a common and frequently reported disease of poultry. In recent years there has been increased interest in pox in wildlife and caged birds, which are being submitted to diagnostic laboratories in increasing frequencies.

ETIOLOGY

1. Fowl pox is caused by a large DNA Avipoxvirus of the family Poxviridae. Many strains of virus are recognized and naturally infect the species given in their name. Some common examples are:

fowl poxvirus (type species)	quail poxvirus
turkey poxvirus	mynah poxvirus
pigeon poxvirus	psittacine poxvirus
canary poxvirus	

2. Poxviruses appear to be closely related, however, strong host specificity is found with most poxvirus strains. In some instances, exposure to one of the viruses in the group engenders immunity to that virus and one or more of the other viruses in the group. Poxvirus isolates from Hawaiian forest birds (alala and apapane poxvirus strains) are more related to each other than to fowl poxvirus indicating genetically distinct poxviruses in this region. Perhaps all strains are host-modified variants of what was once a single virus.

3. The various strains of avian poxvirus are morphologically identical. Strain classification has traditionally depended upon the cross-protection test in birds but these are not practical for routine diagnosis. Restriction endonuclease analysis of DNA has been successful in differentiating strains.

4. Recovery from poxvirus infection usually results in a strong, enduring immunity to later exposure to the same virus. Also, in turkeys and chickens vaccination is usually quite effective in preventing pox. Recently, however, several outbreaks of fowl pox have occurred in vaccinated chickens.

5. Virus is present in lesions and in desquamated scabs. Poxvirus is quite resistant to environmental factors and persists in the environment for many months.

6. Most poxviruses stimulate the formation of inclusion bodies in infected epithelium. Cytoplasmic inclusion bodies (Bollinger bodies) contain elementary bodies (Borrel bodies). Bollinger bodies are quite large and readily identified microscopically.

EPIDEMIOLOGY

1. The virus-containing crusts (scabs) formed on the skin are desquamated into the environment. Virus persists in the environment and may later infect susceptible birds by entering the skin through minor abrasions. Mechanical transmission via cannibalism is thought to play a significant role in some outbreaks. Respiratory tract infection can result from inhalation of aerosolized feathers and scabs containing virus.

2. Certain mosquitoes, and possibly other blood-sucking arthropods, can transmit virus from infected to susceptible birds. Mosquitoes remain infective for several weeks. Mosquito-transmitted outbreaks may result in rapid spread.

3. Poxvirus infection can result from mechanical transmission from toms to turkey hens via artificial insemination.

CLINICAL SIGNS

1. In poultry, onset often is gradual and the disease may go undetected until cutaneous lesions are numerous and obvious in the flock. The disease spreads slowly and severe outbreaks may last many weeks. Turkey pox infection is generally more chronic than fowl pox infection. Canaries can have systemic infection with high mortality. Signs vary somewhat with the two overlapping forms of pox:

   A. Cutaneous form

   This form predominates in most outbreaks. Birds often show few signs other than a mild to moderate reduction in rate of gain, a temporary loss in egg production, or a lack of flock vigor. Mortality is low if the disease is uncomplicated.

   B. Diphtheritic form

   Lesions in the upper respiratory or digestive tract may result in dyspnea or inappetence, respectively.

Lesions in the nasal cavity or conjunctiva lead to nasal or ocular discharge. Mortality is low to moderate and is often due to suffocation or starvation and dehydration.

LESIONS

1. Cutaneous lesions vary in appearance according to whether the papule, vesicle, pustule, or crust (scab) stage is observed. In most outbreaks the terminal reddish brown to black scab stage (Fig. 1) is present on at least some of the birds presented for diagnosis. Papules, the initial lesions, are light-colored nodules in the skin. Vesicles and pustules are raised, usually yellow. Occasionally, small papilloma-like lesions occur. Lesions usually occur on the unfeathered skin of the head (Fig. 2) and neck but may occur around the vent or on the feet or legs (Fig. 3). Cage birds and wild birds often have lesions on the feet or legs and these may appear as horny growths.

2. Diphtheritic lesions are raised, buff to yellow plaques on mucous membranes. They usually predominate in the mouth (Fig. 4) but may be present in the sinuses, nasal cavity, conjunctiva, pharynx, larynx, trachea (Fig. 5), or esophagus. Diphtheritic lesions often accompany cutaneous lesions but may occur alone in some birds.

3. Turkey pox (Fig. 6) has been observed in turkeys previously vaccinated with fowl pox vaccine. Occasional birds develop lesions on the conjunctiva, mouth, and upper digestive tract. Economic loss is often due to poor feed conversion.

4. Microscopically, epithelial hyperplasia (Fig. 7) with eosinophilic cytoplasmic inclusion bodies (Fig. 8) and surrounding inflammation are observed whether the lesion is cutaneous or diphteric.

DIAGNOSIS

1. Typical skin lesions are very suggestive of the disease. The diagnosis can be confirmed by demonstrating intracytoplasmic inclusion bodies in stained sections or in scrapings of the lesions.

2. Typical skin lesions can be reproduced in a susceptible bird of the same species. Ground lesion material should be inoculated into scarified skin or empty feather follicles and should produce a typical pox "take" at the application site in about 5-7 days.

3. Virus-containing lesion material will produce pocks on the dropped chorioallantoic membrane of embryonated chicken eggs. The lesions contain typical intracytoplasmic inclusion bodies.

4. Some poxvirus strains, particularly turkey pox, may not have demonstrable inclusion bodies in tissue sections. Electron microscopy may be helpful in these cases.

CONTROL

1. Pox can be prevented in chickens, turkeys, pigeons, canaries, and quail by vaccination. Vaccination is usually done when the birds are 4 weeks of age but can be done at any age if necessary. Pullets should be vaccinated 1-2 months before production begins.

2. Chickens and pigeons usually are vaccinated by the wing web-stick method. An applicator with two slotted needles is dipped in vaccine and thrust through the wing web. Turkeys may be vaccinated by the wing web route but lesions may be transferred to the head from the vaccination site. Vaccination by a drumstick-stab method when the birds are 2-3 months old is the recommended route. Turkeys retained as breeders should be revaccinated.

3. Pigeon pox vaccine is now widely used in chickens either alone or in combination with fowl pox vaccine. Chickens purchased as replacements for layers should be revaccinated if the initial vaccination occurs prior to 10 weeks of age. Pigeon pox vaccine can cause severe reactions in pigeons if not applied properly.

4. Turkeys are usually vaccinated with fowl pox vaccine. Turkey pox, quail pox, and canary pox vaccines are commercially available when circumstances indicate that these strains are the causative agents. Fowl and pigeon pox vaccines are not cross-protective with these strains. Fowl pox vaccine should not be used to vaccinate pigeons.

5. Vaccination produces a small lesion ("take") at the site of vaccination. A generous sample of the birds should be examined for vaccination lesions about 5-7 days after vaccination. Takes caused by turkey pox vaccine generally appear later (8-10 days after vaccination) than those caused by fowl pox. A large percent of those birds should have takes or revaccination is necessary.

6. Broilers are not vaccinated unless there is pox in the area. Broilers may be vaccinated with a mild tissue culture fowl pox vaccine administered subcutaneously at 1 day of age. This vaccine does not produce a visible take, but may result in a small number of birds that exhibit central nervous system (CNS) signs at 4-12 days postvaccination. *In ovo* injection of this vaccine may magnify the number of chicks exhibiting CNS reactions.

7. Control cannibalism with proper beak trimming and reduced environmental light intensity.

8. Fowl pox is currently being employed as a vector for recombinant vaccines.

TREATMENT

There is no satisfactory treatment for pox.

FOWL POX

Fig. 1
Terminal reddish brown to black scab stage of a fowl pox infection in a broiler breeder chicken.

Fig. 2
Pox lesions on the unfeathered skin of the head of a hen.

Fig. 3
Cutaneous pox lesion (foot) in an experimentally infected bird.

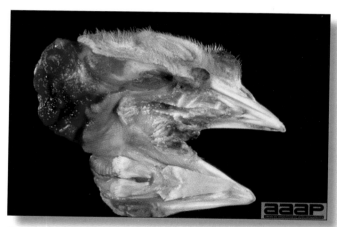

Fig. 4
Diphtheritic pox lesions in the mouth of a naturally infected chicken.

Fig. 5
Diphtheritic pox lesions in a naturally infected hen showing a tracheal plug.

Fig. 6
Young turkey affected with dry pox.

FOWL POX

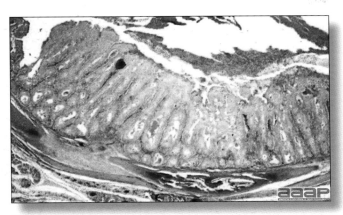

Fig. 7
Microscopic lesions of the trachea showing epithelial hyperplasia and inflammation. Note the presence of necrotic cells in the lumen.

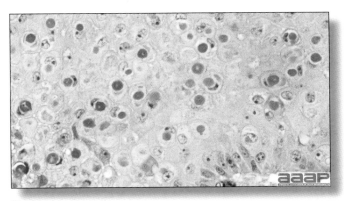

Fig. 8
Microscopic lesions showing ballooning degeneration and characteristic eosinophilic cytoplasmic inclusion bodies in infected cells.

HEPATITIS E VIRUS
(HEV or Hepatitis-Splenomegaly Syndrome in Chickens)

DEFINITION

Hepatitis E Virus causes a disease known as Hepatitis-splenomegaly (HS) syndrome in both layer and broiler-type chickens. It is characterized by increased mortality and decreased egg production. Dead birds have hemorrhagic livers, with clotted blood around the liver or abdominal cavity and splenomegaly. The disease has been seen in the USA, Australia, Canada, Europe, and China and is also probably present in other parts of the world.

OCCURRENCE

HS syndrome was first reported in western Canada in 1991, and since then has been recognized in the United States, Australia and Europe. The disease has been called by many names in the US and Canada; weeping liver disease, necrohemorrhagic hepatitis, necrotic hemorrhagic hepatosplenomegalic syndrome, chronic fulminating cholangiohepatitis and necrotic hemorrhagic hepatitis splenomegaly syndrome. In Australia, the disease is called Big Liver and Spleen (BLS) disease.

ETIOLOGY

HS is primarily caused by Hepatitis E virus (HEV) distantly related (58 to 61 % with the helicase gene) to human and swine Hepatitis E viruses. Hepatitis E virus is a spherical, non-enveloped, symmetrical virus of about 32-34 nm in diameter. It is a single-stranded, positive sense RNA virus that has been placed in a new family Hepeviridae and genus *Hepevirus*. There are genetic differences among various isolates of Avian HEV isolated from different geographic regions such as Australia, the USA and Europe. Avian HEV has also been isolated apparently from clinically normal chickens.

It has been determined that there is a 79 % nucleotide sequence (in the helicase gene) similarity between avian Hepatitis E viruses that cause HS and BLS. The syndrome is most common in laying hens between 30 and 72 weeks of age, with the highest incidence occurring between 40 and 50 weeks of age. Leghorn hens in cages are typically affected and on some farms HS frequently reoccurs. The disease is endemic in chicken flocks in the US. Serological studies in the US revealed that 71 % of the flocks and 30 % of chickens are positive for avian HEV antibodies. About 17 % of chickens less than 18 weeks of age and about 36 % of adult chickens are positive for avian HEV antibodies. Antibodies to BLS have also been demonstrated in chickens in the US.

EPIDEMIOLOGY

1. Transmission of Avian HEV appears to be by fecal-oral, but experimentally the disease has been reproduced by oronasal route of inoculation.

2. In the field, transmission occurs readily within and between chicken flocks.

3. Embryonic chicken eggs can be infected by the intravenous route.

CLINICAL SIGNS

1. Clinical signs due to HS are non specific and consist of anorexia, depression, pale combs and wattles and soiled vents.

2. Some birds can die suddenly without exhibiting any clinical signs.

3. The morbidity and mortality in the field can be low. Mortality can be 1 % per week lasting for 3 to 4 weeks.

4. Egg production drops are above normal, but can be significant in some affected flocks, as high as 20 %. In broiler chickens small eggs with thin and poorly pigmented shells can be observed.

LESIONS

1. Reported gross lesions include hemorrhages in the abdominal cavity and/or on the livers (Fig. 1), as well as red fluid within the abdominal cavity.

2. Livers can be enlarged, friable, and stippled with mottled white, red or tan foci. Subcapsular hematomas can be seen occasionally in the liver.

3. Spleens can be severely enlarged and mottled white (Fig. 2). Ovaries are often regressing.

4. Microscopically, liver lesions range from multifocal to extensive hepatic necrosis and hemorrhage (Fig. 3), with infiltration of mononuclear inflammatory cells around portal triads. Infiltration of lymphocytes in and around the blood vessels in the liver is a characteristic lesion of this syndrome. Microscopic lesions in the spleen include lymphoid depletion, hyperplasia of mononuclear phagocytic system cells and the accumulation of eosinophilic material in and around small arteries and in the interstitium of the vascular sinuses. Similar eosinophilic material can also be present in the interstitium of the liver (Fig. 4). This material is usually positive for amyloid using Congo red stain (Fig. 5).

DIAGNOSIS

1. A presumptive diagnosis can be made based on clinical signs, mortality patterns combined with gross and microscopic lesions. However, gross lesions of HS can appear similar to hemorrhagic fatty liver syndrome (HFLS) of chickens. With HS syndrome the livers tend not to be fatty both grossly and microscopically and amyloidosis is not seen with HFLS.

2. Virus can be isolated in chicken egg embryos by inoculation through intravenous route but it is not practical as this method is difficult and many embryos may die by this technique. Negative stain electron microscopy to detect 30 to 35 nm virus particles in the bile or the feces in chickens suffering from HS syndrome can also be used.

3. Immunohistochemistry (IHC) on tissues can also be used for diagnosis.

4. Serology by ELISA and AGID are other methods that can be used for diagnosis of HS.

5. Currently, the diagnosis of avian HEV is made primarily based on the detection of virus RNA by RT-PCR either in the feces or liver samples.

CONTROL AND TREATMENT

Biosecurity implementation may help limit the spread of the virus. Currently there is no treatment available to control HS. One study suggested that immunization of chickens with avian HEV recombinant ORF2 capsid protein with aluminum as adjuvant can induce protective immunity against avian HEV infection.

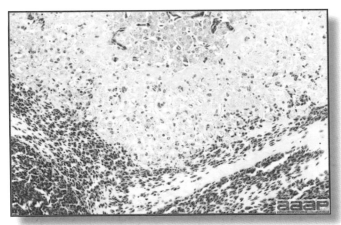

Fig. 3
Acute hemorrhage with architectural disruption of hepatocellular cords and hepatic sinusoids (H&E).

Fig. 1
Enlarged and hemorrhagic liver from a 63-week-old chicken with hepatitis E virus infection.

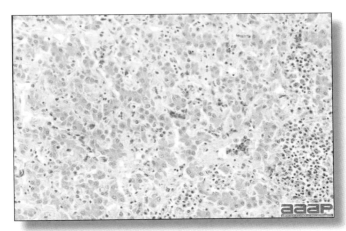

Fig. 4
Photomicrograph of a liver from a chicken infected with HEV showing accumulation of homogeneous eosinophilic material, amyloid, in the interstitium stained with H&E.

Fig. 2
Two enlarged and mottled white spleens from 56-week-old chickens with hepatitis E virus infection. the spleen on the left is of normal size.

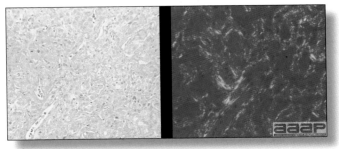

Fig. 5
Congo red stain positive shows orange colored amyloid (on the left) and apple green birefringence property of amyloid under polarizing filter (on the right).

INFECTIOUS BRONCHITIS
(IB)

DEFINITION

Infectious bronchitis (IB) is an acute, highly contagious viral disease of chickens characterized by respiratory signs (gasping, sneezing, coughing, and nasal discharge), severe renal disease associated with nephrotropic strains, and a marked decrease in egg production.

OCCURRENCE

1. IB occurs naturally only in chickens. All ages are susceptible, assuming they have not had prior exposure to the virus or are not passively immune.

2. The disease is present in all countries where chickens are raised in large numbers. In the United States the disease occurs frequently and throughout the year, even in vaccinated flocks.

HISTORICAL INFORMATION

1. In 1930, IB was first observed in young chicks. By the 1940's, IB was a serious disease of laying flocks causing marked loss in egg production. Nephropathogenic IB was first observed in the 1960s.

2. The virus was first isolated by Beach and Schalm in 1936 and multiple serotypes were first reported in 1956.

3. Vaccination became commercially available in the 1950's and is currently practiced worldwide.

ETIOLOGY

1. IB is caused by a coronavirus. The virus is fairly labile and can be destroyed by many common disinfectants.

2. Most IB virus (IBV) without enzyme treatment do not hemagglutinate erythrocytes as do Newcastle and influenza viruses.

3. There is considerable antigenic variation among IBV strains and many serotypes of the virus have been identified. Common serotypes (Connecticut, Massachusetts, Arkansas 99, DEO72 and GA98) are used in US vaccine preparation. There is little or no cross-protection among different serotypes.

4. Some IBV strains have a distinct predilection for renal tissue and these nephrotropic strains can induce significant mortality.

5. IBV has a high mutation rate, making diagnosis and control very difficult.

EPIZOOTIOLOGY

1. Transmission of IBV is by inhalation of virus-containing droplets expelled by infected chickens. Aerosol transmission apparently can occur over considerable distance. Spread of infection throughout a flock is explosively rapid.

2. The virus may persist on contaminated premises for approximately 4 weeks or longer under favorable conditions. Susceptible birds brought on the premises during that interval may contract the disease.

3. A few birds may periodically shed the virus for months after infection. Intermittent virus shedding can contaminate the environment and be a source of infection for susceptible chickens.

4. Vertical transmission has not been documented.

CLINICAL SIGNS

Baby chicks

1. Signs include coughing, sneezing, rales, and nasal and ocular discharge (Fig. 1). Morbidity is virtually 100%, although severity of signs varies. Signs can develop within 48 hours postinfection.

2. There is weakness, depression, and huddling near heat sources.

3. Mortality in young chicks is usually negligible unless the disease is complicated by other infectious agents. Nephrogenic strains may cause high mortality.

Laying chickens and broilers

1. Coughing, sneezing, and rales are common. Seldom is there nasal or ocular discharge.

2. Egg production drops markedly (up to 50%). Effects on production can last 6-8 weeks or longer. Eggs are often soft-shelled or misshapened (Fig. 2). Egg albumin may be watery. Low egg quality and shell irregularities may persist long after an outbreak of IB. Chickens that had IB or a severe reaction to IB vaccine when less than 2 weeks of age may suffer permanent damage to the oviduct resulting in poor-to-no egg-laying capacity.

3. Chickens that have IB or a severe reaction to IB vaccination may develop airsacculitis, due to an increased susceptibility to secondary infectious agents (especially *Escherichia coli* or *Mycoplasma gallisepticum*). This complication can be very severe and may accentuate respiratory signs, especially in young chickens.

4. Mortality associated with swollen pale kidneys and urolithiasis (Fig. 3) is induced by nephrotropic IBV strains in pullets and even in mature birds.

LESIONS

1. There is mild to moderate inflammation of the upper respiratory tract (Fig. 4). There may or may not be airsacculitis (Fig. 5). Severe airsacculitis is manifested as a marked thickening and opacity of the air sac membranes and often is accompanied by exudate in the air sacs. Airsacculitis can result in high mortality in young, growing birds, especially if husbandry is poor. Older birds are usually more resistant.

2. The kidneys sometimes are swollen and the ureters and tubules contain uric acid crystals, especially in young birds, including broilers.

3. Yolk material frequently is present throughout the peritoneal cavity and the ovarian follicles appear flaccid. These lesions are not specific for IB but accompany many acute diseases of layers.

4. In layers that had IB or a severe vaccination reaction while less than 2 week old, there may be abnormalities of the oviducts (particularly the middle third) in some birds. Oviducts may be hypoplastic or cystic and such birds may deposit yolks or fully formed eggs in the abdominal cavity and are referred to as internal layers.

5. Histologically, tracheitis is characterized by an edematous mucosa, cilia loss, rounding and sloughing of epithelial cells and presence of inflammatory cells (Fig. 6). Kidney lesions are those of an interstitial nephritis (Fig. 7).

DIAGNOSIS

1. Tests of paired acute and convalescent serum can be very useful in demonstrating a specific immune response. Several procedures including serum-virus neutralization (VN), enzyme-linked immunosorbent assay (ELISA) and modified hemagglutination inhibition (HI) are available, but only VN and to some extent HI tests (due to cross-reactions) are serotype specific.

2. For diagnosis it is necessary to isolate and identify the IBV type. Isolation is usually is done in 9-12-day-old embryonated eggs. Trachea, lungs, air sacs, and kidneys are good sources of virus. In infections beyond 1 week duration, cecal tonsils and cloacal swabs can sometimes be productive. The type of the virus can be determined by VN testing, HI tests, monoclonal antibodies, and RT-PCR and sequencing.

3. Nine to 12-day-old embryonated eggs inoculated with supernatant containing IBV develop lesions that are useful in diagnosis. The mesonephros of living embryos surviving 5-7 days postinoculation contains excessive urates. IBV causes dwarfing and stunting of some inoculated chick embryos. Also, the amnion and allantois are thickened and closely invest the embryo. After initial isolation it may be necessary to passage the virus three to five times to obtain embryo lesions. Similar embryo lesions can be observed with some lentogenic strains of Newcastle virus.

4. All most all IBV isolates do not hemagglutinate erythrocytes naturally, but will hemagglutinate if treated with neuraminidase. Newcastle virus and avian influenza virus can hemagglutinate erythrocytes without any prior treatment.

5. The fluorescent antibody technique or electron microscopy can be used on tracheal samples for rapid diagnosis of IB but do not differentiate the serotype.

CONTROL

1. Modified live IBV vaccines are used in young chickens for prevention. Vaccines are effective only if they contain the homologous serotypes in a given area. Typically a prime at one-day of age followed by a boost at around 2 weeks of age are given, particularly in birds with maternal immunity. Polyvalent bronchitis vaccines are typically used but can cause more severe vaccine reactions in naive chicks. IBV vaccine is often combined with Newcastle vaccine in the same vial but can cause interference with the Newcastle vaccine if not commercially prepared as a combination vaccine. Vaccines are generally applied via the drinking water or by spray. Utmost care needs to be taken to preserve the vaccine integrity as the vaccine virus can be prone to inactivation under adverse conditions.

2. Vaccinated birds should be watched carefully for possible onset of airsacculitis following vaccination. If signs or lesions of airsacculitis are detected, broad-spectrum antibiotics added to the feed or water will usually minimize the airsacculitis and reaction.

3. Killed virus vaccines (oil emulsion base) are now widely used. They are administered by injection (subcutaneous or intramuscular) to breeders or layer replacement pullets from 14 to 18 weeks of age. They induce high and sustained antibody levels.

TREATMENT

4. No effective treatment of IB is known although broad-spectrum antibiotics may control the complications. If there are no complications of IBV infection or vaccination, medication following vaccination or infection is not recommended.

5. For baby chicks with IBV, it may be helpful to increase the room temperature, encourage the birds to eat by using a warm moist mash, and correct any apparent management deficiencies.

Fig. 1
Chick with ocular discharge.

Fig. 2
Misshapened and softshelled eggs from IB positive broiler breeder hens.

Fig. 3
Kidney lesions associated with IB caused by a nephrogenic strain.

Fig. 4
Mild inflammation of the upper respiratory tract, note the foamy exudate in the laryngeal area.

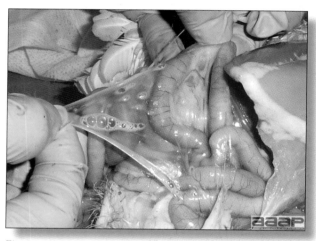

Fig. 5
Mild and acute airsacculitis in a broiler chicken. Note the foamy exudate in the abdominal airsac.

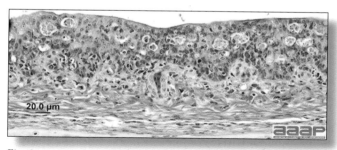

Fig. 6
Viral tracheitis characterized by cilia loss, mucous gland depletion, and mucosal epithelial degenerative changes including hyperplasia, and inflammatory cell infiltration.

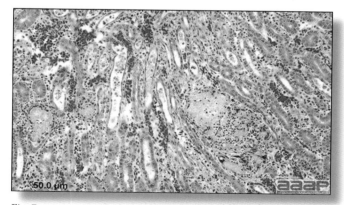

Fig. 7
Tubulointerstitial nephritis in a chicken infected with a nephrogenic IBV strain.

INFECTIOUS BURSAL DISEASE
(IBD; Gumboro Disease)

DEFINITION
Infectious bursal disease (IBD) is an acute, contagious, viral disease of young chickens characterized by inflammation followed by atrophy of the Bursa of Fabricius and variable degrees of immunosuppression.

OCCURRENCE
IBD occurs in all of the major poultry-producing countries of the world. Clinical signs are variable and are generally more severe in birds 3-6 weeks old. However, IBD may occur in chickens as long as they have a functional bursa of Fabricius (1-16-weeks of age). Birds infected at less than 3 weeks of age do not have clinical signs. However, destruction of the bursa results in immunosuppression. The younger the bird at the time of infection, the more severe the immunosuppression, resulting in a high degree of susceptibility to subsequent pathogens. Once a premise has been contaminated with IBD virus, the disease tends to recur, usually as a subclinical infection.

In turkeys, subclinical infection with IBD virus occurs without immunosuppression. However, there is no known disease associated with IBD viral infection. IBD viruses from turkeys are serologically distinct from those isolated from chickens. Ducks can also be subclinically infected with no apparent immunosuppression.

HISTORICAL INFORMATION
A disease caused by infectious bursal disease virus (IBDV) was first described in Gumboro, Delaware in 1962, hence the IBDV-related clinical problems and associated conditions are often referred to as "Gumboro disease". The immunosuppressive effects of IBD were first reported by Allan in 1972. For years IBD was successfully controlled with "classic" vaccines based on early IBD isolates. Variant strains of serotype 1 IBD were found in the Delmarva region in the 1980's. In the United States the disease is a persistent problem in the broiler industry despite vaccination. Very virulent strains of IBD have been reported in The Netherlands, Africa, Asia, South America and the United States.

ETIOLOGY
1. IBD is caused by a double-stranded RNA virus belonging to the genus Avibirnavirus of the family Birnaviridae. The viral genome has two double-stranded RNA segments. The virus may be propagated in chicken embryos or chicken embryo cell cultures. Two serotypes exist, with only serotype 1 being pathogenic.

2. The virus is very resistant to environmental factors and many disinfectants. It can persist for months in contaminated houses and for weeks in water, feed, and droppings. It can be transmitted by fomites. It has some susceptibility to formalin and iodide disinfectants. Invert soaps with 0.05% sodium hydroxide may kill IBD virus.

EPIZOOTIOLOGY
1. The virus spreads rapidly from infected chicks and from contaminated premises or fomites to susceptible chicks. The disease is highly contagious.

2. Vertical transmission of IBD virus has not been documented and there is no evidence for a carrier state.

3. The lesser meal worm (*Alphitobius diaperinus*) harbors the virus for weeks after an outbreak and may transmit it to susceptible birds. This worm lives in poultry litter.

4. The incubation period is very short with clinical signs evident 2-3 days post exposure.

5. Subclinical infection (before 3 weeks of age) is economically important due to suppression of humoral immunity and subsequent secondary infections. The virus is lymphocidal (immunoglobulin-bearing lymphocytes) and severely damages the bursa of Fabricius. The thymus, spleen, and cecal tonsils are also damaged but less severely.

6. It has been demonstrated that IBD can severely damage the humoral responsiveness of susceptible chicks when they are infected at less than 3 weeks of age. Those chicks then do not respond properly when vaccinated against other diseases. There is evidence that inclusion body hepatitis and gangrenous dermatitis occur frequently in such flocks. Some live vaccines may have a similar potential for damage as field infections.

7. The passive transfer of maternal antibodies to baby chicks is very important for the prevention of early infections with the virus. Breeder flocks must receive vaccines or field exposure to the virus followed by booster vaccinations to stimulate high levels of maternal antibody. Progeny from well-immunized breeder flocks may resist infection for 2-3 weeks. Passive immunity will interfere with vaccinations and it is necessary to vaccinate chickens after maternal immunity has fallen to a point that the vaccine will overcome the lower levels of maternal antibody.

CLINICAL SIGNS
1. Clinical disease is observed only in birds infected after 3 weeks of age. There is a sudden onset, particularly with the first outbreak. There may be tremor or unsteadiness. There is depression, anorexia, ruffled feathers, and a droopy appearance (Fig. 1) that resembles coccidiosis.

2. Diarrhea and dehydration are usually present. Occasionally there is voiding of blood and straining

during defecation. Vent picking is common and may be self-inflicted.

3. Morbidity is very high. Mortality is usually low although it can be substantial (approaching 30%) if husbandry is poor or if strains are particularly virulent. Mortality in a flock has usually peaked and receded within a week of onset. IBD tends to be more severe in leghorn strains than in broiler stock.

LESIONS

1. In the acute phase, the bursa is very enlarged with subserosal edema (Fig. 2) and mucosal to transmural petechial (Fig. 3) to ecchymotic hemorrhage. Caseous exudate may be found in the lumen of the bursa as a result of the extensive necrosis and inflammation of the bursal follicles during the acute phase of the disease (Fig. 4). The swelling recedes by the 5th day and the bursa atrophies rapidly until 8-10 days post infection (Fig. 5). There is increased mucus in the intestine.

2. Petechial and echymotic hemorrhages are common in thigh (Fig. 6) and pectoral muscles and, sometimes at the junction of the proventriculus and gizzard.

3. Kidneys may be swollen and the ureters may contain urates. The spleen can be slightly enlarged and contain small pale foci.

4. Necrotic lesions/atrophy may also be found in other lymphoid tissues such as the thymus, Harderian gland, cecal tonsils and Peyer's patches, particularly with highly virulent IBD strains.

5. Microscopically, in the bursa there is marked lymphoid follicle necrosis with heterophil rich cellular infiltrates, edema and hyperemia (Fig. 7) followed by atrophy and interfollicular fibroplasia (Fig. 8). Transient lymphoid necrosis occurs in the spleen, thymus, cecal tonsils and Harderian gland. The renal lesions are non-specific with tubular casts of protein and sometimes heterophils and are likely secondary to dehydration.

6. Some variant strains of the virus cause few clinical signs and minimal gross acute changes in the bursa. However, these variant strains may induce follicular lymphoid necrosis without the inflammatory component and rapid bursal atrophy and severe immunosuppression.

7. IBD infection results in immunosuppression, so birds are more susceptible to secondary infections such as gangrenous dermatitis, IBH, coccidiosis, etc. Historically, IBH was preceded by an immunosuppressive infection such as IBD but recently IBH has been recognized as a primary disease.

DIAGNOSIS

1. In an acute outbreak in susceptible chicks, the short course and bursal lesions are very suggestive of IBD. Signs and lesions can be less apparent in subsequent outbreaks and in chicks with parental antibody.

2. PCR, paired serologic testing with rising titers using the ELISA, agar-gel precipitin, or virus neutralization can be used to confirm the diagnosis.

3. PCR and various molecular typing assays, antigen-capture enzyme immunoassay with monoclonal antibodies and virus neutralization assay can also be used to differentiate among serotype 1 subtypes.

CONTROL

1. Vaccination of breeders to confer immunity to progeny is an effective method of reducing the disease in young chicks. Vaccination programs typically include "priming" with live vaccines and "boosting" with inactivated oil-emulsion vaccines to produce high and long-lasting levels of antibody in breeders.

2. Chicks can be vaccinated against the disease but timing the vaccination in maternally immune chicks can be difficult. When maternal antibodies wane, use of "hot" vaccines in nonimmune chicks may result in bursal atrophy. Vaccination with milder vaccines will not be effective in birds with high levels of maternal antibody. Therefore, knowledge of passive antibody levels and correct timing are necessary for successful vaccination.

3. An *in ovo* immune complex vaccine is available that results in decreased vaccine pathogenicity without loss of immunogenicity.

4. Sanitation programs are rarely successful due to the highly resistant nature of the virus.

TREATMENT
Good husbandry may reduce the severity of the disease.

ZOONOTIC POTENTIAL
None reported.

INFECTIOUS BURSAL DISEASE

Fig. 1
Broiler chicken showing depression, ruffled feathers, and a droopy appearance.

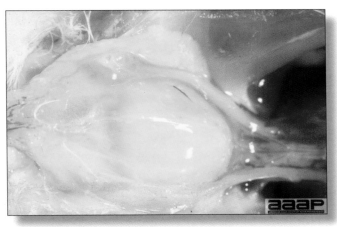

Fig. 2
Swollen and edematous bursa. By the 2nd or 3rd day postinfection, the bursa is covered with a gelatinous yellowish transudate.

Fig. 3
Enlarged bursa with mucosal petechiation.

Fig. 4
Caseous exudate in the lumen of the bursa as a result of the extensive necrosis and inflammation of the bursal follicles during the acute phase of the disease.

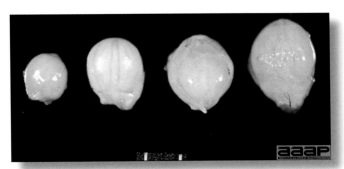

Fig. 5
Bursas of Fabricius showing edema (on the right) and atrophy (on the left). The middle bursas are showing some degrees of subserosal edema.

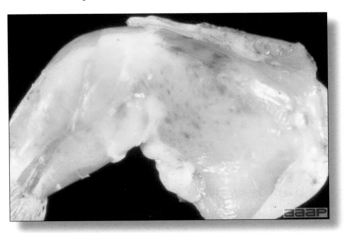

Fig. 6
Petechial and echymotic hemorrhages in thigh muscles.

INFECTIOUS BURSAL DISEASE

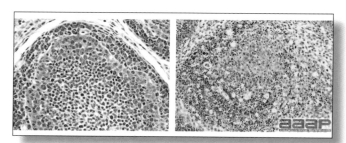

Fig. 7
Microscopically there is marked lymphoid follicle necrosis with heterophilic infiltration, edema and hyperemia of bursa (on the right). The bursa on the left is normal.

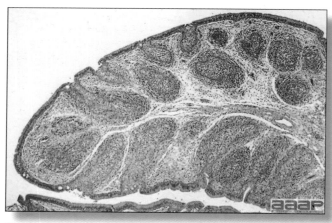

Fig. 8
Microscopic lesions showing atrophy and interfollicular fibroplasia of the bursa.

INFECTIOUS LARYNGOTRACHEITIS

DEFINITION
Infectious laryngotracheitis (ILT) is an acute viral disease of chickens, and, rarely, pheasants and peafowl characterized by marked dyspnea, coughing, gasping, and expectoration of bloody exudate.

OCCURRENCE
ILT is worldwide in distribution. Most outbreaks in chickens occur in broilers more than 4 weeks of age or in mature or nearly mature chickens, although all age groups are susceptible.

ETIOLOGY
ILT is caused by a DNA virus belonging to the genus *Iltovirus* of the family Herpesviridae. The virus is sensitive to many disinfectants and is not highly resistant outside of the host. There appears to be only one immunologic strain, although strains vary considerably in pathogenicity.

EPIZOOTIOLOGY
Some recovered chickens and vaccinated chickens become carriers and will shed virus for long periods of time or much later can shed virus following stress-induced reactivation of latent infections, thus exposing other susceptible birds. Mechanical transmission of virus via fomites also is possible. The disease spreads horizontally via the respiratory tract after it has been introduced. However, spread is often less rapid than with other viral respiratory diseases of chickens. There is no evidence of vertical transmission.

CLINICAL SIGNS

Signs of markedly pathogenic ILT
5. There is marked dyspnea (Fig. 1), often with loud gasping sounds and coughing. Severely affected chickens often raise and extend their head and neck during inspiration (Fig. 2) and make loud wheezing sounds.

6. Expectoration of bloody mucus may occur as a consequence of coughing and head shaking. Beaks, faces, or feathers of occasional birds may be bloody.

7. High morbidity and considerable mortality are common. Morbidity as high as 50-70% has been reported but mortality usually is in the 10-20% range. There is also lowered egg production. The disease often persists for as long as 2-4 weeks in the flock, a course longer than that of most viral respiratory diseases of chickens.

Signs of ILT of low pathogenicity
1. Signs often include hemorrhagic conjunctivitis with watery eyes, lacrimation, persistent nasal discharge, swollen infraorbital sinuses, generalized unthriftiness and lowered egg production.

LESIONS
1. Infected birds often have blood exuding from the nostrils, mouth or staining the feathers of the face and neck. With milder strains of ILT virus, only swollen eyelids (Fig. 3) and ocular and nasal discharge may be seen.

2. Lesions are most common in the nasal turbinates, sinuses, conjunctiva, larynx and trachea. Lesions are variable and depend upon the virulence of the virus. There may only be edema and congestion of the conjunctiva, nasal and sinus epithelium and reddening of the tracheal mucosa with serous or mucoid exudates. With more pathogenic strains of ILT virus, the tracheal mucosa will be congested, hemorrhagic and roughened with mucoid or mucohemorrhagic or fibrinonecrotic exudates (Fig. 4) sometimes with the formation of tracheal casts (Fig. 5) that can result in tracheal occlusion and death from suffocation. Inflammation may extend into the primary bronchi and airsacs.

3. Microscopical lesions are those of epithelial erosion and ulceration of the conjunctiva, nasal turbinates and sinuses, larynx, trachea and primary bronchi with formation of multinucleated syncytial cells containing the characteristic eosinophilic intranuclear herpesvirus inclusions (Fig. 6). Luminal exudate includes mucus, proteinaceous fluid, heterophils, macrophages, red blood cells, sloughed syncytia of epithelial cells bearing intranuclear herpesvirus inclusions (Fig. 7). The lamina propria is generally inflamed, congested and sometimes focally hemorrhagic.

DIAGNOSIS
The signs and lesions of the pathogenic type of ILT are distinctive enough to incite suspicion of ILT. However, there may be few signs and lesions with ILT of low pathogenicity. ILT can usually be confirmed by one or more of the following steps:

1. Demonstration of the characteristic multinucleated epithelial syncytial cells bearing eosinophilic intranuclear herpesvirus inclusions in tissues including conjunctiva, nasal turbinates, larynx and trachea histologically during early stages of the disease.

2. Demonstration of viral antigen or DNA in clinical samples, usually tracheal epithelium, by the use of the fluorescent antibody, immunoperoxidase, electron microscopy, DNA hybridization techniques, antigen capture ELISA and PCR.

3. Growth of the virus on the chorioallantoic membrane of embryonating chicken eggs. Typical plaques are produced and inclusion bodies can be demonstrated in them by histologic means and by the fluorescent antibody technique.

4. Exposure of known-immune and known-susceptible chickens to virus.

CONTROL

1. Avoid adding vaccinated, recovered, or exposed birds to a susceptible flock because these birds may include recovered carrier birds with latent infections. Better yet, raise susceptible flocks in strict quarantine and never add birds of any kind.

2. Premises contaminated with laryngotracheitis virus should be depopulated, cleaned, disinfected, and left vacant for 4-6 weeks before being used again. Due to the heat-labile nature of the virus, virus destruction is enhanced by heating the contaminated poultry house (38°C for 72 hours).

3. In areas where ILT is endemic, vaccination of layers is frequently practiced and is quite effective. Attenuated vaccines are available and can be administered by eyedrop, in the drinking water, or by aerosol spray. Drinking water vaccination may not be reliable because it depends upon the vaccine contacting nasal epithelium with high virus titers. Birds vaccinated prior to 10 weeks of age should be revaccinated at 10 weeks of age or older to confer lifelong immunity. Vaccination of broilers, when indicated, should be done before 4 weeks of age to minimize losses from severe vaccine reaction. Do not mix ILT vaccines with other vaccines.

4. Rarely, clinical ILT, indistinguishable from the natural disease, may occur 1-4 weeks after vaccination. These vaccine-related episodes are usually characterized by low morbidity and mortality.

5. ILT is a reportable disease in some states and provinces.

TREATMENT

There is no effective treatment. However, vaccination of unaffected birds and those in other houses on an infected farm may provide protection and stop the outbreak.

ZOONOTIC POTENTIAL

None reported.

Fig. 1
Broiler chicken with marked dyspnea.

Fig. 2
Severely affected chicken raising and extending its head and neck during inspiration.

Fig. 3
Chicken with swollen eyelids.

INFECTIOUS LARYNGOTRACHEITIS

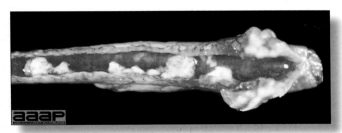

Fig. 4
Hemorrhagic tracheal mucosa with fibrinonecrotic exudate.

Fig. 5
Formation of tracheal casts that resulted in tracheal occlusion and death from suffocation.

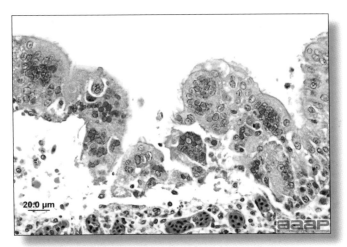

Fig. 6
Trachea with formation of multinucleated syncytial cells containing the characteristic eosinophilic intranuclear herpesvirus inclusions.

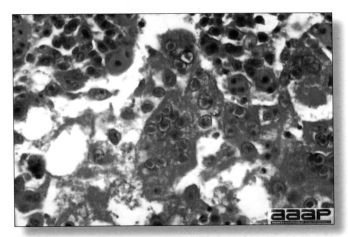

Fig. 7
Micrograph of the tracheal lumen debris from a case of laryngotracheitis. Note the numerous intranuclear inclusion bodies in sloughed epithelial cells and synctia formation (40X).

NEWCASTLE DISEASE

DEFINITION

Newcastle disease (ND) is a viral disease of many kinds of poultry, wild and cage birds characterized by marked variation in morbidity, mortality, signs and lesions. For the purposes of policies, control measures and international trade, ND is currently defined as: "an infection of birds caused by a virus of avian paramyxovirus 1 (APMV-1) that meets one of the following criteria for virulence:

a. The virus has an intracerebral pathogenicity index (ICPI) in day-old chicks (Gallus gallus) of 0.7 or greater, or

b. Multiple basic amino acids have been demonstrated in the virus at the C-terminus of the F2 protein and phenylalanine at residue 117, which is the N-terminus of the F1 protein."

OCCURRENCE

The disease usually occurs in chickens or (less often) in turkeys, although most poultry and many wild and cage birds are susceptible. All age groups are susceptible. The disease occurs in all poultry-raising countries. Natural or experimental infection has been demonstrated in at least 241 bird species.

HISTORICAL INFORMATION

1. ND first appeared in Java, Indonesia and Newcastle-upon-Tyne, England in 1926. Within 10 years it had spread to many countries throughout the world. The disease persists in many countries and in its velogenic form is one of the most devastating diseases of poultry, with mortality up to 100% in chickens.

2. APMV-1 strains of low to moderate virulence have been present in the United States since about 1940. In chickens these forms are well controlled by often-repeated vaccinations. These forms have seldom been a major problem except in chickens.

3. Velogenic ND, the most pathogenic type, occurred in the United States in 1941, 1946, and 1951 but was quickly eradicated. Extensive outbreaks began in California (and other locations) in 1971 and 2002 and were eradicated at great expense ($52M and $170M, respectively).

4. It has become apparent that velogenic ND is usually introduced by imported cage birds or fighting cocks, in many instances by illegally introduced birds.

ETIOLOGY

The causative agent of ND is APMV-1, a single-stranded RNA virus belonging to the genus *Avulavirus* of the family Paramyxoviridae. The many known strains of APMV-1 vary greatly in pathogenicity and are often referred to as:

a. Lentogenic—these are mildly pathogenic (examples: B-1, F, LaSota).

b. Mesogenic—these are moderately pathogenic.

c. Velogenic—these are markedly pathogenic (examples: Milano, Herts, Texas GB).

Most enzootic APMV-1 strains are lentogenic or mesogenic. Vaccines prepared from lentogenic strains tend to produce immunity that is weak and of short duration so that frequent revaccination is necessary to maintain immunity. Conversely, more pathogenic (mesogenic) strains used for vaccine tend to produce longer, stronger immunity but may produce mortality in unthrifty chickens and are not used in the United States. All velogenic strains are classified as select agents in the US.

EPIZOOTIOLOGY

1. Virus-containing excretions from infected birds, including aerosols and feces, can contaminate feed, water, footwear, clothing, tools, equipment, and the environment. Exposure of susceptible birds to any of these sources of virus can result in transmission via inhalation or ingestion. Also, infected poultry may spread the virus if their tissues are used without proper processing in rendered products.

2. Eggs laid by infected hens may contain virus. Such eggs seldom hatch and few are laid due to cessation of production caused by NDV infection. If they are accidentally broken in the hatcher, the entire hatch of chicks may be exposed. The exposed, apparently normal chicks may then be divided into small lots of birds and widely disseminated before the disease becomes apparent.

3. Live-virus vaccines may constitute a reservoir of APMV-1. Chickens often shed the vaccine virus. There is no evidence that attenuated viruses regain their virulence through passage.

4. At various times APMV-1 has been isolated from sparrows, pigeons, doves, crows, owls, and waterfowl. Recent experience suggests these birds do not play a significant role in the spread of ND.

CLINICAL SIGNS

Based on clinical signs in infected chickens strains of APMV-1 have been grouped into five pathotypes:

1. Asymptomatic enteric: subclinical enteric infection;

2. Lentogenic or respiratory: mild respiratory infection;

3. Mesogenic: respiratory signs and occasional nervous signs with low mortality;

4. Neurotropic velogenic: respiratory and nervous signs with high mortality;

5. Viscerotropic velogenic: hemorrhagic intestinal lesions are frequently seen with high mortality.

Adult chickens - Lentogenic APMV-1 infection
A. May not produce any clinical signs at all, or may produce mild respiratory signs and decreased egg production in laying flocks. A few eggs may be soft-shelled, roughened, or deformed.

Adult chickens - Mesogenic APMV-1 infection
A. Sudden onset with mild depression and anorexia. Respiratory signs usually occur but may be in a mild or inapparent form. Mortality is low or absent.
B. Signs suggestive of CNS disease may occur in a few birds but often do not appear.
C. In layers, production almost completely ceases within a few days. Eggs laid are of low quality and may be soft-shelled, roughened, or deformed (Fig. 1). Production is resumed slowly, or not at all, depending on the stage of lay at the time of infection.

Adult chickens - Velogenic APMV-1 infection
A. Signs vary according to tropism of the virus. Dyspnea often is marked. There is violent diarrhea, conjunctivitis, paralysis, and death in 2-3 days in many chickens. There may be swelling and darkening of tissues about the eyes with sticky ocular and nasal discharge. Some birds that survive a few days may exhibit signs of CNS involvement (e.g., tremors, twisting of the head and neck, circling, paresis, paralysis, terminal clonic spasms). Morbidity and mortality are high—up to 100%.

Young chicks - Lentogenic APMV-1 infection
A. Broilers may show sudden onset of respiratory signs including gasping (Fig. 2), sneezing, coughing, rales, and nasal and lacrimal discharge. Some birds may have swollen heads. Even mild B1 strain vaccines may cause these signs in broilers with low immunity.

Young chicks - Mesogenic APMV-1 infection
A. Sudden onset with marked depression and prostration. Marked respiratory signs that include gasping, coughing, hoarse chirping, and nasal discharge.
B. Signs of CNS disease may accompany or closely follow the onset of respiratory signs. Abnormal positions of the head and neck ("star gazers") are common. Usually only a modest number (0-25%) show CNS signs.
C. Eventually there is paralysis, prostration, trampling by pen-mates, and death. Mortality can be very high (up to 50%) regardless of whether CNS signs occur.

Young chicks - Velogenic APMV-1 infection
A. Signs are similar to those induced with mesogenic strains in young birds but mortality is very high (50-100%) and the course is more acute.
B. In wild and cage birds, APMV-1 infection often is inapparent. Signs, when apparent, are variable but often include gasping respiration, diarrhea, and later signs of CNS involvement. Sudden deaths are often the first indication of ND.

LESIONS
Lentogenic and mesogenic APMV-1 infection
A. Usually gross lesions are minimal in young or old birds although there may be mild conjunctivitis, rhinitis, tracheitis (Fig. 3) and airsacculitis that can be complicated with secondary bacterial infections such as *Escherichia coli* resulting in colisepticemia, pneumonia and airsacculitis. With mesogenic strains, a drop in egg production may be reported.
B. Microscopically, there is deciliation, necrosis, attenuation of the respiratory epithelium with infiltration of the lamina propria by a mixture of mononuclear cells with fewer heterophils. Luminal exudate is composed of exfoliating epithelial cells, some mucus and a mixture of inflammatory cells,

Velogenic APMV-1 infection
A. Ocular and respiratory lesions include conjunctival hemorrhage (Fig. 4), edema, hemorrhage, congestion and necrosis of the tracheal epithelium (Fig. 5) with paratracheal edema most often observed near the thoracic inlet and inflammation of the air sacs, with catarrhal or fibrinoheterophilic exudates if complicated with a secondary bacterial infection such as *E. coli*.
B. Facial edema (Fig. 6) with hemorrhage and epidermal necrosis and congestion and petechial hemorrhage of the comb and wattles has been described.
C. Focal to locally extensive hemorrhage and/or necrosis in the oral-pharyngeal and esophageal mucosa can be prominent (Fig. 7).
D. Hemorrhages occasionally occur in the mucosal surface of the proventriculus (Fig. 8) or in the gizzard, Peyer's patches, of the small (Fig. 9) and large intestine and the cecal tonsils (Fig. 10). There can be focal necrosis of the spleen. Mature birds in production can have egg yolk in the abdominal cavity and ovarian follicles are regressing with hemorrhagic stigmata.
E. With peracute mortality, gross lesions can be minimal.
F. Microscopic lesions in the CNS include nonsuppurative encephalomyelitis with

neuronal degeneration, perivascular cuffing with lymphocytic cells, multifocal gliosis and endothelial hypertrophy. Vascular lesions include medial degeneration, microvascular hyalinization and thrombosis, endothelial necrosis with widespread congestion, hemorrhage and edema. Primary and secondary lymphoid organs show marked lymphocyte necrosis sometimes associated with hemorrhage. In the respiratory system, there is widespread decilation of the tracheal epithelium with mucosal congestion, edema and hemorrhage and marked lymphohistiocytic cellular infiltrates in the lamina propria. Multifocal lymphoid infiltrates in the pancreas has been reported.

DIAGNOSIS

Clinical diagnosis based on history, signs, and lesions may establish a strong index of suspicion once ND has been positively identified in an area, but laboratory confirmation should always be pursued in order to identify the strain.

Laboratory diagnosis typically includes:

1. Virus isolation and subsequent pathogenicity testing,

2. RT-PCR – demonstration of viral RNA and subsequent typing,

3. Serology - demonstration of increasing titer of Newcastle antibody in the flock from onset to convalescence by ELISA or HI.

APMV-1 hemagglutinates the erythrocytes of many species, including those of many birds. Hemagglutination and hemagglutination inhibition tests are helpful in virus identification.

CONTROL

1. Chickens and turkeys can be immunized against ND by proper vaccination. The method of vaccine administration has considerable influence on the immune response. Low-virulence live-virus vaccines are administered by a variety of routes and schedules from hatching through grow-out. Killed-virus oil-emulsion vaccines are administered parentally as a final vaccine prior to the onset of egg production. Although proper vaccination protects the birds from serious clinical disease it does not prevent virus replication and shed, which could be a source of infection to other flocks.

2. Stringent laws are in effect pertaining to the importation of poultry, poultry products, and cage birds. However, enforcement is often a difficult and nebulous problem.

3. ND is a reportable disease. All suspected outbreaks of ND must be reported to animal health authorities immediately.

TREATMENT

No treatment available.

ZOONOTIC POTENTIAL

Humans who come in close contact with APMV-1 may develop a temporary, localized eye infection (conjunctivitis). There were suggestions that a more generalized infection resulting in chills, headaches and fever may sometimes occur, with or without conjunctivitis. Human-to-human spread has not been described.

NEWCASTLE DISEASE

Fig. 1
Soft-shelled, roughened and deformed eggs.

Fig. 2
Severe dyspnea and gasping in a young broiler chicken affected with Newcastle disease.

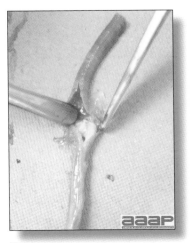

Fig. 3
Severe tracheitis in a broiler chicken affected with lentogenic Newcastle disease. Note the tracheal cast in the lumen.

Fig. 4
Conjunctival hemorrhage.

Fig. 5
Diphteritic laryngotracheitis.

Fig. 6
Facial edema in a young broiler.

NEWCASTLE DISEASE

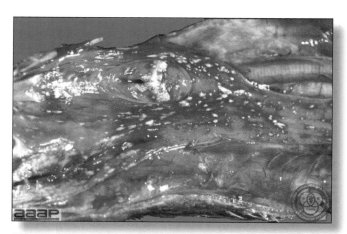

Fig. 7
Diphtheritic oro-pharyngo-esophagitis.

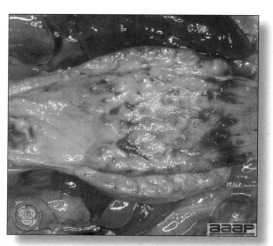

Fig. 8
Proventricular hemorrhages.

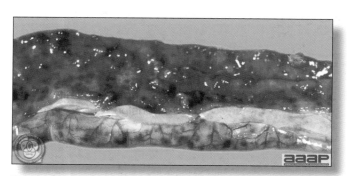

Fig. 9
Small intestine hemorrhage.

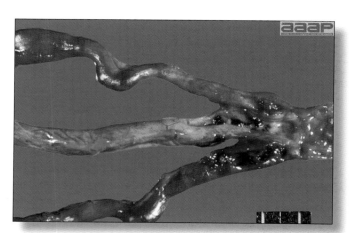

Fig. 10
Cecal tonsil necrosis.

VIRAL ARTHRITIS

DEFINITION
Avian reoviruses have been associated with several poultry disease conditions including enteric and respiratory syndromes, hepatitis and so-called stunting/malabsorption syndrome. However, a direct link between the virus and a disease has only been demonstrated for viral arthritis.

Viral arthritis is a reovirus infection primarily of meat-type chickens characterized by arthritis and tenosynovitis (primarily of the tarsus and metatarsus) and, occasionally, by rupture of the gastrocnemius tendon(s).

OCCURRENCE
Viral arthritis occurs primarily in meat-type chickens with rare reports of the disease in egg-type chickens and turkeys.

HISTORICAL INFORMATION
Viral arthritis was first reported in 1957. Since then there have been numerous reports on the disease and much has been learned about it and the virus that causes it. Viral arthritis is of special importance to the broiler industry because broilers frequently are infected.

Numerous turkey flocks in North Central United States have been experiencing turkey viral arthritis since 2009.

ETIOLOGY
Avian reoviruses are double-stranded RNA viruses belonging to the genus *Orthoreovirus* of the family Reoviridae. The reovirus is quite resistant to many environmental factors.

EPIZOOTIOLOGY
Reovirus is discharged in the feces of infected chickens and may contaminate eggshells. It is transmitted laterally to susceptible chickens. Egg transmission has been demonstrated. Reovirus is known to persist in infected birds for at least 289 days. Age-associated resistance has been described.

CLINICAL SIGNS
1. Lameness and swelling of the tendon sheaths of the shanks and of the gastrocnemius tendon above the hock are early signs (Fig. 1). The shanks of affected chickens are enlarged. Many infected birds are in good condition but some are unthrifty and stunted. Mortality usually is quite low.

2. If the gastrocnemius tendon has been ruptured, the affected foot cannot be extended and the bird cannot bear weight on the affected leg. If both tendons are ruptured, the bird is immobilized (Fig. 2). There usually is swelling and discoloration of the skin over the site of tendon rupture (Fig. 3).

LESIONS
1. In chickens, in the acute phase of the disease there is typically bilateral swelling and inflammation of the tendons and tendon sheaths just above the hock and along the posterior aspect of the shanks. The tendon sheaths are edematous and the hock joints may contain small amounts of yellow coloured or blood tinged exudates (Fig. 4). The articular cartilages of the hock may be eroded and the synovial membranes of the hock and tendons are thickened, edematous and may contain hemorrhages (Fig. 5).

2. Chickens that cannot extend the hock often have rupture of the gastrocnemius tendon, usually just above or over the hock joint with green discolouration of the skin and subcutaneous tissues as a result of the hemorrhage. This is especially true of older, heavier birds.

3. In chronic cases there is usually less inflammatory exudate and more fibrosis of affected tendons and tendon sheaths with fusion.

4. Turkeys with viral arthritis show lameness, swollen hock joints and ruptured tendon.

5. Microscopically, in experimental infections during the acute phase, there is thickening of the tendon sheaths, with synoviocyte proliferation, infiltration of lymphocytes and macrophages. The synovial spaces contain heterophils, macrophages, exfoliating synoviocytes and there is periostitis. In the chronic phase, there is synovial villous hyperplasia, formation of subsynovial lymphoid nodules, increase in fibrous connective tissue and marked infiltration of lymphocytes, macrophages and plasma cells with fibrosis of tendons and tendon sheaths leading to eventual fusion (Fig. 6). Myocardial lesions include multifocal infiltrates of heterophils, sometimes with populations of mononuclear cells between myofibres.

DIAGNOSIS
1. A tentative diagnosis often can be made on the basis of history and bilateral enlargement of the tendon sheaths of the shanks and histological confirmation of the inflammation of the tendon sheaths and tendons and myocardial infiltration of heterophils sometimes accompanied by mononuclear cells.

2. The presence of the virus can be confirmed by isolation of the virus in chicken liver and kidney cell cultures or chicken embryos, by RT-PCR and by direct fluorescent antibody test.

3. If acute and convalescent sera can be obtained it may be possible to demonstrate seroconversion reovirus using the agar-gel precipitin test or ELISA. Antibody may disappear as early as 4 weeks post-infection in some birds but it persists in birds with joint involvement.

4. It is essential to exclude other causes of lameness including mycoplasmosis (especially infectious synovitis), staphylococcal and other bacterial arthritides such as salmonellosis, and pasteurellosis and bone deformities, and certain nutritional diseases. It should be remembered that dual infections can exist.

CONTROL

1. Vaccination of breeder flocks with live and inactivated vaccines results in protection of 1- day-old chicks.

2. Due to the age-associated resistance to disease after 2 weeks, the ubiquitous distribution of the virus in poultry-raising areas, and the resistance of the virus to inactivation, prevention of this disease should be directed at preventing early infection.

TREATMENT
No treatment is available.

ZOONOTIC POTENTIAL
None reported.

VIRAL ARTHRITIS

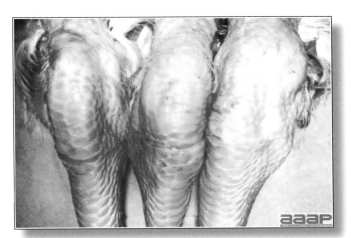

Fig. 1
The tendon sheaths of the shank and above the hock are markedly swollen.

Fig. 2
Chickens with clinical signs of viral arthritis "tendinitis"/"tenosynovitis" tend to sit and are reluctant to move.

Fig. 3
Swelling and discoloration of the skin over the site of tendon rupture.

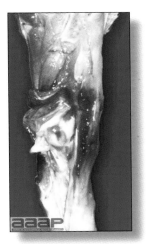

Fig. 4
Ruptured gastrocnemius tendon with hemorrhage.

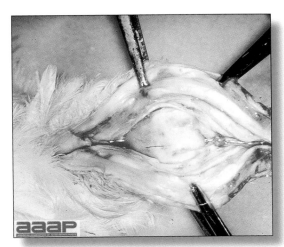

Fig. 5
Starting at about 42 days post-infection, erosions appear in the cartilage of the posterior and distal tibia.

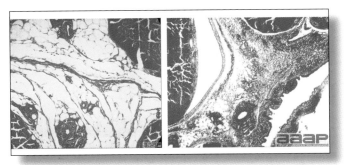

Fig. 6
On the right; infected tendon sheaths from the digital flexor tendons of the shank of a chicken with viral arthritis showing thickened synovium due to hyperplasia of synovial cells and connective tissue with extensive accumulation of lymphocytes, plasma cells and heterophils in the synovial cavity. Normal tendon sheath on the left (H&E, 30X).

TURKEY CORONAVIRUS ENTERITIS

DEFINITION
Enteritis caused by a turkey coronavirus (TCV) is an acute, highly infectious disease of turkeys, especially poults, characterized by anorexia, diarrhea, dehydration, and variable mortality.

OCCURRENCE
The disease occurs throughout the year in turkeys of all ages, but is seen more frequently in young turkeys. It is recognized in the United States, Canada, Australia, Brazil, Italy and United Kingdom.

HISTORICAL INFORMATION
Coronavirus enteritis was first reported in 1951 in Washington and shortly thereafter in Minnesota. Historically it was also called "blue comb" and "mud fever".

ETIOLOGY
1. Turkey coronavirus belongs to family Coronaviridae. Viral particles are enveloped, pleomorphic but often spherical and 80-160 nm in diameter. Genetic material of the virus is a single-stranded positive sense RNA.

2. Under experimental conditions the virus is readily destroyed in batteries and cages. Destruction of the virus is more difficult under natural conditions because it survives well in frozen feces.

3. A number of other enteric viruses including rotavirus, reovirus, astrovirus, enterovirus, and calicivirus have been identified from turkey feces. Their role in enteritis is still not fully understood.

EPIZOOTIOLOGY
The virus spreads by contact of susceptible birds with infected birds or their feces. Once introduced into a flock, the disease spreads rapidly among susceptible birds. The virus is shed in the feces of recovered birds for several weeks. Further, the virus persists in frozen feces for several months. There is no evidence that the virus is transmitted vertically.

CLINICAL SIGNS
1. In young poults the signs appear suddenly after an incubation period of 1-5 days. Signs include anorexia, depression, frothy diarrhea, subnormal temperatures, darkening of the head and skin, and loss of weight. Birds tend to huddle around heat sources. Spread is rapid and morbidity is close to 100%.

2. The signs seen in young poults may also be observed in laying turkeys but usually are less marked. Moreover, there is a sudden decrease in egg production and some eggshells are chalky.

3. Good husbandry and supplemental heat tend to suppress mortality. Mortality varies with the age of the birds affected and can range from 5 to 50% in natural infections.

4. The course of the disease in a flock is around 2 weeks. Recovery may be prolonged, particularly in males, and the flock may become uneven in size.

LESIONS
1. Marked dehydration and emaciation may be seen.

2. The small intestine is thin-walled and distended with foamy yellow fluid content (Fig. 1).

3. The bursa of Fabricius is often small.

4. Histologically, within the small intestine, there is prominent villus atrophy and sloughing of intestinal villi with cryptal hyperplasia and moderate lymphoid and heterophilic infiltration of the superficial and deep lamina propria.

5. There is necrosis of the superficial bursal epithelium with heterophils within and beneath the epithelium and moderate depletion of lymphoid follicles.

DIAGNOSIS
1. The history, signs, and lesions are suggestive of the diagnosis.

2. The virus can be isolated in embryonating eggs and can be identified by electron microscopy (EM). EM can also be used to detect virus particles in intestinal contents. Immunohistochemistry can detect TCV antigens (Fig. 2).

3. RT-PCR has been described as a sensitive and specific diagnostic test for detection of viral nucleic acid.

4. ELISAs have been used to detect antibodies against TCV.

5. In differential diagnosis one should consider the following diseases:

Young Poults (Less than 7 weeks)	Growing and Mature Turkeys
Salmonellosis	Erysipelas
Hexamitiasis	Trichomoniasis
Starve outs	Hemorrhagic enteritis
(very young birds only)	
Coccidiosis	

CONTROL
1. No licensed vaccine is available for immunization; therefore prevention is the preferred method of control.

2. Turkeys should be reared under the all-in, all-out system. Use quarantine and a high standard of sanitation to prevent introduction of the virus.

3. If the disease has been present in prior broods, thoroughly clean and disinfect the premises after

complete depopulation. Leave the premises empty for at least a month, perhaps longer during the winter season.

TREATMENT
No treatment is available.

ZOONOTIC POTENTIAL
None reported.

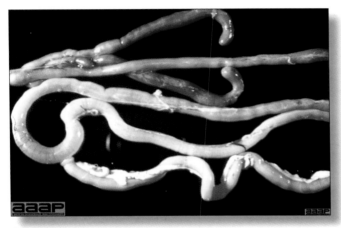

Fig. 1
The small intestine is thin-walled and distended with fluid content.

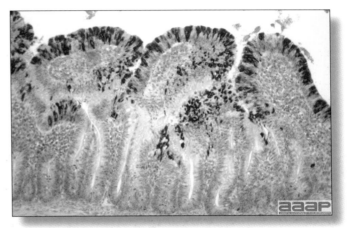

Fig. 2
Immunohistochemistry of the intestines.

TURKEY VIRAL HEPATITIS

DEFINITION
Turkey viral hepatitis is a contagious, often subclinical disease of turkey poults characterized by lesions in the liver and pancreas.

OCCURRENCE
The clinical disease is seen only in turkey poults less than 5 weeks old. Other birds and mammals appear not to be affected. The disease has been observed in most turkey-producing areas of the world.

HISTORICAL INFORMATION
Turkey viral hepatitis was first reported in 1959. Subsequently, the disease has been reported in a number of states and countries but has seldom been associated with severe mortality. Incidence and distribution is difficult to evaluate due to the subclinical nature of the disease.

ETIOLOGY
The etiologic agent is a virus that has been suggested to be a picornavirus based on its molecular, immunologic and morphologic features. It can be grown in the yolk sac of 5-7-day-old chicken or turkey embryos. Embryo mortality occurs in 4-10 days. The virus rarely achieves high titers despite repeated embryo passages. Young turkey poults can be infected following parenteral inoculation of infective tissue suspensions.

EPIZOOTIOLOGY
Up to 28 days after infection, the virus can be isolated consistently from feces and liver of infected poults. Transmission readily occurs by direct or indirect contact of susceptible poults. There is clinical evidence suggestive of egg transmission of virus.

CLINICAL SIGNS
The disease is usually subclinical. It may be that signs are apparent only if there are other concurrent diseases or stresses on the poults. Infected flocks in the early stages of the disease show variable depression with sporadic deaths of well-fleshed birds. Morbidity varies greatly. Mortality is usually very low (>5%) but has been as high as 25% during a 7-10 day period in occasional flocks.

LESIONS
1. In the liver, there are focal gray areas 1 mm or more in diameter that may coalesce (Fig. 1). Lesions are often slightly depressed and can be concealed by congestion and focal hemorrhages. Bile staining may be apparent.

2. In the pancreas, lesions are less consistent and appear as focal gray to pink areas (Fig. 2). They are usually more evident later in the course of the disease and on the dorsal side of the pancreas.

3. Microscopically, inclusion bodies have not been identified. The lesions observed grossly are found to be focal areas of necrosis, which subsequently become infiltrated with mixed inflammatory cells in which lymphocytes and reticular cells predominate; heterophils are present but not in high numbers. Syncytial cells arising from hepatocytes can often be found along lesion margins.

DIAGNOSIS
1. Typical lesions in both liver and pancreas are diagnostic. If turkey viral hepatitis is suspected, both liver and pancreas need to be submitted for histopathologic examination even if there are no visible lesions in the latter; often microscopic lesions can be found in the pancreas in the absence of gross lesions. If lesions are present only in the liver, they will have to be differentiated from those of blackhead and systemic bacterial infections.

2. Isolation and identification of the virus can be used for confirmation. An agar-gel precipitin test using rabbit antisera to the virus has been developed but is not in general use.

CONTROL
1. A high standard of sanitation along with proper nutrition and good husbandry should minimize the effects of the disease. There is no vaccine for prevention. Treatment of concurrent diseases is important.

2. Eggs from infected flocks should not be used for hatching because there may be transovarian transmission of virus.

TREATMENT
There is no proven effective treatment. Fortunately, most well cared for flocks recover in a few weeks.

TURKEY VIRAL HEPATITIS

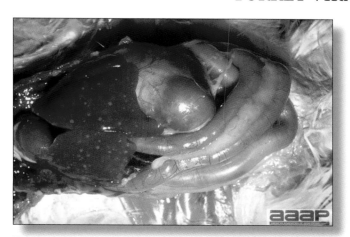

Fig. 1
Focal gray areas 1 mm or more in diameter in the liver.

Fig. 2
In the pancreas, focal gray to pink areas of necrosis.

BACTERIAL DISEASES

Revised by Richard M. Fulton, new sections on *Campylobacter* and *E. cecorum* by Martine Boulianne

AVIAN CHLAMYDOPHILOSIS/ AVIAN CHLAMYDIOSIS
(Psittacosis; Ornithosis)

DEFINITION
Avian chlamydophilosis is a reportable, acute or chronic infectious disease of poultry, many caged birds and wild and migratory birds. In clinically ill birds, the disease is characterized by systemic, pulmonary, or enteric signs and lesions. The latent, inapparent chlamydial infection has long been recognized as the predominant and most important state in the zoonotic relationship between chlamydial agent, birds and humans.

In the Psittacidae (parrots, parakeets, cockatoos, macaws, etc.) and humans, avian chlamydiosis is called psittacosis. Historically, avian chlamydiosis has been called ornithosis in other avian species.

OCCURRENCE
Chlamydophilosis occurs in many kinds and ages of birds. Most acute outbreaks are in young birds. Parrots, parakeets, cockatiels, and pigeons frequently are infected. In poultry, occasional outbreaks occur in turkeys. Chickens seldom are affected. Severe outbreaks occasionally have been reported in shorebirds and migratory birds. Important outbreaks of psittacosis occasionally occur in humans, usually following exposure in poultry processing plants. The presence of wild pigeons in some cities in North America, Europe, and Asia poses a major problem in efforts to control human chlamydiosis.

HISTORICAL INFORMATION
1. Psittacosis is a significant public health concern. The incidence in humans in the US appears to be on the decline since between 1987 through 1996 831 cases of psittacosis were reported to the United States Center for Disease Control while there has been fewer than 50 cases reported per year since 1996. In the United States most human cases were related to exposure of people to infected cage birds, especially parrots, or to sick turkeys in processing plants.

2. Some of the first recognized outbreaks of avian chlamydiosis were in pigeons. Important outbreaks soon were recognized in turkeys and ducks.

3. Interest in chlamydiosis waned somewhat with the introduction of antibiotics that control mortality in people with the infection.

4. During the last decade *Chlamydophila psittaci* outbreaks have occurred relatively infrequently among birds and humans. However, in 1974-1975 there were at least 11 outbreaks in turkey flocks, mostly in Texas. People were infected in at least seven of the outbreaks. More recently, outbreaks in ducks have been described. These outbreaks again have focused attention on the public health aspects of avian chlamydiosis.

5. During the past few years chlamydiosis has been recognized as a common and major problem in imported and domestic exotic birds.

ETIOLOGY
1. In birds the etiologic agent is *Chlamydophila psittaci* (*Chlamydia psittaci*). Chlamydia is closely related to rickettsia.

2. *C. psittaci* can be grown in chicken embryos, cell culture, mice, and guinea pigs. Chlamydia form intracytoplasmic inclusion bodies in many kinds of cells, including epithelial and macrophages, and inclusions can be demonstrated in stained smears (Fig. 1) and histologic sections. All chlamydia are highly susceptible to tetracyclines. They are obligate intracellular gram negative bacteria and cannot be grown in artificial media.

3. Isolates of *C. psittaci* vary greatly in pathogenicity. Concurrent infection, especially with *Salmonella*, sometimes enhances the pathogenicity of *C. psittaci*. Younger birds are more susceptible. Crowding or otherwise unfavorable environmental conditions and stress from shipping, racing, and handling contribute to the severity of the disease.

4. A common group-specific antigen is present in all chlamydia. Antibody to that antigen can be demonstrated in the sera of exposed or sick birds after an appropriate interval of time has elapsed.

EPIDEMIOLOGY
1. It is believed that wild birds that are carriers (and cage birds) transmit chlamydia to their nestlings and some surviving nestlings in turn become carriers. A delicate host-parasite relationship is established so that stressed carriers intermittently shed chlamydia in their secretions and excretions thus exposing other susceptible birds.

2. Chlamydiosis may become epizootic when large numbers of birds, including disseminating carriers, are in close contact. Transmission is primarily by inhalation of chlamydia in fecal dust but also can result from ingestion of *C. psittaci*.

3. It is suspected that wild birds may transmit chlamydia to poultry. Pigeons are strongly suspected of being important disseminators. Wild migratory birds such as gulls, egrets, and ducks are known to excrete chlamydia under certain conditions. Little is known of

possible transmission of chlamydia between infected mammals and poultry but the chlamydia found in mammals are believed to be distinct from those in birds.

CLINICAL SIGNS

1. Mild outbreaks of avian chlamydophilosis result in few signs and may go unrecognized. Alternatively, mild respiratory signs or diarrhea may be noted.

2. In more pathogenic outbreaks in turkeys there is depression, weakness, inappetence, and loss of weight and there may be nasal discharge and respiratory distress. Frequently there is marked, yellowish-green diarrhea. Similar signs occur in many other kinds of birds such as ducks, geese, and pigeons and may reflect systemic, pulmonary, or enteric involvement. A watery diarrhea is noted in ducks, geese, and pigeons. An unbalanced gait has been reported in ducks, geese, and pigeons.

3. In pigeons, conjunctivitis often occurs and should lead the diagnostician to suspect chlamydiosis. Other signs include depression, anorexia, diarrhea, or rales. The latter signs resemble those seen in many cage birds with psittacosis.

LESIONS

1. The basic lesions in chlamydophilosis are characterized by fibrinous response grossly with pleocellular infiltrate and necrosis histologically. These basic tissue responses may lead to pneumonia, airsacculitis, hepatitis, myo- and epicarditis, nephritis, peritonitis, and splenitis.

2. In turkeys the severity of lesions is in proportion to the pathogenicity of the strain of *C. psittaci*. In turkeys that succumb, there is wasting, vascular congestion, fibrinous pericarditis (Fig. 2), fibrinous airsacculitis, and perhaps fibrinous perihepatitis. The lungs are congested and often there is a fibrinous pneumonia. The spleen is enlarged and congested, and may be the only lesion.

3. In pigeons there is conjunctivitis with encrusted, swollen eyelids. There may be hepatomegaly, airsacculitis and enteritis.

4. In cage birds that succumb, the spleen frequently is enlarged and contains white foci. Often there is hepatomegaly with focal necrosis and yellowish discoloration, airsacculitis, pericarditis, and congestion of the intestinal tract.

DIAGNOSIS

1. A tentative diagnosis can be made on history, signs, and lesions and the demonstration of intracytoplasmic inclusions on impression smears (Fig. 3) made from fresh exudates (monocytic cells) from the surface of the air sac (epithelial cells), spleen, liver, lung, serous surface, and pericardium. Where available, the fluorescent antibody technique may be useful in demonstrating chlamydia. A definitive diagnosis is usually obtained by isolation and identification of chlamydia organisms, genetic detection or by a fourfold rise in antibody titer to chlamydial group antigen.

2. Every precaution should be taken to avoid self-infection. Psittacosis is highly contagious and many laboratory workers have been infected while handling infected birds or their tissues. Dead birds should be completely immersed in effective disinfectant solutions because the highly infectious nasal and fecal secretions dried on feathers can give rise to infectious aerosols.

3. Chlamydophilosis in turkeys must be differentiated carefully from *Mycoplasma gallisepticum* infection, influenza, aspergillosis, septicemic colibacillosis and cholera. Lesions in turkeys closely resemble those of *M. gallisepticum* infection. In general, diagnostic specimens from birds for chlamydia diagnosis should also be cultured for species of *Salmonella, Pasteurella, Mycoplasma*, and other bacteria as well as viruses.

4. Spleen, liver, lung, fibrinous exudate, air sacs, nasal washings and mucosa, fecal samples, or intestinal loops should be used for microbiologic or pathologic evaluation.

CONTROL

1. Because there is no effective vaccine against chlamydophilosis, prevention of the disease in poultry depends upon avoidance of exposure. Facilities should be cleaned and disinfected prior to use. Flocks should be started and raised as units and no birds should be added to a started flock.

2. Poultry should not be exposed to other birds, especially wild birds, animals, or their excreta. A preventive level of tetracycline can be added to poultry rations if exposure of the flock is suspected or anticipated. Poultry farm workers should not own any pet birds or poultry.

3. Federal law specifies how commercial birds imported for resale, research, breeding, or public display must be handled. During the quarantine period, all exotic birds of the psittacine family are treated with chlortetracycline as a precautionary measure against psittacosis. The quarantine period, designed to prevent the introduction of velogenic Newcastle disease, is 30 days and an effective treatment for chlamydophilosis requires 45 days.

TREATMENT

Avian chlamydophilosis is a reportable disease in most states and must be reported to the state veterinarian or other designated officials. Flocks should be treated only under supervision. Infected turkey flocks often have been treated with chlortetracyclines and slaughtered under supervision without human infection.

AVIAN CHLAMYDIOSIS

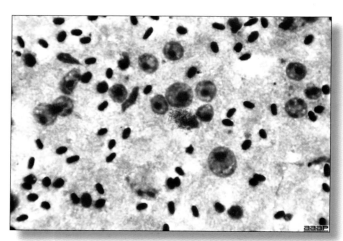

Fig. 1
Inclusions can be demonstrated in stained smears (Giemsa stain elementary bodies).

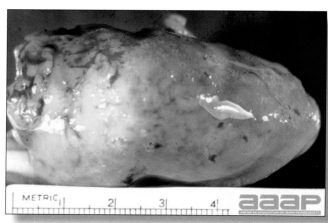

Fig. 2
Fibrinous pericarditis in a turkey.

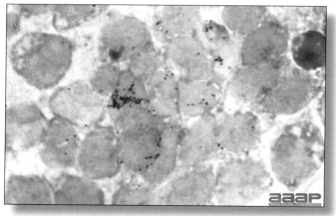

Fig. 3
Giemsa stain elementary bodies in smear.

AVIAN TUBERCULOSIS
(Avian TB; TB)

DEFINITION
Avian tuberculosis is a slow-spreading, usually chronic, granulomatous infection of semimature or mature birds, characterized by progressive weight loss and, ultimately, by emaciation and death.

OCCURRENCE
Avian tuberculosis occurs in many kinds of birds, including poultry, game birds, cage birds, wild birds, and zoo birds. Most outbreaks are encountered in old backyard chickens. Avian tuberculosis also occurs in mammals, including swine, sheep, mink, cattle, and, rarely, humans. Among mammals, swine are more frequently infected. Avian tuberculosis is worldwide in distribution.

HISTORICAL INFORMATION
1. Tuberculosis in chickens was first recognized as a separate disease about 1884. Once identified, it soon was recognized in many countries. Avian tuberculosis eventually was found to be transmissible to certain other birds and mammals, especially swine, and was shown to sensitize cattle to tuberculin and johnin.

2. In the United States there once were many farm flocks of chickens and a farm flock often was kept for years. In old flocks avian tuberculosis was a very common disease. Later, farm flocks largely were replaced by large commercial flocks, which are sold after one laying cycle, a practice that greatly restricts the spread of avian tuberculosis.

3. Avian tuberculosis is seldom seen today in poultry species. However, there is a tendency toward the reestablishment of small farm flocks, which probably will be kept long enough for tuberculosis to develop.

4. Avian tuberculosis commonly infects swine, interferes with the eradication of bovine tuberculosis, and sometimes infects humans. However, there presently is no formal eradication program for this disease.

ETIOLOGY
1. The etiologic agent is *Mycobacterium avium* subspecies *avium* (referred herein as *M. avium*), a highly resistant, acid-fast bacillus. It resists heat, cold, water, dryness, pH changes, and many disinfectants and survives in soil for years.

2. Destruction by disinfection is impractical on most poultry farms. *M. avium* is distinct from the bacilli that cause human and bovine types of tuberculosis, although all three types share many similar characteristics.

3. *M. avium* is present in large numbers in avian tubercles. Stained impression smears or histologic sections made from the centers of tubercles readily reveal the acid-fast bacilli. Their demonstration permits a strong presumptive diagnosis of tuberculosis.

EPIDEMIOLOGY
1. In chickens (and many other birds) small round to oval nodules (tubercles) develop as diverticuli along the intestine. These tubercles discharge viable tubercle bacilli into the intestine. Infectious feces and other excretions contaminate feed, water, litter, and soil and survive in the environment for months to years. Transmission of the organism is predominantly through ingestion of contaminated feed, water, litter, and soil.

2. During intermittent periods of bacteremia, tubercle bacilli spread from the intestine to most other organs and tissues. If bacteremic or dead infected poultry are cannibalized or consumed by other susceptible poultry or mammals (e.g., swine) the bacillus can be transmitted.

3. Other sources of dissemination of the bacilli include offal from infected chickens, excretions from wild birds (pigeons, sparrows, starlings, etc.), contaminated shoes or equipment, and the feces of infected mammals, especially swine.

4. With the increased popularity of exotic birds as pets, *M. avium* has become increasingly important as a potential zoonotic agent. Although *M. avium* of human origin and *M. avium* of avian origin differ in their genetic makeup, there is concern that *M. avium* can cause disseminated disease in humans with immunosuppressive disease conditions (e.g., AIDS).

CLINICAL SIGNS
1. In chickens there is progressive wasting leading to emaciation, although the appetite is usually maintained. Diarrhea is common and there may be lameness in occasional birds. The skin of the face, wattles, and comb often appears pale.

2. The course in the individual bird and in the flock is prolonged. Total morbidity and total mortality are high, although both are spread over a period of many months and hence are misleading unless records are kept.

LESIONS
1. A bird with advanced tuberculosis is very light in weight and there is marked emaciation. Few other diseases result in such extreme emaciation. These features are unique enough that they should alert the prosector to the possibility of avian tuberculosis.

2. In chickens, gray to yellow nodules (tubercles) often are attached to and scattered along the periphery of the intestine (Fig. 1). Smaller, discrete granulomas usually are present in parenchymatous organs, especially the liver and spleen (Fig. 2). In advanced cases few organs are spared and tubercles often can be demonstrated in the bone marrow of the femur. The lung often has few or no gross lesions.

DIAGNOSIS

1. A history of a chronic disease and persistent mortality in an old flock is suggestive of tuberculosis. Diagnosis often can be confirmed by the postmortem demonstration of typical gross lesions and the demonstration of acid-fast bacilli in impression smears or sections of tubercles (Fig. 3 and 4). The tubercle bacillus should be cultured and identified.

2. Tuberculin testing was once utilized as a flock test and is still available but has fallen into disuse. Chickens are tested by inoculating avian tuberculin into one wattle. The other wattle is used as a control. Turkeys are tested by wing web inoculation. Skin tests are read in 48 hours. The tuberculin test has been used in other avian species with some success.

3. An enzyme-linked immunosorbent assay (ELISA) has been developed for the detection of mycobacterial antibodies in serum and this test has greater promise in the detection of avian tuberculosis in individual exotic birds and aviaries.

CONTROL

1. All poultry should be maintained in single-age groups. This will help control the disease by eliminating infected birds that might be disseminators. Thoroughly clean and disinfect buildings between flocks. Maintain a high standard of sanitation at all times.

2. Young birds should be raised away from old birds on clean premises. Insofar as is possible, raise them in quarantine, thus avoiding exposure to all possible carriers including wild birds.

3. The use of tuberculin testing or ELISA serology monitoring may be of value in aviaries to identify and remove infected birds before widespread dissemination of the disease occurs.

TREATMENT

Avian tuberculosis is a reportable disease in some states and appropriate authorities should be notified. Treatment of avian tuberculosis is not recommended because of the zoonotic potential. Furthermore, *M. avium* is resistant to many of the drugs used in treating other types of tuberculosis.

AVIAN TUBERCULOSIS

Fig. 1
Nodules (tubercles) are attached to and scattered along the periphery of the intestine of a chicken.

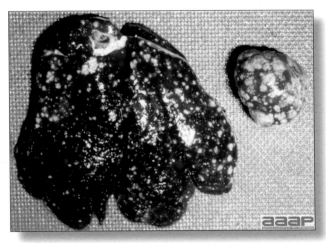

Fig. 2
Smaller, discrete granulomas in parenchymatous organs, especially the liver and spleen.

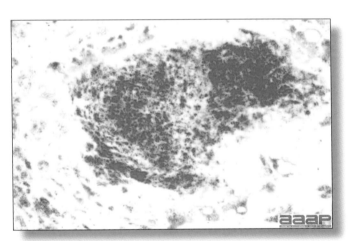

Fig. 3
Demonstration of acid-fast bacilli in section of a tubercle.

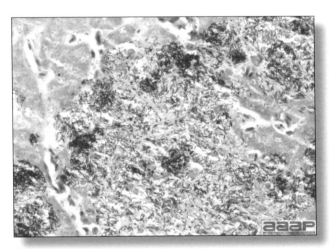

Fig. 4
Acid-fast bacilli in liver of a finch.

BORDETELLOSIS
(Turkey Coryza; Bordetella avium)

DEFINITION

Bordetellosis is an acute, persistent, contagious upper respiratory disease of turkeys characterized by ocular exudation and rhinitis in young turkeys and tracheitis in older turkeys caused by *Bordetella avium*.

OCCURRENCE

1. Bordetellosis occurs most commonly in turkeys 1-6 weeks of age. All ages of turkeys are susceptible, including breeders.

2. Outbreaks of the disease occur in most turkey-producing areas of the United States. Similar diseases have been reported from Canada, Australia, Germany, France, England, Italy, Israel, and South Africa.

3. Farms with continuous confinement production and multiage flocks have the greatest problems with bordetellosis. Bordetellosis occurs most commonly in the summer and fall.

4. *B. avium* has been recovered from chickens and occasionally other avian species. Presence of *B. avium* has been associated with increased severity of respiratory disease in broilers, especially when flocks are concurrently infected with infectious bronchitis virus, but its role as a primary pathogen in chickens is less obvious than in turkeys.

HISTORICAL INFORMATION

1. The term turkey coryza (TC) was first used in Canada in 1967 to describe a clinically distinct, acute respiratory disease of turkeys. TC was recognized in Iowa in 1971. Following greater awareness of TC, others recalled similar disease outbreaks that occurred in turkey- producing areas for at least the last three decades. As the number of turkeys being reared in confinement has increased, TC has been identified as an increasingly important respiratory disease and cause of economic loss.

2. The terms alcaligenes rhinotracheitis and turkey bordetellosis were introduced following preliminary identification of the causative agent as *Alcaligenes faecalis* or *Bordetella bronchiseptica*-like, respectively.

ETIOLOGY

1. *B. avium* has been identified as the cause of bordetellosis. *B. avium* can be distinguished from other species of *Bordetella* and nonfermenting, gram-negative bacteria. Hemagglutination of guinea pig erythrocytes is associated with pathogenicity and is useful in distinguishing *B. avium* from *B. hinzii* (formally *B. avium*-like). More recently, some strains of *B. hinzii* have been shown to cause clinical signs and lesions of bordetellosis in turkeys but not chickens.

2. Strains vary greatly in virulence but virulence does not appear to be related to the presence or absence of plasmids.

3. *B. avium* produces hemagglutinin, heat-stable and heat-labile toxins, a dermonecrotic toxin, a tracheal cytotoxin, and an osteotoxin.

4. Presence of other infectious agents, notably Newcastle virus, other paramyxoviruses, *Mycoplasma gallisepticum*, *Pasteurella* and *Escherichia coli* increases the severity of bordetellosis.

EPIDEMIOLOGY

1. *B. avium* is susceptible to most disinfectants and environmental conditions, especially drying.

2. Although carrier state has not been confirmed, older flocks are thought to serve as carriers and the most important source of infection for younger susceptible flocks on multiage farms. Transmission between flocks occurs as a result of human activity. There is no evidence of egg transmission.

3. Litter and contaminated water have been shown to be sources of infection. The organism has been found to persist for at least 6 months in moist litter but not dry litter. Contaminated water can remain in water lines and be a source of infection for new flocks.

4. Infection of flocks less than 10 days of age strongly suggests the environment as the source of the organism. Infection between 2 and 4 weeks may result either from the environment if poults had substantial maternal immunity or introduction from an outside source. Outbreaks in flocks over 4 weeks of age result from introduction of *B. avium*.

CLINICAL SIGNS

1. Onset is abrupt 4-7 days after exposure, with high morbidity and low mortality. Growth rate is decreased.

2. In young turkeys, initial clinical signs are clear, mucoid, nasal discharge and frothy ocular exudate accompanied by sneezing, "snicking", and flicking of the head. Activity is reduced and heat sources are sought out.

3. Exudates become progressively thicker with pasting of nostrils and matting of eyelids. The palpebral opening often assumes an almond shape. There are voice changes or loss in more severely affected birds, accompanied by tracheal rales. Birds show mouth breathing. The intermandibular tissue tends to balloon giving the profile a baggy appearance (Fig. 1). Poults may scratch at matted eyes causing trauma to eyelids. Dried exudate is commonly found on the upper wings and lower neck where the bird wipes off nasal-ocular exudates. Swollen infraorbital nasal sinuses are not typical of bordetellosis but are occasionally seen in a few birds.

4. Tracheal rales persist for several weeks after apparent recovery. Turkeys have been found to be culturally positive for at least 4 months after infection.

5. In uncomplicated outbreaks, mortality remains low. In bordetellosis outbreaks complicated by other respiratory disease agents mortality usually begins 10-14 days after onset of clinical signs and may be high (10-60%). *Escherichia coli* is the most common cause of mortality. Flocks in poor environments, especially if ammonia levels are high, have higher mortality and greater production losses.

6. In older turkeys, nasal and ocular exudation does not occur. Typically the only sign observed in these birds is tracheal rales.

LESIONS

1. Epiphora, serous to catarrhal rhinitis, sinusitis, and tracheitis with hyperemia of the trachea are the only consistent lesions. In severely affected birds, there is distortion of tracheal rings in proximal segments of the trachea, which leads to narrowing of the tracheal lumen and retraction of the larynx. Cross sections through an affected segment will reveal the characteristic flattening or dorsal infolding of the trachea (Fig. 2). Death occurs by suffocation from an obstructed trachea.

2. A variety of other lesions can be found in complicated outbreaks, depending upon the etiologic agents present.

3. *B. avium* attaches readily to ciliated epithelial cells of the upper respiratory tract (Fig. 3). This leads to deciliation, altered mucus production, impairment of mucociliary clearance, and mucus accumulation. Inflammatory changes are not pronounced but are chronic, which leads to distortion of tracheal rings and hyperplastic bronchial-associated lymphoid tissue.

4. Infection with *B. avium* has been shown to interfere with vaccination for fowl cholera but the mechanism is unknown.

DIAGNOSIS

1. The bacterium is readily isolated from the trachea. Typical nonfermenter colonies occur on MacConkey agar in 48-72 hours.

2. If high populations of fermenting organisms are present on the plate, *B. avium* may be inhibited. This situation often occurs when the disease has been going on for several weeks. Early in the outbreak, almost pure, dense growths of *B. avium* are readily obtained.

3. *B. avium* should be looked for in any respiratory disease of turkeys even if another cause is identified because it is a significant predisposing factor to severe respiratory disease outbreaks.

4. A variety of serological tests including rapid plate agglutination, microagglutination, and enzyme-linked immunosorbent assay (ELISA) tests have been developed to detect antibodies to *B. avium*. The microagglutination and ELISA tests are commonly used for diagnostic purposes.

CONTROL

1. Clean and disinfect the brooder house and all equipment between flocks. Make sure house and equipment are thoroughly dry. Depopulate problem farms.

2. Flush water lines with disinfectant between flocks. Use halogens or similar substances to treat drinking water.

3. Control traffic patterns. Traffic should always move from younger to older flocks without backtracking. Ideally only one person who has no other contact with poultry should care for a single brooder house (isolation brooding).

4. Prevent contact between wild birds and young turkeys.

5. An oil-emulsion bacterin is available for use in breeder hens. This will provide poults with maternal immunity for up to 4 weeks, the interval when infection generally results in a more severe disease.

6. A live vaccine prepared from a temperature-sensitive mutant of *B. avium* is available for use in poults. Two doses are recommended, the first given via spray cabinet in the hatchery with a booster administered through the drinking water at 2-3 weeks of age.

TREATMENT

Although *B. avium* is susceptible to most antibiotics on sensitivity tests, treatment with antibiotics is generally ineffective. This is thought to be due to failure of the antibiotic to reach effective levels in the trachea where the organism is located. Aerosol administration of oxytetracycline is effective in reducing clinical signs during the treatment period but has little long-term benefit. The best management for an infected flock is to move the birds to range if possible. If not, increase ventilation, increase house temperature, and frequently stimulate the flock to move thus encouraging them to eat and drink. Higher density "stress" rations and use of vitamins and electrolytes in water are useful adjuncts to general support of sick birds.

BORDETELLOSIS

Fig. 1
Young turkey showing mouth breathing, frothy ocular exudate and the intermandibular edema (bottlejaw).

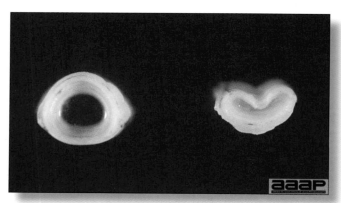

Fig. 2
Cross-section of the trachea, normal on left and characteristic flattening or dorsal infolding of the trachea (on the right).

Fig. 3
Partial deciliation of the tracheal epithelium. *B. avium* attaches readily to ciliated epithelial cells of the upper respiratory tract.

BOTULISM
(Limberneck; Western Duck Sickness)

DEFINITION
Botulism is an intoxication caused by ingestion of the toxins of *Clostridium botulinum.*

OCCURRENCE
1. In birds, botulism occurs frequently in wild waterfowl, captive pheasants and occasionally in chickens. Except for vultures, most birds are susceptible. Most outbreaks in poultry occur in semimature or mature chicken flocks. Many mammals, including humans, are susceptible.

2. In waterfowl (especially wild ducks) occurrence is related to shallow water conditions in lakes and ponds with much decaying vegetation and alkaline water.

3. In some intense broiler rearing areas there is a recurring form of botulism on certain premises. Outbreaks occur in almost every new flock with a seasonal high incidence in the warmer months.

HISTORICAL INFORMATION
1. The first report of botulism in chickens was made in the United States in 1917. Within 25 years the disease had been reported frequently in chickens and in turkeys and waterfowl.

2. During the first half of the 1900s, humans and chickens sometimes died from eating improperly canned foods containing the toxins of *C. botulinum.* Small farm flocks and home canning are now out of vogue and botulism is seldom seen in farm flocks or humans. However, botulism is still an important disease of wild waterfowl, especially ducks. Botulism seldom occurs in well-managed commercially raised poultry.

ETIOLOGY
1. Botulism is caused by ingestion of the preformed toxins of *C. botulinum* in feeds, foods, dead poultry, or toxin-containing maggots (Fig. 1). Although *C. botulinum* or its spores are not pathogenic and are commonly found in the environment and in the intestinal tract, under ill-defined circumstances it colonizes the intestines, produces toxin and causes botulism.

2. The toxin of *C. botulinum* is extremely potent. The minimum lethal dose (MLD) for guinea pigs is 0.00012 mg/kg subcutaneously (The MLD for cobra venom is 0.002 mg/kg). The toxin is relatively heat stable.

3. Based on specific conditions, there are eight types of toxins produced by *C. botulinum.* Type C is most common in poultry outbreaks although other types have occurred.

4. Botulism should not be confused with pseudobotulism of chickens. Pseudobotulism closely resembles botulism except that affected birds almost invariably recover within 24 hours. Pseudobotulism is now considered to be a transient manifestation of Marek's disease.

EPIDEMIOLOGY
1. *C. botulinum* is ubiquitous in nature and commonly present in feeds. When ideal conditions for growth occur, large amounts of exotoxin may be formed. If adequate toxin is consumed, botulism will develop. Improperly sterilized canned fruits and vegetables, spoiled animal feeds, decaying poultry carcasses and the insects that feed on them can contain enough exotoxin to be lethal, even when taken in small amounts.

2. It is speculated that wild waterfowl contract botulism in the following ways:

    A. The toxin may be consumed in decaying vegetation in shallow, alkaline lakes as they dry up or are created by irrigation during the summer. Alternatively, it may be that the toxin is in larvae or crustaceans in the vegetation. Invertebrates killed by anaerobic conditions contain toxin from growth of *C. botulinum* within them and may be consumed by some waterfowl. This condition may also occur in small ponds via water temperature inversion following summer storms.

    B. Ducks that die from various causes may allow toxin production and dissemination after death by *C. botulinum* normally present in their intestine. Toxins are formed in the cadavers. Ducks that feed on the cadavers or on maggots from the cadavers may be poisoned.

3. A growing body of evidence suggests that *C. botulinum* type C can produce toxin within the intestinal tract of the live broiler chicken. This type of botulism has been termed toxico-infectious botulism.

CLINICAL SIGNS
Signs appear within a few hours to days. Clinical signs include drowsiness, weakness, and progressive loss of control and flaccid paralysis of the legs, wings, neck, hence the term 'limberneck' (Fig. 2 and 3) and eyelids. Paresis soon progresses to paralysis and the recumbent bird closes its eyes and appears to be in a deep coma. Fine tremors of muscles and feathers occur in some birds. Death may occur shortly or may be delayed for a few hours. Most visibly affected birds die.

LESIONS
Most birds with botulism are free of gross lesions. Rarely, in birds that have lived for some time, there may be mild enteritis. The upper digestive tract (especially the crop) may contain putrid ingesta or maggots but is usually empty.

DIAGNOSIS

1. In chickens and turkeys diagnosis is based largely on history, signs, the presence of putrid feed or maggots in the digestive tract, looseness of feathers (chickens only), and the absence of lesions. Finding a decaying cadaver upon which birds have been feeding may assist in diagnosis.

2. Saline gizzard or intestinal washings or blood serum from an affected bird can be tested for toxicity. Either can be injected into mice that have been inoculated with protective antiserum and into mice without antiserum. Results should clarify diagnosis. Culture is meaningless.

3. A group of affected birds can be treated with polyvalent antitoxin. Recovery in a high percent of the birds tends to confirm a diagnosis of botulism. Unfortunately, commercial availability of antitoxins may be a problem.

CONTROL

1. The disease can be avoided by preventing access of poultry to any source of toxin. Sick and dead birds should be picked up regularly and frequently because they are a common source of toxin.

2. Type C toxoid can be used to immunize birds, although this is seldom done.

3. Wild ducks can be baited or frightened away from shallow lakes. Water sometimes can be pumped into shallow lakes to raise the water level and botulism may not occur.

4. On broiler farms where botulism is enzootic, the prophylactic use of selenium and antibiotics has been effective. They also aid in treatment of affected flocks.

TREATMENT

1. Antitoxin can be given to valuable affected birds. Results often are good although this will depend on the specificity of the antisera. Type C antitoxin is usually given. Polyvalent antisera (especially types A and C) are preferred but often difficult to obtain. It is important that the treated birds have access to fresh, clean, nonalkaline water.

2. Because toxico-infectious botulism has not been experimentally induced, treatment of this condition is based solely on the apparent response to therapy during field outbreaks of this condition. Treatment of flocks with sodium selenite and vitamins A, D, and E has been reported to reduce mortality. Treatments with bacitracin, streptomycin, chlortetracycline, and penicillin have also been reported to be efficacious.

Fig. 1
Decaying waterfowl carcass with toxin-containing maggots.

Fig. 2
Flaccid paralysis of the legs, wings, neck in a chicken.

Fig. 3
Flaccid paralysis of the legs, wings, neck in ducks (limberneck).

CAMPYLOBACTER

DEFINITION

Campylobacteriosis is a disease caused by the bacteria *Campylobacter* sp. It occurred as a disease of egg-laying chickens in the 1950's and 60's and was known at that time as vibrionic hepatitis. The disease in chickens has since disappeared. Campylobacteriosis more recently has been recognized as a leading cause of human foodborne illness and is even more common than Salmonellosis. There are many sources of campylobacteriosis in humans, but poultry is often considered as a major source.

This bacteria does not cause disease in poultry but since the handling and consumption of raw or undercooked poultry is the most frequently identified source of sporadic campylobacteriosis in humans, it is important to understand this bacteria to better implement control and preventative measures.

OCCURRENCE

Campylobacter is a common cause of diarrhea in the developing world, particularly in the under five year of age group. Most developed countries, with surveillance systems, have reported an increase in the number of cases over the last 25 years. In European countries, the overall incidence of human campylobacteriosis ranged from 50 to 90 cases per 100,000 people in 2005, while the United States reported a 12.7 per 100,000 the same year. Australia had a similar level to that of Europe while New Zealand was the highest of all industrialized countries with 396 per 100,000 in 2003. It is now agreed that the most common source for campylobacteriosis is the handling or consumption of raw or undercooked poultry products.

HISTORICAL INFORMATION

In poultry, in the 1950s, a possible association was established between sporadic cases of vibrionic hepatitis in laying hens and vibrio-like organisms now known as *Campylobacter jejuni*. Layers during these episodes showed a drop in egg production with increased mortality. Lesions were found in the liver and appeared stellate, asterisk-shaped, or cauliflower-like. This condition mysteriously disappeared in the late 1960s but since then, cases have occasionally been reported in the literature.

Awareness of the public health implications of campylobacteriosis in humans has evolved over the past century. In 1886 Escherich observed organisms similar to *Campylobacter* in fecal samples from children with diarrhea while *Campylobacter* was isolated as *Vibrio fetus* in 1909 from spontaneous abortions in livestock. The development of selective growth media in the 1970's allowed easier culture of *Campylobacter* from human feces. Soon *Campylobacter* species were established as common human pathogens in the occurrence of bacterial gastro-enteritis

ETIOLOGY

Thermophilic *Campylobacter* species *C. jejuni* and *C. coli* can be isolated from the intestines of poultry. It is a microaerophilic, small curved or spiral Gram-negative rod with a rapidly darting motility under a phase-contrast microscope.

Unlike Salmonella, *Campylobacter* does not tolerate exposure to oxygen or drying, and does not multiply on food left at room temperature. Indeed, since the optimal growth temperature of thermotolerant *Campylobacter* lies between 37 and 42°C (very close to the chicken body temperature), they are not able to grow below 30°C i.e., at room temperature.

EPIDEMIOLOGY

Campylobacter is a commensal organism in avian hosts. Colonization is mostly with *C. jejuni* with a much lower proportion of *C. coli* strains. It occurs mostly in the ceca, more specifically in the intestinal mucous layer covering the intestinal crypts. Large percentages of poultry lots at slaughter can be found to be *Campylobacter* positive, but since sampling methodology varies greatly from one study to another, comparison is almost impossible. In Canada and Europe, prevalence estimates range from 18% to 82% in broiler chicken carcasses.

Broilers are typically shown to be free of *Campylobacter* at a day of age and detection in a flock is usually possible at 2 to 3 weeks of age. Maternal immunity is thought to play a protective role in delaying *Campylobacter* infection. Interestingly, experimental studies have demonstrated that day-old chicks are susceptible to colonization and *Campylobacter* has also been isolated from the ovaries and semen of broiler breeders as well as from paper pad liners in hatching box. Vertical transmission is still being debated.

The organism can rapidly spread horizontally in chickens, presumably through fecal contact, a common water source or with flies as vectors. The latter might partly explain the seasonality of campylobacteriosis in poultry flocks with an observed higher prevalence in summertime. Indeed, recent trials using insect screens on chicken houses found that this sole preventative mean reduced flock prevalence by up to 70%. Once infection has entered the broiler house, its spreads rapidly; within 10-14 days post-infection, more than 90% of the birds in a flock will be positive.

When *Campylobacter* positive chickens enter the slaughterhouse with large numbers of *Campylobacter* in the intestines, as well as on the feathers and skin, this inevitably leads to a cross-contamination of the equipment, environment and other processed birds.

CLINICAL SIGNS
Chickens are carriers of *Campylobacter* and typically do not exhibit clinical signs or lesions. Egg drop and increased mortality in egg-laying chickens was observed with vibrionic hepatitis.

LESIONS
No lesions are observed in carrier birds. Hepatitic necrosis consisting of small stellate whitish foci to fairly large diffuse areas has been described in egg-laying chickens with vibrionic hepatitis (Fig. 1).

DIAGNOSIS
1. Culture of the causative organism from egg-laying hens with lesions.

2. Bile is a better source of organism than liver.

3. Samples are best kept at 4°C.

CONTROL
Biosecurity, including rodent and insect control.

TREATMENT
No treatment is currently used.

CONCLUSIONS
Campylobacteriosis is currently only a public health concern and not a poultry health concern. It is important that a thorough knowledge of transmission of and infection with *Campylobacter* be developed in order to provide effective and economical means of controlling its incidence in poultry.

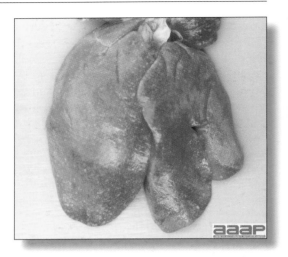

Fig. 1
Hepatic necrosis.

COLIBACILLOSIS
(*Escherichia coli* Infections)

DEFINITION
Avian colibacillosis is an infectious disease of birds in which *Escherichia coli* is the primary or secondary pathogen. Infections include airsacculitis, cellulitis, omphalitis, peritonitis, salpingitis, synovitis, septicemia and coligranuloma.

OCCURRENCE
Colibacillosis occurs in all types and age groups of poultry as well as in other birds and many kinds of mammals. Most reported outbreaks in poultry have been in chickens, turkeys, and ducks. Many outbreaks occur in poultry raised under a low standard of sanitation, poor environmental conditions, or after a respiratory or immunosuppressive disease. Infection is more frequent in young than mature birds. Colibacillosis is common throughout the world.

HISTORICAL INFORMATION
Colibacillosis was first described in chickens in 1894. Since then, there have been numerous reports on colibacillosis in poultry and considerable research on the disease has been completed. Many investigators doubt that *E. coli* is a primary pathogen. Others are convinced that certain serotypes are primary pathogens and their opinion seems to prevail. Most investigators agree that *E. coli* frequently can be isolated from a variety of well-defined syndromes in poultry.

ETIOLOGY
The etiologic agent is *Escherichia coli* (*E. coli*). The O (somatic) antigen serotypes most commonly associated with disease outbreaks are O1, O2, O35, O36, and O78. The K (capsular) antigens most commonly associated with virulence are K1 and K80. In the intestinal tract of normal poultry, nonpathogenic serotypes far outnumber pathogenic serotypes, with 10% to 15% of intestinal coliforms being potential pathogens.

EPIDEMIOLOGY
1. *E. coli* is present in the intestine of birds and mammals and is disseminated widely in feces. Birds are continuously exposed through contaminated feces, water, dust, and environment. Any time a bird's resistance to disease is impaired, pathogenic or facultative pathogenic strains may infect the bird. Sequestered *E. coli* in such sites as the intestine, nasal passages, air sacs, or reproductive tract may be a latent source of infection. Certain pathogenic serotypes may have the ability to infect a normal bird.

2. *E. coli* has been isolated from the eggs of normal hens. Its presence has been attributed to ovarian infection, oviduct infection, and to eggshell contamination followed by penetration. Chicks may hatch with a latent infection; however, active infection will typically only occur if some environmental stress or lesions initiates the disease process.

CLINICAL SIGNS AND LESIONS
A variety of lesions from which *E. coli* has been isolated include:

1. Airsacculitis
Respiratory signs occur and vary in severity. This pathology may be associated with poor environmental conditions such as dusty litter, poor ventilation, high ammonia levels, sudden variation in the barn temperatures, but also with concomitant respiratory (infectious bronchitis virus, Newcastle disease virus, laryngotracheitis virus, mycoplasmas) or immunosuppressive (infectious bursal disease, chicken anemia virus) diseases. In these cases, *E. coli* is a secondary pathogen and will cause the airsacculitis lesions. Air sacs are normally thin, glistening and transparent (Fig. 1) but bacterial infection will cause the air sacs to become thickened, number of blood vessels within the air sac walls increases and exudate will accumulate within the cavity of the air sac. An acute inflammation will be characterized by the presence of mucous exudate (Fig. 2) which will eventually become fibrinous (Fig. 3). Thickened air sacs and caseous exudate in the air sac will be present (Fig. 4) in more severe and chronic cases. There often is an accompanying adhesive pericarditis, fibrinous perihepatitis and peritonitis (hence a polyserositis). Airsacculitis occurs chiefly in 3-7-week-old broilers, probably peaking at 5-6 weeks.

2. Pericarditis
Most serotypes of *E. coli*, after a septicemia, cause a pericarditis (Fig. 5). Opaqueness and thickening of the pericardial sac, an edematous epicardium along with myocarditis typically occurs. Pericarditis can also be caused by other bacteria including *Chlamydophila* sp.

3. Omphalitis and yolk sac infection
E. coli is often isolated in pure culture from organs or the yolk sac of recently hatched birds having depression, septicemia, and variable mortality. With omphalitis the navel is swollen and inflamed (Fig. 6) and the bird feels wet. Abnormal yolk material and peritonitis is typically seen on necropsy of birds with an *E. coli* infection of the yolk sac (Fig. 7).

A great variety of other organisms such as species of *Aerobacter, Proteus, Klebsiella, Pseudomonas, Salmonella, Bacillus, Staphylococcus*, enteric *Streptococcus*, and *Clostridia* are frequently isolated from yolk sacs of embryos and navels of chicks, most likely as mixed infections.

4. Coliform septicemia of ducks (duck septicemia)
E. coli, Salmonella, and *Riemerella (Pasteurella) anatipestifer* produce respiratory signs, airsacculitis,

pericarditis, perihepatitis, and peritonitis. In outbreaks of *R. anatipestifer*, involvement of the air sacs and a dry, thin transparent covering over visceral organs are present. In coliform septicemia (*E. coli*) usually a moist, granular to coagulative exudate of varying thickness is present on abdominal and thoracic viscera and surfaces of air sacs. The spleen and liver are swollen and dark with bile staining of the liver.

5.   Acute septicemia
An acute septicemic disease caused by *E. coli* resembles fowl typhoid and fowl cholera. Birds are in good flesh and have full crops suggesting acuteness of the disease. This can occur in young or mature birds.   There are sudden deaths, and variable morbidity and mortality. Parenchymatous organs are swollen with congested pectoral muscles. Livers are green in color and may have small necrotic foci. There may be petechial hemorrhages, pericarditis, or peritonitis. Acute systemic disease may also be caused by various *Pasteurella, Salmonella, Streptococci*, and other organisms.

6.   Enteritis
Enteritis caused by *E. coli* is considered rare but pathogenic attaching effacing *E. coli* have been reported. Diarrhea and dehydration are noted on clinical examination. At necropsy there is enteritis, often with excessive fluid in the intestines. *E. coli* may be isolated from parenchymatous organs.

7.   Salpingitis
This lesion may occur following entry of coliform bacteria from the vagina in laying hens. It is also likely to develop when the left greater abdominal air sac becomes infected by *E. coli*, causing a chronic salpingitis. Affected birds usually die during first 6 months postinfection and never lay. The oviduct is distended with exudate (Fig. 8) that may be caseous and has a foul odor. No specific signs are noted but there may be an upright (penguin) posture.

8.   Coligranuloma (Hjärre's disease)
Signs vary in this uncommon disease of chickens and turkeys. Nodules (granulomas) occur along the intestinal tract, and mesentery, and in the liver (Fig. 9). The spleen is not involved. The lesions resemble those of tuberculosis. The agent is a mucoid coliform, possibly not *E. coli*. Granulomas of the liver have many causes, some of which would include the anaerobic genera *Eubacterium* and *Bacteroides*.

9.   Synovitis and osteoarthritis
Affected birds are lame or recumbent. There is swelling of one or more tendon sheaths or joints (Fig. 10). Synovitis and/or osteoarthritis are frequently a sequel to a systemic infection. With synovitis many birds will recover in about 1 week. Osteoarthritis is a more severe and chronic condition where the joint is inflamed and the associated bone has osteomyelitis. These severe chronic infections make birds unwilling or unable to walk and necropsy findings often include dehydration and emaciation. Synovitis-arthritis may also be caused by reovirus, or species of *Mycoplasma, Staphylococci*, and *Salmonella*.

10.   Panophthalmitis and meningitis
Occasional birds have a hypopyon and/or hyphema, usually in one eye, which is blind. Likewise, meningitis is a rare sequelae to *E. coli* septicemia.

11.   Cellulitis (Infectious process)
This is one of the most common causes of condemnation at slaughter in broiler chickens in the United States, some European countries, and Canada. It is recognized primarily at post-mortem inspection, with no abnormality having been noted in live birds. The USDA Food Safety and Inspection Service designates cellulitis as "infectious process" or "IP". Gross lesions include variable yellowing and thickening of the skin lateral to the vent (Fig. 11) and extending in severe cases over the ventrocaudal aspect of the breast. On incising the skin a yellow caseous plaque of variable size (Fig. 12) is noted in the subcutis. Histologically there is cellulitis involving both dermis and subcutis. The inflammatory reaction includes edema and heterophil infiltration in active areas, whereas there is accumulation of a walled-off causative sheet of exudate surrounded by a zone of giant cells in more chronic areas of involvement. Coccobacillary bacteria can be seen in microcolonies within the exudate and *E. coli* is recovered quite consistently on culture. This condition may affect up to 8% of entire flocks at slaughter resulting in extensive trim-out, downgrading, or whole-carcass condemnation. Cellulitis is caused by the secondary infection of skin wounds. Risk factors such as certain broiler breeds, poor feathering, sex (males more susceptible), skin scratches, increased stocking density and litter type have been associated with this condition.

DIAGNOSIS
Diagnosis of primary colibacillosis is based on the isolation and typing of a coliform into one of the serotypes recognized as pathogens. Diagnosis based merely on the isolation of *E. coli* is of questionable validity. The possibility of other infections (viruses, bacteria, fungi, *Chlamydophila*, and mycoplasmas) should have been eliminated through culture or other means. When *E. coli* is isolated secondary to some other primary disease, it should be diagnosed as secondary colibacillosis.

CONTROL
1.   Measures should be taken to minimize eggshell contamination of newly laid hatching eggs. Eggs should be disinfected on the farm prior to storage and should be stored under ideal conditions. Scrupulous hatchery sanitation, disinfection, and/or fumigation procedures should be practiced.

2.   A vigorous sanitation program should be followed in raising poultry.

3. Diseases, parasitism, and other stresses on a flock should be minimized as much as possible. Dust should be controlled.

4. Only feeds free of fecal contaminations should be fed to poultry. Pelleted feeds are more likely to be free of contamination.

5. Treatment of water with halogens and related compounds as well as conversion to nipple drinkers has greatly decreased the incidence of septicemic forms of colibacillosis.

TREATMENT

Numerous antimicrobials have been utilized for treatment. These have included tetracyclines, neomycin, sulfa drugs and others but *E. coli* has developed resistance to many of these commonly used antimicrobials. Antibiotic sensitivity testing is therefore strongly suggested as well as record keeping of treatment history by farm.

COLIBACILLOSIS

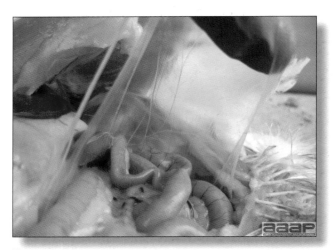

Fig. 1
Air sacs are normally thin, glistening and transparent.

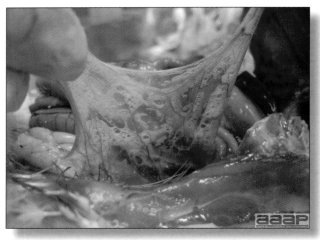

Fig. 2
An acute inflammation will be characterized by the presence of mucous exudate in the air sacs.

Fig. 3
Fibrinous exudate and neovascularization of the air sacs.

Fig. 4
Chronic airsacculitis: caseous exudate and thickened air sacs.

Fig. 5
Pericarditis and perihepatitis.

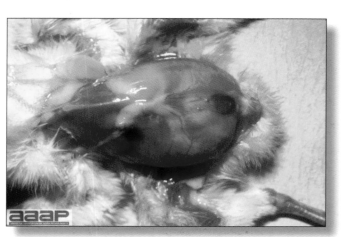

Fig. 6
Swollen and inflamed navel in a case of omphalitis.

COLIBACILLOSIS

Fig. 7
E. coli infection of the yolk sac. Note the periombilical neovascularization.

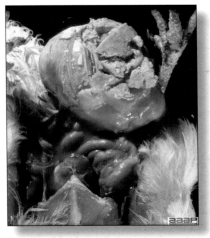

Fig. 8
Oviduct is distended with caseous exudate.

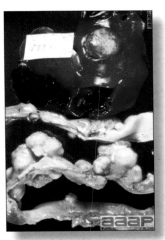

Fig. 9
Nodules (granulomas) along the intestinal tract, and mesentery, and in the liver.

Fig. 10
Arthritis and synovitis.

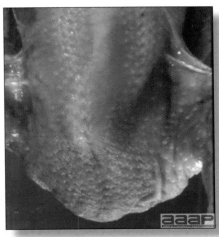

Fig. 11
Cellulitis: yellow discoloration of the skin in the right inguinal area of a broiler chicken carcass at slaughter.

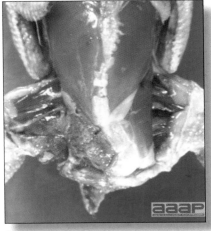

Fig. 12
Subcutaneous caseous exudate in the right inguinal area of a broiler chicken carcass at slaughter.

ENTEROCOCCUS CECORUM

DEFINITION
Recently, *Enterococcus cecorum* has been recognized as an emerging avian pathogen, associated with spondylitis, femoral head necrosis, and osteomyelitis in broiler chicken and broiler breeder flocks in Europe and North America. *E. cecorum* can cause significant economic losses, due to mortality and culling, poor feed conversion and increased condemnations.

OCCURRENCE
E. cecorum arthritis and osteomyelitis have been reported in Europe, United States and Canada in broiler chickens and broiler breeders.

HISTORICAL INFORMATION
Enterococcus cecorum joint and bone infections in broilers were first described in 2002.

ETIOLOGY
Enterococcus spp. are facultative anaerobic, Gram-positive, catalase-negative cocci. Members of this genus are widespread and normal inhabitants of the intestinal tract of mammals and birds.

EPIDEMIOLOGY
Broiler chickens, mostly males, start presenting clinical signs at the age of 7 – 14 days and the morbidity will increase in the following weeks. Mortality at the end of the growing period will reach 2 to 7%. The problem seems to be recurring in the same barn.

The condition can be observed in broiler breeders as early as 3 weeks of age with males being predominantly affected.

The pathogenesis is not well understood, but since *E. cecorum* is a normal inhabitant of the gastrointestinal flora, increase in numbers and invasion into systemic circulation could occur.

CLINICAL SIGNS
Affected birds are lame and reluctant to walk. Some may sit back on their hocks (Fig. 1) and tails with their feet and shanks raised.

LESIONS
The disease appears to have a shorter clinical course in broiler chickens, with macroscopic inflammatory lesions in hock and stifle joints and corresponding tendon sheaths, as well as osteomyelitis of the femur and tibiotarsus, or the thoracic vertebra. In broiler breeders, typical cases present arthritis lesions and osteomyelitis of the fourth thoracic vertebra.

Sagittal section of the vertebral column will show abscess formation (Fig. 2), necrosis of the bone resulting in the dorsal displacement and compression of the overlying spinal cord (Fig. 3).

Histopathology reveals heterophilic to granulomatous inflammation of the synovium and tendon sheath of the affected joints, focal osteomylitis of the affected bones, with Gram-positive coccoid bacteria visible in the lesions.

DIAGNOSIS
Lesions and isolation of *E. cecorum* are diagnostic.

Vertebral osteomyelitis should be differentiated from other causes of spinal cord compression e.g. spondylolisthesis.

CONTROL
Clean, disinfect and fumigate the premises.

Water sanitation has been reported to reduce the incidence.

The pattern of recurrent disease at the flock level might be stopped with amoxicillin and/or tylosin administered in prevention starting at a day of age.

ENTEROCOCCUS CECORUM

Fig. 1
Chicken sitting on its hocks.

Fig. 2
Abscess in the vertebral column of a male broiler chicken.

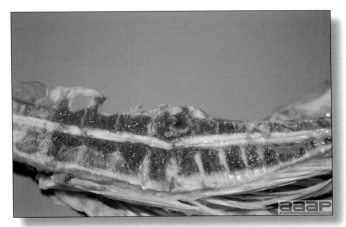

Fig. 3
Necrosis of the bone resulting in the dorsal displacement and compression of the overlying spinal cord.

ERYSIPELAS

DEFINITION

Erysipelas is an acute septicemic disease occurring most commonly in older male turkeys, characterized by serosal, cutaneous, and muscular hemorrhages and splenomegaly. Chronic erysipelas (polyarthritis, endocarditis) occurs occasionally, usually after acute outbreaks.

OCCURRENCE

Erysipelas is of primary importance in turkeys although outbreaks sometimes occur in geese, ducks, pheasants, other game birds, wild birds, and (rarely) in chickens. Sporadic cases have been reported in many wild birds. Erysipelas also occurs in swine, sheep, sea mammals, fish, and many wild animals. In humans, erysipelas is caused by streptococci, whereas *Erysipelothrix* typically causes a localized inflammation designated erysipeloid but may cause septicemia and death. In turkeys erysipelas usually occurs in toms approaching market weight; it seldom occurs in turkeys less than 10 weeks old, although no age or sex resistance has been found experimentally. The peak incidence in tom turkeys roughly coincides with puberty and the disease occurs occasionally in hen turkeys following artificial insemination. Because the bacterium is ubiquitous in nature, erysipelas probably affects poultry and birds throughout the world.

HISTORICAL INFORMATION

The economic significance of erysipelas in turkeys was pointed out in 1939 and the disease was soon recognized as a major disease of turkeys. In the United States, recognition of erysipelas in turkeys as an important disease roughly paralleled the recognition of erysipelas as an important disease of swine. The disease, while still occurring, is not common today because most turkeys are raised in confinement, thus reducing exposure to the organism.

ETIOLOGY

The etiologic agent is *Erysipelothrix rhusiopathiae*. It is a Gram-positive, slender, slightly bent, pleomorphoric rod. Filamentous, beaded forms that tend to decolonize easily are often seen in cultures. It grows well on enriched media especially when incubated in a 5-10% carbon dioxide atmosphere (candle jar). Colonies tend to be quite small and grow slowly, making them easily overgrown by faster growing bacteria. Selective and enrichment media assist in recovering the organism. Production of hydrogen sulfide in iron-containing media is a useful characteristic for presumptive identification of isolates. The organism is quite resistant to many environmental factors and disinfectants and remains viable in favorable (alkaline) soils for months to years.

EPIDEMIOLOGY

1. The organism is shed in the feces of some recovered turkeys for up to 41 days. It also is shed in the feces of infected swine and lambs. Because turkeys can be infected experimentally by the oral route, it is believed that oral exposure is a common route of natural infection. Infection may occur after ingestion of contaminated soil, water, fish meal, meat meal, or following cannibalism of infectious live or dead birds.

2. The organism can also infect turkeys through breaks in the skin or mucous membranes. Cutaneous wounds commonly occur in tom turkeys, which are inclined to fight as they reach puberty. A typical maneuver during fighting is to grab the snood of the opponent and shake him violently causing considerable trauma to this skin appendage. The snood is considered to be a prime site of infection with *Erysipelothrix* for this reason.

3. Stress often precedes outbreaks of erysipelas. A major outbreak of erysipelas occurred in a large goose flock after the birds were plucked. Other stressors include such things as poor sanitation, bad weather, vaccinations, changes in the ration, etc. The role of vectors, if any, is unknown.

4. Important outbreaks of erysipelas have been reported in hen turkeys following artificial insemination. Presumably, infectious semen comes from carriers that shed the organism in their semen.

5. Injecting numerous birds in a flock with the same needle may spread the organism if septicemic birds are present.

6. *Erysipelothrix* can persist for years in the soil. Outbreaks occur most commonly during the late fall and winter following periods of cold, wet weather. Repeated reoccurrence on a farm is common even with flocks in confinement.

7. Feeding dead turkeys to swine has resulted in outbreaks of erysipelas in the pigs. There is clinical evidence indicating that people who move between swine herds and turkey flocks can spread the organism from the pigs to turkeys.

CLINICAL SIGNS

1. The onset of erysipelas in turkeys is usually sudden with a few birds being found dead. At that time careful examinations of the flock often reveals other turkeys that squat on the floor, and appear sleepy and depressed. They can be aroused but have an unsteady gait when forced to move. Occasional birds may exhibit respiratory signs or have yellow-green diarrhea.

2. Within a few days morbidity increases markedly. The course of the disease in affected turkeys is short, often only a few hours or overnight and most visibly sick birds die.

3. Occasionally, infected turkeys have a swollen snood or irregular, dark red skin, and demarcated lesions on their dewlap, face, or head (Fig. 1). In recently inseminated, infected hens there may be perineal congestion and hemorrhage.

4. Crippled birds with swollen joints are seen in chronic infections. These often occur after an acute outbreak.

LESIONS

1. The lesions are those of a septicemia. The carcass is congested and parenchymatous organs (liver, kidney, spleen) are swollen. Splenomegaly is often marked (Fig. 2).

2. Petechial or suffusive hemorrhages often occur in heavy muscle masses, in pericardial fat, on the epicardium, under serous membranes, and in mucous membranes. Hemorrhages vary greatly.

3. There usually is a marked catarrhal enteritis, often more apparent in the duodenum, with excess mucus in the gut.

4. Skin lesions occur occasionally and are more apparent on the face, head, and neck. Inseminated hens may have peritonitis, perineal congestion, and hemorrhage.

5. Purulent arthritis, often in more than one joint, and vegetative valvular endocarditis are seen in chronic cases.

6. In all organs the dominant histopathologic finding is vascular damage as evidenced by generalized congestion, edema, focal hemorrhages, disseminated fibrin thrombi and numerous Gram-positive bacterial aggregates either within fibrin thrombi or engulfed by reticuloendothelial cells (Fig. 3).

DIAGNOSIS

1. History, signs, and lesions may suggest erysipelas but the etiologic agent should be isolated and identified for confirmation. Erysipelas must be differentiated from fowl cholera. Helpful necropsy findings are the markedly enlarged spleen seen in erysipelas but not fowl cholera, and the pneumonia that often occurs in fowl cholera but not erysipelas. Erysipelas also should be differentiated from acute colisepticemia, salmonellosis, streptococcosis, chlamydiosis, and virulent Newcastle disease. Erysipelas can occur concurrently with other diseases including fowl cholera, chlamydiosis, and internal parasitism.

2. Gram-stained impression smears from the cut surface of liver, spleen, and bone marrow will reveal Gram-positive, slightly bent, thin bacilli. Stained smears may be valuable differentiating erysipelas from cholera and colisepticemia. If available, fluorescent antibody technique can be used to identify the organism in smears or tissue sections.

CONTROL

1. Poultry should be raised separately from older turkeys, which may be carriers. Poults should be started and raised as a flock and no birds should be added. Contact should be prevented between the turkeys and animal carriers, especially sheep and swine. Raise turkeys in houses that were cleaned and disinfected and on ranges without a history of outbreaks of erysipelas.

2. If erysipelas is enzootic in the area, turkeys should be vaccinated with bacterin when 8-12 weeks old. Immunity is enhanced if bacterin inoculation is repeated at least once. Breeders should be revaccinated prior to onset of egg production. A live oral erysipelas vaccine is also available for water vaccination.

3. Semen for artificial insemination should come from tom turkeys with no history of erysipelas infection.

4. Formerly, desnooding of males was a common hatchery practice. This practice is declining because outbreaks of erysipelas are not as common and oral infection with acute systemic disease now appears to be more common than skin infection.

5. Selection for genetic resistance may be possible. Strains of turkeys selected for rapid growth have been found to be more susceptible to naturally occurring erysipelas than unselected lines or lines selected for high egg production.

TREATMENT

1. Penicillin and erysipelas bacterin often are inoculated simultaneously into all birds of an infected flock. Sick birds should be inoculated with a fast-acting form of penicillin. It may be necessary to repeat the inoculation. A longer acting form of penicillin can be used in birds not obviously sick.

2. Water-soluble penicillin used at a rate of 1,5 million units/gal is effective, but the disease often resumes after treatment is stopped. Depending on market conditions the cost of treatment may be greater than the value of the commercial birds.

PUBLIC HEALTH

Slaughter plant management should be notified prior to the affected flock being slaughtered since these birds could serve as a source of infection for the slaughter plant employees.

ERYSIPELAS

Fig. 1
Turkey with a markedly swollen snood and dewlap.

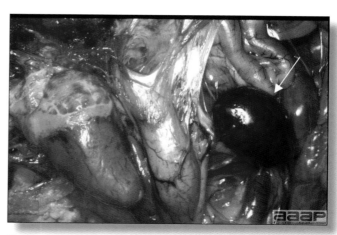

Fig. 2
Splenomegaly.

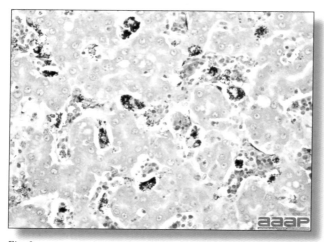

Fig. 3
Numerous Gram-positive bacterial aggregates in the liver of a 22 week-old pheasant. Gram stain.

FOWL CHOLERA
(Cholera; Pasteurellosis)

DEFINITION
Fowl cholera is an infectious disease of poultry, waterfowl, and many other birds, usually appearing in poultry as an acute septicemic disease with high morbidity and mortality. A chronic, localized form, most common form in chickens, occurs in poultry and may follow the acute form, or may occur independently.

OCCURRENCE
Fowl cholera is a disease of many species of birds, including chickens, turkeys, geese, ducks, quail, canaries, and many wild and zoo birds. Perhaps all birds are susceptible under appropriate conditions. In poultry, most outbreaks occur in semimature or mature birds, although there are exceptions. The disease occurs more frequently in turkeys than in chickens. The disease occurs frequently in domesticated waterfowl and often causes extensive losses among wild waterfowl. Geese are highly susceptible. Fowl cholera is more likely to occur in birds that are stressed by such things as poor sanitation, parasitism, malnutrition, and other diseases. Fowl cholera occurs worldwide and is a relatively common disease. There is no relationship between cholera in humans (caused by *Vibrio cholerae*) and fowl cholera.

HISTORICAL INFORMATION
Fowl cholera has been recognized as a disease of poultry for more than 200 years. About 100 years ago, Pasteur isolated the organism and used it in one of the first vaccines. In the United States, Dr. Salmon studied the disease as early as 1880. Fowl cholera was one of four major livestock diseases that stimulated formation of the Veterinary Division of the United States Department of Agriculture. Although fowl cholera has been recognized and studied for almost 200 years, it still remains an important disease of poultry.

ETIOLOGY
1. The etiologic agent is *Pasteurella multocida*, a Gram-negative, bipolar-staining rod that grows readily on blood agar but not on MacConkey agar. Virulence among isolates is highly variable. Encapsulated strains are usually highly virulent; unencapsulated isolates are typically of low virulence.

2. The organism varies greatly in its antigenic makeup, a characteristic responsible for difficulties in producing effective bacterins and vaccines. The gel diffusion precipitin test has been used to describe 16 *P. multocida* serotypes, all of which have been isolated from avian hosts. Serotypes 1, 3, and 3X4 are most commonly isolated from poultry outbreaks.

3. *P. multocida* is easily destroyed by many disinfectants and by sunlight, heat, and drying. However, the organism persists for months in decaying carcasses and moist soil.

EPIDEMIOLOGY
1. Poultry flocks that have recovered from an outbreak of fowl cholera will remain carriers of *P. multocida* and spread the disease to susceptible flocks. These carriers harbor the organism in the choanal cleft and contaminate feed, water, and the environment with oral fluids. Likewise, wild birds may carry the organism and introduce it into the poultry flock if appropriate biosecurity practices are not followed.

2. Several mammalian species are carriers of *P. multocida* and may introduce the organism to poultry flocks. Swine, cats and raccoons have been shown to be carriers of *P. multocida* and those isolated have been shown to be pathogenic in poultry.

3. Birds that die of septicemic cholera have the agent in most of their tissues. Cannibalism of sick or dead birds is an important method of dissemination of the disease.

4. Resistance to cholera is correlated to humoral immunity. Immunosuppression increases susceptibility.

5. *P. multocida* is resistant enough to be readily spread on contaminated crates, feed bags, shoes, equipment, etc.

CLINICAL SIGNS
1. With acute cholera, sudden unexpected deaths occur in the flock. Mortality often increases rapidly. Laying chickens may be found dead on the nest. Geese have been reported to just drop dead while walking across a barnyard. Poisoning is often initially suspected in outbreaks of acute cholera.

2. Sick birds show anorexia, depression, cyanosis, rales, nasal and oral discharge of mucus, and white watery or green mucoid diarrhea. The course of illness is short and often followed by death. Affected chickens often conceal themselves under equipment.

3. Chronic fowl cholera is most common in chickens. Often there is swelling of a joint, wattle (Fig. 1), foot pad, or tendon sheath. Exudate, often caseous, may accumulate in a conjunctival sac or infraorbital sinus. There may be torticollis in a few birds (Fig. 2).

4. Abscesses of the infraorbital sinuses and middle ear infection resulting in torticollis, often occur in turkeys with chronic cholera.

5. In turkey breeders there is a drop in egg production and increased mortality following handling of hens during insemination. Affected toms produce thin, watery, poor quality semen.

LESIONS

1. Lesions may be absent if the disease is very acute. Usually there are petechial and ecchymotic hemorrhages at a few sites, for example, on the heart, under serous membranes, in mucous membranes, on the gizzard, or in abdominal fat. There is often a generalized hyperemia of the upper intestine. Acute lesions develop as a result of disseminated intravascular coagulation. In layers and breeder hens, free yolk in the peritoneal cavity, acute oophoritis with regressing follicles, and acute diffuse peritonitis are frequently seen. These lesions can accompany many other acute diseases.

2. In acute cases of cholera, as with other septicemias, there often is hepatomegaly. If the birds live a few days, there may be a few or many small necrotic foci in the liver (Fig. 3). Consolidation of lungs is a common finding in affected turkeys (Fig. 4 and 5). With time, these lesions become sequestered as necrotic areas in the lungs and these lung lesions often are extensive.

3. In chronic cases there may be localized inflammatory lesions. These often involve a joint, tendon sheath (Fig. 6), wattle, conjunctival sac, infraorbital sinus, the nasal turbinates, the middle ear, or cranial bones at the base of the skull. Caseous exudate in a localized lesion (Fig. 7) should arouse suspicion of cholera.

DIAGNOSIS

1. At necropsy, Gram-stained impression smears of liver, spleen or heart blood from septicemic cases often reveal bipolar-stained, Gram-negative rods suggestive of *P. multocida* (Fig. 8). Use of blood stains or methylene blue readily demonstrates the bipolar morphology of the organism.

2. Although the history, signs, and lesions may strongly suggest fowl cholera, *P. multocida* should be isolated and identified for confirmation. Isolates should be tested for antibiotic susceptibility because of widespread resistance and should be serotyped, especially if routine treatment and vaccination procedures appear ineffective or if vaccination for future prevention is desired.

3. Cholera must be differentiated carefully from erysipelas and acute colibacillosis in turkeys and other birds that are susceptible to both diseases. Erysipelas is caused by a Gram-positive rod. Cholera can be differentiated readily from most septicemic and viremic diseases of poultry by the isolation of *P. multocida*.

4. Cholera always should be suspected if there are epizootic losses in domesticated or wild waterfowl.

5. Related organisms can cause cholera like diseases or complicate other diseases. These include *Pasteurella gallinarum, P. haemolytica, R. anatipestifer, Moraxella osloensis,* and *Yersinia pseudotuberculosis.*

6. Several serological tests have been developed. Currently an enzyme-linked immunosorbent assay (ELISA) is commercially available and widely used. Serology is used primarily to evaluate efficacy of vaccination rather than for diagnosis of a disease outbreak.

CONTROL

1. *P. multocida* is not transmitted through the egg. Obtain clean birds and raise them in quarantine on disease-free premises and away from all birds and mammals that might be carriers. Never add birds to the flock as they may be carriers. Avoid stresses, insofar as is possible, and practice a high standard of sanitation.

2. Pick up and destroy all sick or dead birds before they can be cannibalized. Birds with cholera are teeming with *P. multocida* and are important in the transmission of the agent. Dispose of carcasses by burying or burning to prevent them from being fed on by scavengers (including dogs and cats).

3. Although bacterins are not always effective, in many instances they do a good job of immunizing birds, especially if they can be repeated at least once. They often are given when birds are about 8 and 12 weeks old. Bacterins do not provide good cross-protection between serotypes. Oil-emulsion bacterins are used to immunize breeders prior to production. They can cause serious drops in egg production if given to laying birds. Bacterins should contain the serotype of *P. multocida* that causes the disease at that premise.

4. Live vaccines are given via wing web inoculation to chickens and via drinking water or wing web inoculation to turkeys. In the United States live vaccines are based on the Clemson University (CU) strain of *P. multocida*. This is a naturally occurring low-virulent organism. Since its introduction as a commercial product, two milder mutants of the original CU strain have been produced: PM-1 and M-9 strains. They frequently are given to turkeys at 2-6- week intervals beginning at 6-7 weeks of age in the drinking water. Some turkey breeders are vaccinated via the wing web. Layers and breeders are inoculated by wing web stick at 10-11 weeks of age, and revaccinated in 6-8 weeks. Fowl pox vaccine may be given concurrently in the opposite wing. The live vaccines have been shown to be safe but vaccine reaction problems can occur in the field, presumably because of immunosuppression, concurrent diseases, breed sensitivity, late vaccination, or management stress such as intentional feed restriction. Parenteral administration may result in a localized lesion, or, more seriously, arthritis. Live vaccines confer better resistance than killed bacterins and offer a broad spectrum of protection against most serotypes.

5. Following an outbreak, depopulation should be considered because many surviving birds become

carriers and transmit *P. multocida*. Following depopulation, the premises and equipment should be thoroughly cleaned and disinfected and, if possible, kept free of poultry for a few weeks.

6. Continuous medication programs have been used but are generally more costly than a vaccination program.

7. Reduce rodents, scavengers, and predators in the farm environment and limit their contact with flocks.

8. Differing susceptibilities among genetic lines of turkeys have been shown, suggesting that selection for resistance to fowl cholera may be possible.

TREATMENT

1. Many sulfa drugs and antibiotics will lower the mortality from cholera but mortality may resume when treatment is discontinued. Most medications are given in the feed or water. Sulfaquinoxaline is one of the better treatments but will depress egg production in layers and may throw them completely out of production. Care should be taken to use only those products approved by the Food and Drug Administration for the class of poultry being treated. Drugs and antibiotics in common use include:

Sulfadimethoxine	Tetracyclines
Sulfadimethoxine + ormetoprim	Erythromycin
Sulfaquinoxaline	Streptomycin
Sulfamethazine	Penicillin

2. Moving an infected flock to clean premises or markedly improving sanitation during an outbreak may slow the course of cholera. Use of live vaccine during the early course of an outbreak may be effective.

3. If cholera cannot be controlled, it may be necessary to market the flock early. Be sure to adhere to regulations relating to withdrawal of medication.

Fig. 2
Head tilt (torticolis) due to otitis in a broiler breeder.

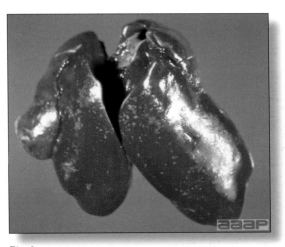

Fig. 3
White foci of necrosis in the liver of a chicken with acute fowl cholera.

Fig. 1
Chronic fowl cholera in a broiler breeder. Severe swelling of the wattles.

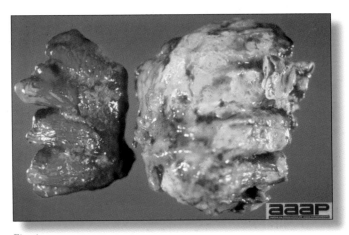

Fig. 4
Subacute to chronic pneumonia in a turkey. Notice the unilateral involvement of the lung.

FOWL CHOLERA

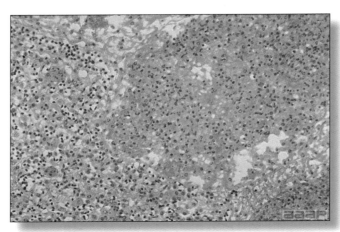

Fig. 5
Histopathology of the lung with acute fowl cholera.
Fibrinosuppurative exudate in the parabronchi extending into the
adjacent parenchyma.

Fig. 6
Subacute to chronic fowl cholera in a broiler breeder with severe
synovitis, cellulitis and tendonitis.

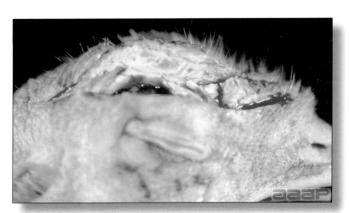

Fig. 7
Facial cellulitis in a turkey with acute fowl cholera.

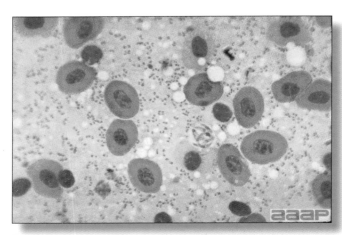

Fig. 8
Gram-stained impression smear of liver with bipolar-stained, Gram-
negative rods suggestive of *P. multocida*. Wright's stain.

GANGRENOUS DERMATITIS
(Necrotic Dermatitis)

DEFINITION
Gangrenous dermatitis is typically a disease of young growing chickens characterized by necrotic areas of the skin and by a severe, underlying, infectious cellulitis.

OCCURRENCE
Most outbreaks have occurred in chickens 4-20 weeks old. Young birds of this age group may be poorly feathered. Outbreaks often occur in excessively warm, humid houses. Gangrenous dermatitis also occurs in turkeys and has recently been a problem for turkey producers.

HISTORICAL INFORMATION
Gangrenous dermatitis was reported first in 1930 although most outbreaks have been reported since 1963. Some of the more recent reports have suggested that affected flocks may be immunologically deficient.

ETIOLOGY
Skin lesions develop because of trauma and growth of *Clostridium septicum*, *C. perfringens* type A, and *Staphylococcus aureus* either singly or in combination.

EPIZOOTIOLOGY
1. Cutaneous wounds probably occur initially as a result of cannibalism, mechanical trauma (from mechanical feeders, etc.), or other trauma. Bacteria either invade or multiply within traumatized skin and underlying tissue and their toxins or metabolites cause cellulitis. Septicemia and toxemia follow, leading to death.

2. Increased susceptibility of affected flocks to infection is an important factor in the pathogenesis. This increased susceptibility is commonly related to immunosuppression secondary to infectious bursal disease or chicken infectious anemia virus. Reticuloendotheliosis virus and adenovirus have also been implicated.

3. Other factors that may enhance susceptibility include aflatoxicosis, nutritional insufficiency or imbalance, or poor poultry house management and sanitation.

CLINICAL SIGNS

A sudden, sharp increase in mortality is often the first indication of onset. When sick birds are observed, they are depressed, and sometimes prostrate or lame. Skin lesions often, moist and crepitant, are apparent in live or dead birds. The course of the illness is often less than 24 hours. Mortality varies but can be quite high.

LESIONS
1. There are scattered patches of darkened, gangrenous skin (Fig. 1), often with cutaneous sloughing or feather loss in affected areas. Marked emphysematous or serosanguineous cellulitis (Fig. 2) underlies some skin lesions, especially with clostridial infections.

2. Swelling and infarction may be apparent in parenchymatous organs with necrotic foci in the liver.

3. Severe atrophy of the bursa of Fabricius and possibly thymus is usually present.

DIAGNOSIS
A tentative diagnosis often can be made on the basis of history and gross lesions. For confirmation, smears or histologic sections of affected tissues will reveal bacteria. Bacteria can be cultured from the area of cellulitis.

CONTROL
1. The cause of skin trauma should be found and eliminated. If cannibalism is a cause, it may be necessary to trim the beaks or improve the quality of previous beak trimming. Mechanical feeders and equipment should be examined carefully as a source of possible trauma.

2. Vaccinate the breeder flock for infectious bursal disease and chicken anemia virus to prevent or reduce possible immunosuppression in the progeny.

3. As much as possible, eliminate all stresses on the birds (e.g., parasitism, malnutrition, coccidiosis, etc.).

4. Improve sanitation in the house, particularly that of the feeders, waterers, and litter. A thorough cleaning and disinfection of the house may be helpful. If litter in the house stays wet, improve moisture control. Repeat problem houses may benefit from salting the floor at clean out. Cheap grade feed salt is used on the soil at a rate of 60-100 lb/1,000 ft^2.

5. Broad-spectrum antibiotics (e.g., penicillin, erythromycin, and tetracyclines) can be added to the ration of the flock and will reduce mortality.

TREATMENT
In addition to adding broad-spectrum antibiotics to the ration, valuable birds can be treated individually with penicillin, tetracyclines, or other fast-acting antibiotics.

GANGRENOUS DERMATITIS

Fig. 1
Darkened, gangrenous skin of a broiler
chicken with gangrenous dermatitis.

Fig. 2
Emphysematous and
serosanguineous cellulitis in
a chicken with gangrenous
dermatitis.

INFECTIOUS CORYZA
(Coryza)

DEFINITION
An acute or subacute disease of chickens, pheasants, and guinea fowl characterized by conjunctivitis, oculonasal discharge, swelling of infraorbital sinuses, edema of the face, sneezing, and sometimes by infection of the lower respiratory tract. Prolonged outbreaks are now believed to be outbreaks complicated by other diseases, especially *Mycoplasma gallisepticum* infection (chronic respiratory disease).

OCCURRENCE
Chickens are primarily affected, although the disease has been reported in pheasants and guinea fowl. All ages of chickens are susceptible although most natural outbreaks occur in chickens that are half grown or older. The disease is seen more frequently on chicken farms where facilities are never emptied of chickens. The disease has a worldwide distribution. Infectious coryza does not occur in turkeys and should not be confused with turkey coryza caused by *Bordetella avium*.

HISTORICAL INFORMATION
Infectious coryza was believed to be a separate disease of chickens as early as 1920 but this was not confirmed until 10-15 years later. The incidence of coryza has varied markedly. Presently coryza is a disease of considerable importance, especially on multiage egg production complexes.

ETIOLOGY
1. The etiologic agent, *Avibacterium paragallinarum* (formerly *Hemophilus paragallinarum* and *H. gallinarum*) is a Gram-negative, bipolar-staining, nonmotile rod with a tendency toward filament formation. *A. paragallinarum* requires V-factor (nicotinamide adenine dinucleotide), which is available in certain enriched medium (i.e., chocolate agar). It grows on blood agar (with a *Staphylococcus aureus* nurse colony) as dewdrop-like satellite colonies in a microaerophilic environment. V-factor independent isolates have been described from South Africa and Mexico.

2. *A. paragallinarum* is not a very resistant organism and will persist outside of the host for only a few days. It is easily destroyed by many disinfectants and by environmental factors. The organism is susceptible *in vitro* to many chemicals and antibiotics, including spectinomycin, neomycin, novobiocin, and tetracycline.

3. *A. paragallinarum* is present in sinus exudate and is easily demonstrated in stained smears.

4. There are several strain classification schemes. The Page scheme recognizes three antigenic types (A, B, C) of *A. paragallinarum*, although all types share certain antigens. Hemagglutinins produced by the organism appear to be important antigens capable of inducing protection against infectious coryza. Bacterins are available that allow limited protection to laying chickens.

EPIDEMIOLOGY
Chronically ill or apparently healthy carrier birds are the major reservoirs of infection and readily transmit the agent to susceptible chickens. Transmission probably occurs by inhalation of infectious aerosol coughed into the air or through ingestion of contaminated feed or water. The etiologic agent can be transmitted by fomites, although it soon perishes outside of the host. Recovered birds are frequently carriers.

CLINICAL SIGNS
1. Usually there is a rapid onset and morbidity is high in the flock. Feed consumption, egg production or growth are reduced noticeably.

2. There is oculonasal discharge, conjunctivitis with some adherence of eyelids, edema of the face (Fig. 1) (occasionally of the wattles) (Fig. 2), respiratory noises, and, perhaps, diarrhea. Later, some of the birds may have swollen infraorbital sinuses and/or exudate in the conjunctival sac. There is considerable variation in the severity and length of course in flock outbreaks.

3. Respiratory signs usually persist for only a few weeks. Persistence of signs occurs when complicated by fowl pox, *M. gallisepticum*, infectious bronchitis, *Pasteurella sp.*, or infectious laryngotracheitis and unthrifty birds will become apparent. Persistence of signs was once attributed entirely to strains of *A. paragallinarum* of low virulence.

GROSS LESIONS
1. There is catarrhal inflammation of nasal passages and sinuses and nasal discharge often is apparent (Fig. 3). One or both infraorbital sinuses may be distended with exudate (similar distension can occur with localized fowl cholera, pox, vitamin A deficiency, and staphylococcal infection).

2. There is conjunctivitis, frequently with adherence of the eyelids or with accumulation of cheesy exudate in the conjunctival sac.

3. There often is edema of the face and, occasionally, of the wattles. In complicated cases there may be tracheitis, pneumonia, or airsacculitis.

DIAGNOSIS
1. Typical history, signs, and lesions are suggestive of coryza, although other respiratory diseases of chickens must be ruled out.

2. A smear of sinus exudate should be made and Gram stained. It should reveal Gram-negative, bipolar-staining rods with a tendency toward filament formation and pleomorphism.

3. Aseptically collect sinus exudate and swab it on blood agar. On the same plate then make an S-shaped streak of *S. aureus* (use a strain that excretes V-factor), which will serve as a feeder colony. Incubate the culture in a candle jar. Tiny dewdrop satellite colonies (Fig. 4) of *A. paragallinarum* will grow adjacent to the feeder colony. The organism can be further identified by biochemical means or by a PCR test specific for *A. paragallinarum*.

4. A nonpathogenic species, *Avibacterium avium*, (*Hemophilus avium*) may be cultured from the sinus, either alone or with *A. paragallinarum*. *A. paragallinarum* is catalase negative and the nonpathogenic species is catalase positive.

5. Put a small amount of sinus exudate in the infraorbital sinus of a few young susceptible chickens. Typical signs and lesions develop in 3-5 days (rarely less).

6. Hemagglutination inhibition and immunodiffusion tests can be used to detect *A. paragallinarum* antibodies in serum. Both tests are serotype specific.

CONTROL

1. Depopulate, if necessary, to eliminate all carrier birds. Leave the premises vacant for two to three weeks after thorough cleaning and disinfection before restocking with 1-day-old or other coryza-free chickens. As much as possible, raise them in quarantine.

2. Commercial bacterins can be used to immunize chickens and protect only for the serotype included in the vaccine. All pullets to be housed on multiage infected farms should receive two injections of the bacterin at 4-week intervals prior to 20 weeks of age. The first vaccination should be given after the birds reach 10 weeks of age.

TREATMENT

Various sulfonamides and antibiotics have been used to alleviate the severity, usually in feed or drinking water. Birds usually respond to treatment but relapses may occur when treatment is discontinued. Erythromycin and oxytetracycline are commonly used in layer operations.

INFECTIOUS CORYZA

Fig. 1
Facial edema in a White Leghorn pullet.

Fig. 2
Male broiler breeder with edema of both face and wattles.

Fig. 3
White Leghorn pullet with slight facial edema and showing nasal discharge.

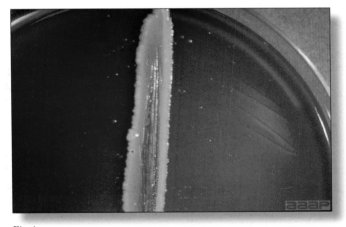

Fig.4
48-hr culture of *A. paragallinarum* on Casman blood agar showing satellite growth of *S. epidermidis* streak.

MYCOPLASMOSIS

PREFACE
Mycoplasmas belong to the Order *Mycoplasmatales*, are the smallest prokaryotic organism (DNA content) cultivatable on artificial medium and are completely devoid of a cell wall. The absence of a cell wall accounts for their pleomorphic shape, fried-egg colony appearance and their resistance to penicillin-like antibiotics. Over twenty-five species have been identified from avian hosts although several isolates are unidentifiable. Generally, mycoplasmas have a narrow host range. Four mycoplasma species are considered pathogenic to commercial poultry; *Mycoplasma gallisepticum, M. synoviae, M. meleagridis* and *M. iowae*. Pathogenic species generally infect the respiratory system but other systems may be involved. Transmission is generally by direct contact although egg transmission, carrier birds and fomites are of importance. The cultivation of *Mycoplasma sp.* is somewhat demanding and requires specialized media containing 10 – 15% serum, yeast-derived components and unique factors for certain species. Colony morphology is variable but is characterized by a "fried-egg" appearance. In general, colony morphology, cultural characteristics or carbohydrate fermentation are not useful for speciation. Identification to the species level is usually based on immunologic tests utilizing species-specific antisera or amplified DNA type tests. Inoculation of 5-7 day-old embryos is an alternative isolation procedure utilized when artificial media is unrewarding. Alternatively, clinical material can be inoculated into young chickens or turkeys and a comparison of pre-inoculated sera with 3-5 week post-inoculated sera by the immunologic tests may provide clues as to which *Mycoplasma sp.* is involved.

I. MYCOPLASMA GALLISEPTICUM INFECTION

(MG; Chronic Respiratory Disease (CRD);
Infectious Sinusitis of Turkeys)
(See Table on Page 116)

DEFINITION
A mycoplasma infection characterized by respiratory signs and lesions, a prolonged course in the flock and primarily affecting chickens and turkeys. In turkeys the disease is frequently manifested by swelling of the infraorbital sinus(es) and called infectious sinusitis.

OCCURRENCE
Mycoplasma gallisepticum (MG) occurs primarily in chickens and turkeys but also has been reported in partridge, pheasants, peafowl, quail, guinea fowl, ducks, geese, and pigeons. All ages of chickens and turkeys can have the disease although the very young are seldom submitted for the disease. Since 1994, a serious *M. gallisepticum* infection of free-ranging house finches has

been shown to cause periorbital swelling, conjunctivitis and mortality.

HISTORICAL INFORMATION
In the United States the disease was first described in turkeys in 1905, then in chickens in 1935. MG became of major importance as the poultry industry expanded over the last 25 years. The first of a series of national conferences on mycoplasmosis in poultry was held in 1962 and recognized the importance of MG. Considerable progress has been made in the control and eradication of MG, especially in turkeys, but the disease is still of major importance.

MG is one of the more costly poultry diseases, sharing that distinction in the United States with Marek's disease and Newcastle disease. A few years ago the annual loss from MG was estimated at 125 million dollars.

ETIOLOGY
1. *M. gallisepticum* is the etiologic agent. In poultry flocks, other infecting organisms complicate or increase the pathogenicity of MG infections. Typical complicating organisms include: infectious bronchitis virus, Newcastle disease virus, *Escherichia coli*, *Pasteurella multocida*, and *Avibacterium paragallinarum*.

2. *M. gallisepticum* seldom survives for more than a few days outside of the host. Carrier birds are essential for its survival.

3. In chickens the organism may be present and cause no disease until triggered by stress, such as changes in housing, management, nutrition, or weather; vaccination against or infection with infectious bronchitis or Newcastle disease; or increased levels of dust or ammonia in the environment.

4. *M. gallisepticum* strain variability exists and accounts for the variability of host susceptibility, clinical presentation and immunologic response.

EPIDEMIOLOGY
M. gallisepticum is transmitted in some of the eggs (transovarian transmission) laid by inapparent carriers. Infected progeny then transmit the agent horizontally, probably through infectious aerosols coughed into the air and through contamination of feed, water, and the environment. The agent probably can be transmitted by other species of birds, domestic or wild. In addition the agent can be transmitted mechanically on shoes, feed sacks, crates, etc.

CLINICAL SIGNS
Signs usually develop slowly in the flock. They vary in severity depending on strain of organism, and may persist for weeks or months. Signs are the same as those observed with many other avian respiratory diseases. They include coughing, sneezing, snicks, rales, ocular and

nasal discharge, and, in turkeys, swelling of the infraorbital sinus(es) in occasional birds. Additional signs are listed below.

1. Adult layers- Drop in feed consumption and egg production. Egg production continues at a lower level. Mortality is low but there may be many unthrifty birds.

2. Broilers (4-8 weeks old) - Signs are more pronounced than in adult birds and the disease is more severe. Feed intake and growth rate are reduced. Mortality is variable but may be high, particularly if poor husbandry, exposure, or other stress factors are present.

3. Turkeys- In addition to swelling of one or both infraorbital sinuses (Fig. 1), infected turkeys may have nasal exudate wiped on their wings. Alternatively, if the air sacs and lungs are primarily involved, mortality from pneumonia and airsacculitis may be very high although few birds have swollen sinuses.

LESIONS

1. Poor physical condition and loss of weight are usually apparent and suggest the presence of a chronic disease.

2. There is marked catarrhal inflammation of the nasal passages, sinuses, trachea, and bronchi and air sacs (Fig. 2). Air sacs often are thickened (Fig. 3) and opaque and may contain hyperplastic lymphoid follicles in their wall (Fig. 4). Recent vaccination against Newcastle disease or infectious bronchitis may enhance opacity of air sacs. Air sacs often contain mucoid or caseous exudate.

3. The following classic triad of lesions is often the basis of extensive condemnation of infected birds at slaughter: airsacculitis, fibrinous perihepatitis, and adhesive pericarditis (Fig. 5). These lesions are not pathognomonic and may occur with chlamydiosis or septicemia.

4. In infectious sinusitis of turkeys, lesions may be restricted to swelling of the infraorbital sinuses. Conversely, sinusitis may be absent although rhinitis, tracheitis, and airsacculitis occur and there may be a fibrinous pneumonia. Occasional turkeys and chickens may have the oviduct distended with exudate (salpingitis).

DIAGNOSIS

1. A history of chronic respiratory disease accompanied by lowered feed consumption, poor weight gains, or lowered egg production is suggestive of MG. Typical gross lesions are very suggestive.

2. Positive serum plate (Fig. 6) or tube agglutination tests for MG on sera from a few birds in the flock strengthen the diagnosis. Because sera from birds with *Mycoplasma synoviae* may cross-react, it is a good plan to confirm some of the agglutination tests with the hemagglutination inhibition (HI) test (Fig. 7) or

MG ELISA test. Cross-reactions usually do not occur when the HI or ELISA test is used. Flocks recently vaccinated with oil-based vaccines may also produce false-positive agglutination tests.

3. A commercial PCR test specific for *M. gallisepticum* is available. Tracheal swabs from a number of birds can be tested.

4. Isolation and identification of *M. gallisepticum* can be cultured from exudate, trachea, sinuses, air sacs, or lungs on artificial media (Fig. 8) or in chick embryos. MG identified isolates can be compared by molecular techniques for epidemiologic purposes.

5. In many instances it will be necessary to differentiate MG from other respiratory diseases of poultry, usually by culture or serologic tests. Pulmonary and air sac lesions may be confused with colibacillosis and aspergillosis. In turkeys, fowl cholera is a frequent and important complication and may be accompanied by a fibrinous pneumonia. Sinusitis in turkeys can also be caused by avian influenza, *M. synoviae* infection, cryptosporidiosis, and avian metapneumovirus.

CONTROL

1. Depopulation of infected premises should precede establishment of a "clean" flock. Thoroughly clean and disinfect the houses and leave them vacant for a few weeks.

2. Prevention is based largely on obtaining chicks or poults hatched from eggs from MG-free breeder flocks. The MG-free progeny are then raised in quarantine. MG-free breeder flocks have been established for both chickens and turkeys as part of the National Poultry Improvement Plan. They are monitored by serologic testing to assure that they and their eggs are free of *M. gallisepticum*. Quarantine measures must be strictly enforced and good management and sanitation must be practiced to keep a flock free of infection.

3. The vaccination of replacement pullets scheduled to enter MG positive layer complexes is practiced in most states. Three live commercial vaccines (F-strain, TS 11 and 6/85) are available and are administered via fine spray or eye-drop to birds during the growing period to protect them from clinical disease during the laying period. The live F-strain MG vaccine is pathogenic in turkeys. An oil-emulsion- based bacterin is also available for use in replacement pullets destined for multiage egg production complexes where MG is established in older hens.

4. Many MG-free breeder flocks were established initially by identifying small flocks that were not infected and using that flock as a nucleus. In addition egg dipping (in antibiotic solutions), heat sterilization, and antibiotic treatment of hatching eggs have all been used in attempts to obtain disease-free progeny from infected breeder flocks. All of the latter three methods reduce the number of infected progeny that hatch from eggs

from infected flocks. No antibiotic or drug given to infected breeders will prevent them from laying some infected eggs.

TREATMENT

1. Marketing an infected flock with a low incidence of the disease may be more economical than treatment because treatment is very expensive and doesn't clear the infection. Consider this possibility initially.

2. Improve the management, husbandry, or nutrition if possible. In particular, try to reduce the dust to which the birds are exposed if dust is excessive. Remove accumulated manure and improve ventilation if ammonia levels are excessive. Eliminate all possible sources of stress.

3. Many broad-spectrum antibiotics have been used for treatment and will suppress losses. However, relapses often occur when treatment is discontinued. Most antibiotics are given in feed or water, preferably in water. Tylosin and tetracyclines have been used extensively for treatment. Injectable antibiotics may be more effective if the disease is advanced and if the flock is small enough to be treated individually. Be sure to follow recommended antibiotic withdrawal times to prevent residues in meat.

II. MYCOPLASMA MELEAGRIDIS INFECTION

(MM Infection)
(See Table on Page 116)

DEFINITION
An egg-transmitted mycoplasmosis of turkeys characterized by inapparent venereal infection in breeder turkeys, airsacculitis in recently hatched poults and late embryo mortality.

OCCURRENCE
Mycoplasma meleagridis (MM) infection is confined to turkeys and occurs in all age groups. The disease is usually inapparent except in nonhatching turkey embryos or recently hatched poults in which it causes airsacculitis. Most large turkey breeder flocks are now free of MM infection.

HISTORICAL INFORMATION
1. Airsacculitis in newly hatched poults was first noted in 1958 but was not considered a major cause of loss because the lesions usually regressed. Airsacculitis at time of slaughter was usually attributed to infection with *Mycoplasma gallisepticum* (MG) or *Mycoplasma synoviae* (MS).

2. After MG and MS infections were brought under control, airsacculitis still remained as a significant cause of condemnation at slaughter. MM infection was found to be the cause.

3. *M. meleagridis* has been eliminated from the major primary turkey breeders. Infection in commercial flocks is not uncommon.

4. Skeletal abnormalities have been referred to as TS-65 (Turkey Syndrome 65) and crooked necks abnormalities have been referred to as wryneck.

ETIOLOGY
1. The etiologic agent is *M. meleagridis*. The organism is fastidious in its growth requirements and, presumably, easily destroyed by environmental factors and most disinfectants.

2. Concurrent infection with other mycoplasmas can occur and increases the severity of lesions.

EPIDEMIOLOGY
1. Infected breeder hens lay some eggs containing *M. meleagridis*, transmitting mycoplasma to some of their progeny. Organisms are spread horizontally to many hatchmates via aerosols from the respiratory tract or to the vent on contaminated hands during vent-sexing.

2. In some progeny, the organism later spreads to and localizes in the reproductive tract. In male turkeys localization is often in the cloaca and/or phallus and their semen may contain the agent.

3. Artificial insemination of turkey hens with pooled infected semen is an important method of MM spread.

4. Most tom and hen turkeys overcome MM infection after one breeding season. By then the infection has already been transmitted to many of their progeny.

5. MM has a distinct predilection for the bursa of Fabricius and can cause immunosuppression, which may explain why infected turkeys are more susceptible to other infections, especially colibacillosis.

CLINICAL SIGNS
Most of the following signs are mild or inapparent on casual examination and go unobserved.

1. Often there is impaired hatchability, due to late term embryo death, in eggs from infected flocks. Embryo mortality is highest after eggs are transferred to the hatcher and at pipping.

2. Poults hatched from infected lots of eggs may have a high incidence of starve outs and may make poor weight gains.

3. Young growing poults may show mild respiratory signs and occasional poults may have sinusitis.

4. Small numbers of poults may have skeletal abnormalities associated with a deforming osteomyelitis in cervical vertebrae (crooked neck) or leg deformities.

5. Adult breeders usually show no signs (Fig. 1) of venereal or respiratory infection.

LESIONS

1. Infected, pipped, unhatched embryos and recently hatched poults have a variable degree of airsacculitis (Fig. 2) manifested by thickening of air sac membranes and possibly by the presence of small amounts of yellow exudate in the air sacs (Fig. 3). In uncomplicated MM, lesions regress and disappear in most turkeys by marketing time.

2. Poults with wryneck may have a cervical airsacculitis and osteomyelitis of adjacent vertebrae that can be demonstrated microscopically.

3. Adult breeders are free of gross lesions of the genitalia. However, outbreaks of airsacculitis, synovitis, and sinusitis have been observed in mature and semimature turkeys from which only MM could be isolated.

4. MM can also cause a generalized skeletal disorder historically known as turkey syndrome 65 (TS 65) characterized by chondrodystrophy, or unilateral or bilateral varus deformities (Fig. 4) and perosis.

DIAGNOSIS

1. Monitor pipped embryos and weak, cull poults for the presence of air sac lesions.

2. Diagnosis can be made by the isolation and identification of MM from infected tissues or exudates. The organism is fastidious and requires special media. Isolates must be differentiated from MG, MS, and other mycoplasmas by serologic methods or fluorescent antibody techniques.

3. Turkeys infected with MM develop antibodies in 3-5 weeks. These antibodies can be demonstrated by plate and tube agglutination tests. Positive reactors can be confirmed by hemagglutination inhibition tests. These tests alone are not adequate for eradication of infection but can indicate infection in a flock. Test antigens are commercially available.

CONTROL

1. Poults should be obtained from MM-free breeder flocks.

2. A breeder flock free of MM infection can be established by inoculating all of the turkey eggs for hatching with 0.6 mg of gentamicin sulfate and 2.4 mg of tylosin in a 0.2-ml volume. The newly hatched flock is monitored by culturing pipped (unhatched) eggs and 1-day-old cull poults and by using the plate agglutination test on sera from the flock.

3. Dipping the eggs from infected breeder flocks into solutions of tylosin or gentamicin will substantially reduce the incidence of infection in the progeny. Dipping is often combined with temperature- or pressure-differential techniques.

4. Repeated serologic testing alone has not been successful in establishing clean flocks. However, both egg dipping and testing programs have value.

5. Poults are often injected with an antibiotic during servicing in the hatchery, which probably aids in reducing MM infections.

6. Treatment of semen with antibiotics results in an unacceptable decrease in sperm viability.

TREATMENT

Because MM is sensitive to tylosin and tetracyclines, they may be of value in controlling airsacculitis in infected turkeys. It is unlikely they would be effective in controlling venereal infection. Use of lincomycin/spectinomycin in the drinking water (2 g/gal) for the first 5-10 days of life reduced the incidence of airsacculitis and improved weight gains.

III. MYCOPLASMA SYNOVIAE INFECTION

(MS, Infectious Synovitis; Tenovaginitis)
(See Table on Page 116)

DEFINITION

Predominately a subclinical upper respiratory infection of chickens and turkeys. Systemic infection results in an acute or chronic condition of chickens and turkeys characterized by inflammation of synovial membranes and, usually, by exudate in the joints and tendon sheaths of many infected birds. Synovial involvement is referred to as infectious synovitis.

OCCURRENCE

Respiratory *M. synoviae* infections are common in multi-age commercial layer flocks and are predominately subclinical. The Infectious Synovitis form occurs in chickens, especially in broilers, but the disease also occurs in turkeys. The disease is usually seen in young (4-12-week-old) chickens or young (10-12-week-old) turkeys. It has been seen in adult layers and in chicks as young as 6 days of age. Synovitis occurs throughout the year but is more severe during the cold, damp seasons or whenever the litter is wet. The disease is probably worldwide in distribution. The incidence of clinical disease has greatly decreased in recent years in the United States.

HISTORICAL INFORMATION

Infectious synovitis was first described in chickens in 1954 and in turkeys in 1955. Although the disease was uncommon initially, it is now well established. It is said to be less commonly encountered today than a decade ago.

ETIOLOGY

1. The etiologic agent is *M. synoviae* (MS), a fastidious organism that requires nicotinamide adenine dinucleotide for growth. There appears to be only one

serotype of the organism, although isolates vary in pathogenicity.

2. *M. synoviae* is a fastidious organism. It can usually be grown in 5-7-day-old embryonating chicken eggs or in special mycoplasma media. A commercially available PCR test can rapidly identify a MS positive flock.

3. Convalescent sera from birds with *M. synoviae* will agglutinate commercially available *M. synoviae* plate antigen. During the early stages of synovitis the sera may also agglutinate *Mycoplasma gallisepticum* plate antigen. Cross-reactions usually do not occur when hemagglutination inhibition (HI) or ELISA tests are used to separate *M. synoviae* from *M. gallisepticum* infection.

EPIDEMIOLOGY

1. Transovarian transmission is an important means of spread of the infectious agent. Only a small number of eggs from reactor birds carry *M. synoviae* and most of them are laid during the earlier stages of infection.

2. Infection also spreads horizontally via the respiratory tract. Such spread is slow and only part of the infected birds develop joint lesions.

3. The organism has a predilection to localize in synovial-lined structures such as joints, tendon sheaths and bursas (breast blisters). It also localizes in the ovary and, occasionally, in the air sacs or sinuses.

CLINICAL SIGNS

1. Lameness in many birds and a tendency of affected birds to rest on the floor are prominent early signs. Many affected birds have pale head parts and swollen hocks or foot pads. The feces of acutely affected birds often are green. Eventually, affected birds become dehydrated and thin because of failure to eat and drink regularly.

2. Morbidity is usually low to moderate but may be high if there is damp, cold weather or the litter is wet. Mortality is usually less than 10% unless there are other diseases present or the husbandry is poor.

3. A slight, transient egg production drop maybe observed in acutely infected layer flocks.

4. Respiratory tract infections are usually asymptomatic.

LESIONS

1. In the early phase of synovitis most synovial-lined structures (joints, tendon sheaths) contain a sticky, viscid, grey to yellow exudate (Fig. 1). This is usually more voluminous in swollen hock or wing joints or under swollen foot pads (Fig. 2).

2. In later stages of the disease the birds may be emaciated or thin and there may be no lesions in internal organs. Exudate in joints and tendon sheaths may be inspissated or joint surfaces may be stained

orange or yellow. Breast blisters often are present secondary to trauma from resting on the floor.

3. Respiratory lesions may be absent or consist of a mild mucoid tracheitis, airsacculitis (Fig. 3), or sinusitis, lesions that are usually associated with *M. gallisepticum* infection (chronic respiratory disease). Such birds may not have the usual lesions of synovitis described above.

DIAGNOSIS

1. Typical signs and gross lesions, especially epizootic lameness and characteristic exudate in swollen joints or tendon sheaths, are suggestive of synovitis. The diagnosis can be strengthened by obtaining positive plate agglutination tests for synovitis on sera from birds in the flock (Fig. 4). Three to 5 weeks are required for antibody formation to have occurred.

2. *M. synoviae* can be isolated on special media or in 5-7-day embryonating chicken eggs. Trachea (Fig. 5), sinuses, air sacs or synovial exudate are preferable for culture. The isolated *Mycoplasma* can be identified by direct fluorescent antibody techniques applied to colony imprints (Fig. 6). Alternatively, exudate may be inoculated into the foot pad of chickens and turkeys to reproduce typical lesions. Later, their preinoculation and convalescent sera may be tested against *M. synoviae* antigen and *M. gallisepticum* antigen. The HI test can be used to confirm agglutination test results.

3. A commercial PCR test specific for *M. synoviae* is available. Tracheal swabs from a number of birds can be tested.

4. Synovitis must be differentiated from arthritis caused by *Staphylococci*, fowl typhoid, pullorum disease, and viral arthritis. Agents of the first three are easily cultured. Viral arthritis should infect experimentally inoculated chickens but not turkeys.

CONTROL

1. In most areas it is now possible to get chicks or poults that were hatched from eggs from MS-free flocks. If possible, start with such chicks or poults.

2. As far as possible, raise the birds in quarantine under the all-in, all-out system.

3. Synovitis can usually be prevented by continuously giving the birds a low-level antibiotic in the feed. This is an expensive procedure. Many antibiotics used for treatment can be used for prevention but are fed at a lower level.

4. Commercial *M. synoviae* vaccine is available and maybe beneficial in certain management situations.

TREATMENT

Treatment of lame birds with well-established synovitis is usually not very satisfactory. Relatively high levels of antibiotics are required and may be given in feed or water. Tetracyclines have been widely used. Streptomycin has been used intramuscularly in small groups of birds where they could be handled individually.

IV. OTHER MYCOPLASMA INFECTIONS

Many other species of mycoplasmas are isolated from poultry, especially hobbyist and household flocks. Few if any cause disease. The only other mycoplasma of note that occasionally causes clinical disease in commercial poultry flocks is *M. iowae*. *M. iowae*, like *M. meleagridis*, is spread venereally and typically causes reduced hatchability and late term embryo mortality in turkeys.

MYCOPLASMA GALLISEPTICUM

Fig. 1
Swollen sinuses in a MG infected turkey.

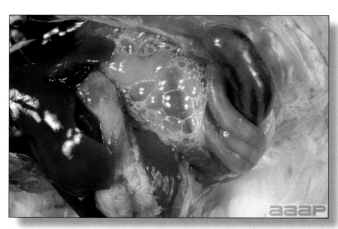

Fig. 2
Acute MG airsacculitis in the abdominal air sac of a broiler chicken.

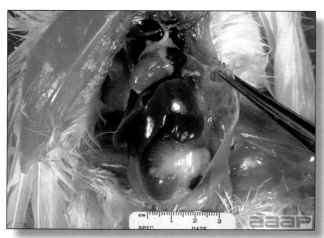

Fig. 3
Thickened posterior thoracic air sac in a chicken.

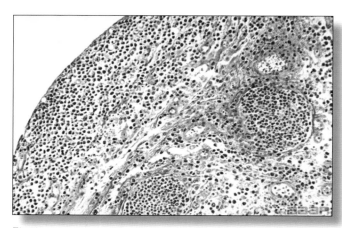

Fig. 4
Lymphoid nodules are prominent in this air sac. Increased air sac thickness is due also to a diffuse infiltration of lymphocytes, macrophages and plasma cells.

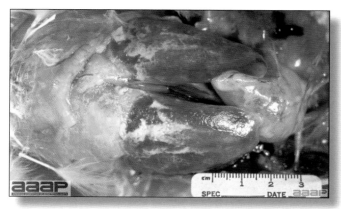

Fig. 5
Classic triad of lesions; adhesive pericarditis, fibrinous perihepatitis and airsacculitis.

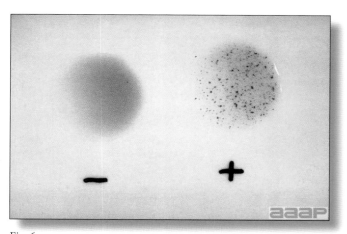

Fig. 6
Serum-plate-agglutination test with MG antigen. The serum sample on the right is positive, and the sample on the left is negative.

MYCOPLASMA GALLISEPTICUM

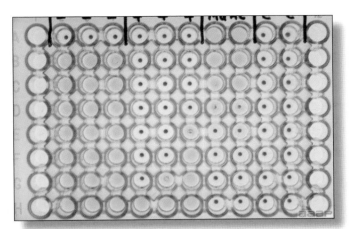

Fig. 7
HI test with 3 negative and 3 positive sera. Sera are diluted from top-to bottom, beginning at 1:10. Sera in rows 2, 3, and 4 are from negative control birds (the top well is a serum control). Sera in rows 5, 6, and 7 are from MG-infected chickens. Their titers are 1:640, 1:320 and 1:80 respectively. Rows 8 and 9 are antigen controls and rows 10 and 11 are cell controls. Rows 1 and 12 are empty.

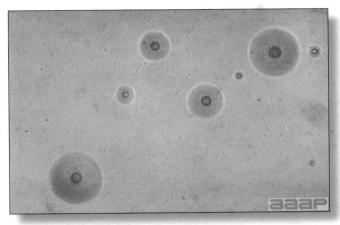

Fig. 8
Typical MG colonies as observed by microscopic examination of agar medium at 35X.

MYCOPLASMA MELEAGRIDIS

Fig. 1
Flock of MM positive turkey hens with no clinical signs of infection.

Fig. 2
Mild airsacculitis.

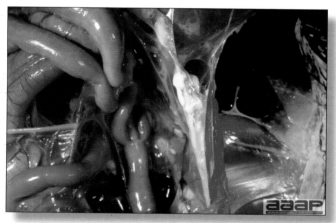

Fig. 3
MM-caused airsacculitis in a 4 week-old poult; the lesion has progressed from the thoracic air sacs to the abdominal air sacs by this age. If the disease is uncomplicated, the lesions will regress within 16 weeks.

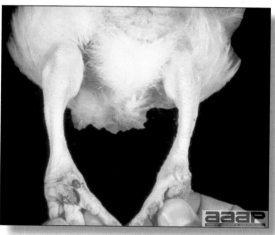

Fig. 4
Bowing of the tarsometatarsal bones of a 3-week-old poult infected during embryonic development with the pathogenic RY-39 strain of MM.

MYCOPLASMA SYNOVIAE INFECTION

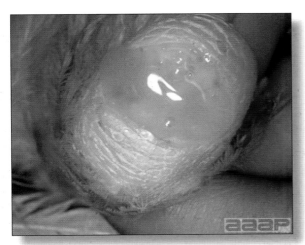

Fig. 1
Incised swollen hock joint of chicken with MS synovitis.

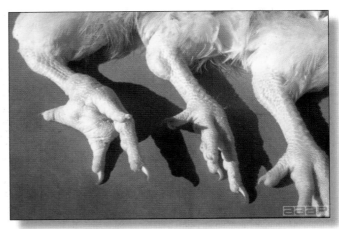

Fig. 2
Enlarged footpads typical of MS infectious synovitis.

Fig. 3
Chronic airsacculitis: thickening of the air sac membrane with large masses of caseous exudate.

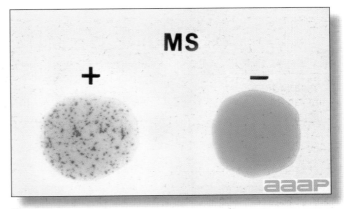

Fig. 4
Rapid serum-plate test to detect for the presence of antibodies for MS in the sera of infected chickens or turkeys. Agglutination of the rose-bengal-stained antigen is noted by aggregates or clumps of MS organism by specific antibodies produced by the infected birds.

Fig. 5
Chicken with its tongue pulled aside to position the larynx for insertion of a cotton swab to obtain tracheal exudate for culture of most avian mycoplasma, including MS. Tracheal infection tends to persist for several weeks to months.

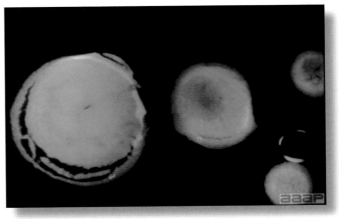

Fig. 6
Greenish glow of colonies MS positive in the fluorescent-antibody (FA) test.

THE MYCOPLASMOSES[A]

Name(s) of the Disease	Etiologic Agent	Nature of the Disease	Major Lesions
Mycoplasma gallisepticum infection; chronic respiratory disease	*Mycoplasma gallisepticum*	A respiratory disease.	Airsacculitis, adhesive pericarditis, fibrinous perihepatitis. Occasionally causes synovitis or salpingitis.
Infectious sinusitis	*Mycoplasma gallisepticum*	Unilateral or bilateral sinusitis. May spread to or occur initially in the lower respiratory system.	Swollen infraorbital sinus(es) may or may not be followed by airsacculitis, pericarditis, and perihepatitis.
Infectious synovitis	*Mycoplasma synoviae*	Involves synovial lining of joints, tendon sheaths. Results in lameness, debility.	Swollen joints and tendon sheaths. Feet, shanks, hocks more obviously affected. Occasionally causes airsacculitis in broilers and turkeys.
Mycoplasma meleagridis infection; MM infection	*Mycoplasma meleagridis*	A venereal infection of turkeys, usually transmitted by infected semen. Produces airsacculitis in many progeny.	Airsacculitis in nonhatching or newly hatched poults. May spread horizontally to other young poults as an airsacculitis. May lead to airsacculitis in market birds.
Mycoplasma iowae infection	*Mycoplasma iowae*	Associated with reduced hatchability and embryo mortality in turkeys. May spread venereally.	Infected turkey embryos die from about 18 days of incubation with lesions of stunning and congestion with hepatitis, edema and splenomegaly.

[A] More than 20 serotypes of *Mycoplasma* have been identified in chickens, turkeys, and ducks. These are the most significant pathogens.

NECROTIC ENTERITIS

DEFINITION
Necrotic enteritis is an acute bacterial infection primarily of chickens and turkeys, although other avian species can be affected. The disease is characterized by sudden death, distended intestines with necrosis of the intestinal mucosal.

OCCURRENCE
Chickens 2-10 weeks of age raised on litter are most frequently involved. Turkeys that are 7-12 weeks of age are also affected. Both species usually have some predisposing enteric condition.

HISTORICAL INFORMATION
Necrotic enteritis was first reported in 1961 and has been reported to occur where ever poultry are produced.

ETIOLOGY
1. *Clostridium perfringens* (mostly type A but also C) and their toxins are the cause of necrotic enteritis. These bacteria are anaerobic Gram-positive rods and produce double-zoned hemolysis on blood-agar plates.

2. Alpha toxin is produced by *C. perfringens* type A and C and beta toxin is produced by *C. perfringens* type C and is responsible for the mucosal necrosis.

3. Recent research on the pathogenesis of necrotic enteritis has focused on a gene known as NetB which is responsible for producing a pore forming compound.

4. *C. perfringens* are ubiquitous and are normal inhabitants of the intestinal tract.

5. Slower peristaltism or intestinal mucosal damage is necessary for the *Clostridia* to attach, proliferate and produce sufficient toxin. Coccidiosis, ascarid migration, hemorrhagic enteritis in turkeys, and severe salmonella infection are predisposing conditions for mucosal damage.

EPIDEMIOLOGY
Necrotic enteritis often develops as an acute terminal complication of other primary intestinal diseases or in situations where the intestinal microflora is disturbed or the host is severely immunosuppressed. A disturbed intestinal microflora can result from sudden changes in feed formulation such as addition of high levels of fish meal or wheat. Immunosuppression from infectious bursal disease or hemorrhagic enteritis frequently precedes necrotic enteritis.

CLINICAL SIGNS
The acute onset of depressed, ruffled birds occurs; these birds rapidly progress to death. There is a rapid increase in mortality.

GROSS LESIONS
1. Lesions are usually found in the mid-small intestines, which are distended and friable.

2. The intestine is typically distended, the mucosa is covered by a brownish diphtheritic membrane (Fig. 1) and contents consist of foul-smelling brown fluid (Fig. 2).

3. Severe dehydration with darkening of the breast muscle and swelling and congestion of the liver may also be present (Fig. 3, normal and congested livers).

DIAGNOSIS
1. Intestinal mucosal appearance and typical history of acute and dramatic increase in mortality is strongly suggestive of necrotic enteritis.

2. Histologically, there is heavy clostridial colonization of the villous epithelium accompanied by coagulative necrosis of the mucosa (Fig. 4).

3. Identification of a predisposing factor is necessary for successful treatment and prevention.

CONTROL
1. Good management practices of cleaning and disinfection of poultry houses prior to bird placement are essential. Repeat problem houses may benefit from salting the floor at cleanout. Cheap grade feed salt is used on the soil at a rate of 60-63 lb/1,000 ft^2.

2. All predisposing factors must be controlled and the coccidia prevention program verified.

3. Administration of appropriate feed medication may be warranted.

TREATMENT
Determination of predisposing condition will dictate specific medication. The clostridial component of this disease usually responds well to the same antibiotics specified for ulcerative enteritis (i.e., bacitracin, penicillin, and lincomycin).

NECROTIC ENTERITIS

Fig. 1
In a case of necrotic enteritis, the intestine are typically distended with the mucosa covered by a brownish diphtheritic membrane.

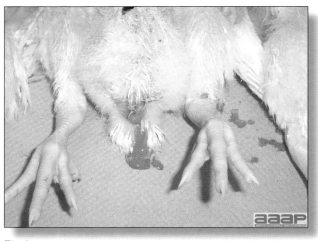

Fig. 2
NE in a broiler chicken: Intestinal contents consist of foul-smelling brown fluid.

Fig. 3
Severe dehydration with darkening of the breast muscle and swelling and congestion of the liver may also be present (chicken dead with necrotic enteritis on the right, normal chicken on the left).

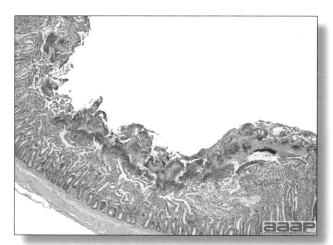

Fig. 4
Histologically, there is heavy clostridial colonization of the villous epithelium accompanied by coagulative necrosis of the mucosa.

ORNITHOBACTERIUM RHINOTRACHEALE INFECTION
(OR, ORT)

DEFINITION
Ornithobacterium rhinotracheale (OR) is a recently encountered bacterium which has been associated with respiratory disease in poultry. Certain bacteriologic and pathologic aspects of the organism and disease have only recently been investigated.

OCCURRENCE
O. rhinotracheale has recently been found with an increasing frequency in broiler and turkey operations experiencing respiratory problems. OR is frequently isolated with other respiratory agents (i.e., *Escherichia coli*, *Bordetella avium*, *Mycoplasma sp.* and respiratory viruses). Broiler and meat turkey operations see primarily birds with respiratory signs, with no consistent mortality. Condemnation and decreased feed efficiency have been reported. Egg production can be decreased in layer or breeder flocks. Mortality appears more severe in turkey breeder operations.

HISTORICAL INFORMATION
O. rhinotracheale was first isolated in 1981 in Germany and in 1986 in the United States. In 1994, the bacterial organism was named by Vandamme.

ETIOLOGY
1. *O. rhinotracheale* is a pleomorphic Gram negative rod (Fig. 1) which grows well (but slowly) on blood agar plates. After 24 hours of incubation at 37 C in 7.5% CO_2, OR colonies are pinpoint in size and show no hemolysis. No growth is observed on MacConkey agar plates.

2. Key biochemical tests include the Gram's stain reaction and morphology (short plump rods, club shaped rods, or long filamentous rods), positive oxidase test, positive β-galactosidase (ONPG) test, and negative catalase test. No reaction is observed in most carbohydrates.

3. The api-ZYM system (Biomerieux, France) is most useful. This system gives fourteen positive reactions and five negative reactions (lipase, β-glucuronidase, β-glucosidase, α-mannosidase, and α-fucosidase).

EPIDEMIOLOGY
1. *O. rhinotracheale* has been isolated from broiler and layer chickens, turkey and chicken breeders, meat turkeys, duck, goose, gull, guinea fowl, pigeons, ostrich, quail, pheasants, partridges, and chukers. Isolates are most frequently obtained from respiratory sites such as the trachea, sinuses, and lungs. Occasionally, systemic involvement is indicated by isolations from the heart, spleen, liver, bone, and joint.

2. In most disease situations, other primary respiratory agents can be demonstrated (*Bordetella avium*, *Mycoplasma sp.*, *Pasteurella sp.*, *Escherichia coli*, paramyxovirus, and infectious bronchitis virus).

CLINICAL SIGNS
Mild respiratory signs are most frequently observed with only a slight increase in mortality. Older birds may experience more severe respiratory signs with gasping, marked respiratory effort and an increase in mortality.

LESIONS
Mild sinusitis, tracheitis, or unilateral or bilateral lung consolidation may be observed. Turkeys frequently have blood-stained mucous in the mouth. Serofibrinous pleuropneumonia (Fig. 2) and inflammation of the air sacs (Fig. 3 and 4) are noted macroscopically and histopathologically (Fig. 5 and 6).

DIAGNOSIS
Bacterial culture is required to demonstrate *O. rhinotracheale*'s involvement in respiratory disease. Care must be taken to prevent its overgrowth by other bacteria. In turkeys, differentiation from *Pasteurella multocida*, *Riemerella anatipestifer* and/or *Escherichia coli* requires bacterial culture.

CONTROL
Little is known on the prevention of *O. rhinotracheale*. It is frequently present in consecutive flocks on the same ranch. Currently, there is no commercial vaccine available but autogenous bacterins have been used with some apparent benefit.

TREATMENT
Treatments with tetracycline and amoxicillin have been reported in Europe. Limited success has been reported with enrofloxacin and trimethoprim/sulfa. Most isolates in the United States have been reported as being susceptible to ampicillin, erythromycin, penicillin, spectinomycin and tylosin but sensitivity testing of the isolate is often necessary.

ORNITHOBACTERIUM RHINOTRACHEALE INFECTION

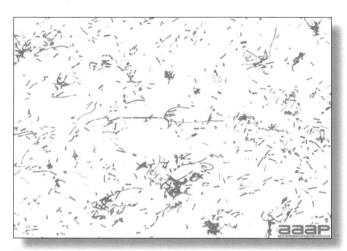

Fig. 1
Ornithobacterium rhinotracheale is a pleomorphic Gram negative rod.

Fig. 2
Serofibrinous pleuropneumonia in a turkey.

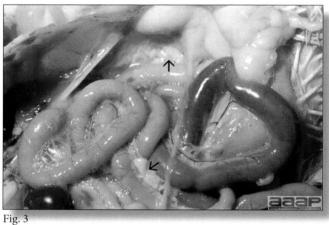

Fig. 3
Airsacculitis in a 28-day-old broiler chicken caused by ORT.

Fig. 4
Airsacculitis in a 35 day-old broiler chicken caused by ORT infection.

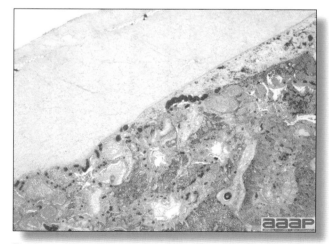

Fig. 5
Pneumonia and pleuritis in a 55 week-old turkey breeder hen.

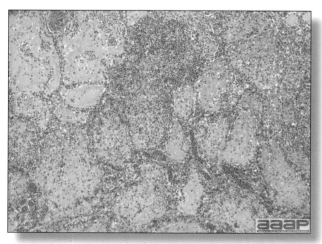

Fig. 6
Severe fibrinoheterophilic inflammation of lung associated with *Ornithobacterium rhinotracheale* infection in a 55-week-old turkey breeder hen.

PSEUDOMONAS INFECTION

DEFINITION

Pseudomonas can cause localized or generalized disease in chickens and turkeys of all ages. In poultry *Pseudomonas* is most commonly associated with hatchery and incubation problems and yolk sac infections. *Pseudomonas* can also cause infection in other species of birds, such as ducks, geese, pheasants, ostriches, pet and captive birds.

OCCURRENCE

Young birds as well as birds stressed and immunodeficient are very susceptible to *Pseudomonas*. Severe outbreaks have also occurred due to the use of contaminated vaccines and antibiotics as a result of poor hygienic conditions during mixing and handling of these products.

ETIOLOGY

Pseudomonas aeruginosa is the most common species that causes infections in poultry and other birds. *Pseudomonas* is a motile, Gram-negative, non-spore forming aerobic rod-shaped bacteria. Other species such as *P. fluorescence* has been associated with death of turkey embryos and *P. stutzeri* has been isolated from chickens with respiratory disease. *Pseudomonas* (*Burkholderia*) *pseudomallei* infections have been reported from Australia in psittacines that had septicemic lesions.

EPIDEMIOLOGY

Pseudomonas organisms are ubiquitous in nature and are most commonly found in contaminated water and soil. *Pseudomonas* is generally considered an opportunistic bacterium that can cause various clinical signs and pathology. Other factors such as concurrent infections with viruses and bacteria can influence infection. *Pseudomonas* is one of several bacteria that are commonly isolated from dead embryos, newly hatched chicks, poults, ducklings and others. Contact with infected birds, continuous intense management of broilers and turkeys with different ages and without periodic change of litter and cleaning and disinfection influences the spread of bacteria. *Pseudomonas aeruginosa* has also been isolated from the surface of eggs and skin of processed broiler chickens.

CLINICAL SIGNS

Clinical signs due to *Pseudomonas* in poultry depend on whether the disease is localized or systemic. These include ruffled feathers, anorexia, stunting, depression, weakness, respiratory signs, swelling of the head, swollen joints or foot pads, lameness, opisthotonus, diarrhea, corneal opacity and swollen conjunctiva. Sudden death without any apparent clinical signs is also common. Morbidity and mortality can vary from 2 -10 % but can be much higher depending upon management factors and concurrent diseases.

LESIONS

Gross lesions due to *Pseudomonas* are not specific and may include yolk sacs which may have watery yellow or caseous exudate, swollen joints with fibrinous exudate, edema and fibrin in the subcutis, fibrinous exudate in the anterior chamber of the eyes, pericardium, air sac, and capsule of the liver; necrotic foci in the liver, spleen, kidneys and occasionally in the brain. Nasal gland adenitis associated with *Pseudomonas aeruginosa* has been reported in ducks. Histologically, the lesions of a *Pseudomonas* infection generally consist of mild to severe fibrinosuppurative or fibrinoheterophilic inflammation mixed with large numbers of rod-shaped Gram-negative bacteria.

DIAGNOSIS

Tentative diagnosis of *Pseudomonas* infections can be made based on careful analysis of the history in combination with clinical signs, gross and microscopic lesions. Demonstration of Gram-negative rods in the smears from lesions can provide a tentative and quick diagnosis. Definitive diagnosis can be made by isolation on suitable media and identification of the organisms. *Pseudomonas* sp. can be isolated from various lesions such as yolk sac, pericardium, air sacs, joints, liver, lungs, skin and other organs.

CONTROL AND TREATMENT

Steps should be taken to identify and eliminate the source of *Pseudomonas*. Cleaning and disinfection of the incubators, hatchers, equipment and the environment are fundamental to the control and prevention of *Pseudomonas*.

Due to the antimicrobial resistance of *Pseudomonas* spp. sensitivity tests should be performed frequently. Some of the antibiotics that may be helpful in reducing losses include gentamicin, streptomycin, amikacin, enrofloxacin.

SALMONELLOSIS

PREFACE

Bacteria of the genus *Salmonella* have long presented serious challenges to the poultry and other food animal industries and are responsible for significant health problems. In this section salmonella infections are presented in four parts covering pullorum disease, fowl typhoid, arizonosis, and paratyphoid. The essential background information on each disease is provided within these parts. It is noteworthy that although the host-specific salmonellae (*Salmonella pullorum* and *Salmonella gallinarum*) literally prevented intensive large-scale poultry production prior to the evolution of practical testing and eradication programs in breeders, it is now the paratyphoid infections that threaten public acceptance of poultry products by virtue of concern for food-borne infection.

Paratyphoid salmonella infections are relatively common in poultry and all reasonable steps should be taken to minimize contamination of the finished product. The poultry industry is justifiably proud of its efficient production systems which provide a wide spectrum of economical and appealing products. It is our duty as professionals to make every effort to protect the industry from either implied unwholesomeness or true food safety problems.

With the advent of molecular technology, molecular bacteriologists have re-examined old bacterial classification and naming schemes to make them more scientifically accurate. For instance, paratyphoid salmonella (motile, not host adapted to the avian species in contrast to the non-motile host adapted *S. pullorum* and *S. gallinarum*) have been subdivided biochemically into 5 distinct subgenera. Further subdivision based on genetic analysis has yielded only 2 species. One, namely *S. enterica,* contains more than 2,500 previously named, motile, non-host adapted salmonella. This species contains *S. enterica* subspecies *enterica* serovar Enteritidis more commonly referred to as *S.* Enteritidis and *S. enterica* subspecies *enterica* serovar Typhimurium also referred to as *S.* Typhimurium. In addition, the bacterium previously named *S. pullorum*, the agent of pullorum disease and *S. gallinarum*, the agent of fowl typhoid have been reclassified and renamed *S. enterica* subspecies *enterica* serovar Pullorum and *S. enterica* subspecies *enterica* serovar Gallinarum. As is evident, this development has led to a longer naming schema and a potential for miscommunication and misunderstanding for the scientist and the student. For the purposes of discussion in this section, the shortened naming scheme provides more concise and more easily understood nomenclature and will be used. Thus, salmonella such as *S. enterica* subspecies *enterica* serovar Pullorum, and others, will be referred to in their shortened form such as *S.* Pullorum.

I. PULLORUM DISEASE

DEFINITION

Pullorum disease is an infectious, egg-transmitted disease of poultry, especially chicks and turkey poults, often characterized by white diarrhea and high mortality in young birds and by asymptomatic adult carriers.

OCCURRENCE

Pullorum disease occurs primarily in young chicks and turkey poults. Many other species can be infected naturally but they usually play an insignificant role in the epidemiology of this disease. Pullorum disease occurs in all age groups of chickens and turkeys but causes greatest loss in those less than 4 weeks old and is worldwide in distribution.

HISTORICAL INFORMATION

1. The bacillus that causes pullorum disease was first described in 1899. Within a few years pullorum disease was recognized as a common, worldwide, egg-borne disease of chickens. A tube agglutination test that would detect carriers was developed in 1913 and a whole blood test was developed in 1931. These tests permitted development of eradication programs.

2. Losses from pullorum disease were once so severe that they impaired expansion of the poultry industry. Pullorum disease sometimes was spread through hatchery-infected chicks. Extensive losses from pullorum disease and fowl typhoid were partly responsible for stimulating the development of the National Poultry Improvement Plan; the plan contains measures for the control of hatchery-disseminated diseases.

3. Through the application of control measures now detailed in the voluntary National Poultry Improvement Plan, pullorum disease has been eliminated from commercial poultry in the United States. The disease still persists in small backyard flocks. It probably could be eradicated if proven control measures could be enforced for all poultry and exotic birds.

4. Pullorum disease still causes catastrophic losses when no effort is made to control it. This occurs repeatedly in developing countries trying to establish a poultry industry.

ETIOLOGY

1. The etiologic agent is *S.* Pullorum, a nonmotile, Gram-negative bacillus adapted to poultry. This organism, like many other *Salmonella* spp., tends to infect young birds more frequently than older individuals and to establish a bacteremia. *S.* Pullorum closely resembles *S.* Gallinarum, the cause of fowl typhoid. They share certain antigens and usually cross-agglutinate on serologic tests.

2. The organism is rather resistant under moderate climatic conditions and can survive for months. However, it can be destroyed by thorough cleaning followed by disinfection. The organism can be killed by formaldehyde gas, which may be used in fumigation of fertile eggs and hatchers.

EPIDEMIOLOGY

1. *S.* Pullorum is primarily spread vertically through occasional infected eggs laid by infected carrier hens. Many of the infected chicks hatch and then transmit the organism horizontally to other birds in the hatch through the digestive and respiratory systems. Sale of exposed but apparently healthy birds to many different purchasers can result in widespread dissemination of the etiologic agent.

2. Adult carriers also shed the organism in their feces. Slow horizontal spread to other adults is possible through contamination of feed, water, and the environment. Also, contamination of nests and eggs therein can result in eggshell penetration and infection of chicks that hatch from those eggs.

3. Cannibalism of infected bacteremic birds can result in transmission.

CLINICAL SIGNS

Adults
Usually there are no signs. The infected adult may or may not appear unthrifty. An infected hen may or may not be a productive layer.

Young chicks and poults

1. In a setting of fertile eggs with a few infected embryos, there may be reduced hatchability. A few of the newly hatched birds appear weak or soon die. In others that develop bacteremia sudden death may occur. Mortality may be low during the first few days if only a few of the eggs contained the organism.

2. Morbidity and mortality begin to increase around the 4th or 5th day. Sick birds appear sleepy and weak. There is anorexia, white adherent diarrhea with pasting of the vent area, huddling near heat sources and shrill chirping. A few days later there may be respiratory signs in birds that inhaled the organism in the hatcher. Losses usually peak during the 2nd or 3rd week and then diminish. Survivors often are irregular in size and some are unthrifty, stunted, or poorly feathered. Many remain carriers and disseminators of the etiologic agent.

3. Mortality varies greatly but often is very high and can approach 100%. Mortality is increased by shipping, chilling, or poor husbandry. Conversely, mortality may be surprisingly low and the disease may go unrecognized.

LESIONS

Adults
Often there are no lesions. Occasionally there is a nodular myocarditis, pericarditis, or abnormal gonads. An abnormal ovary may have hemorrhagic, atrophic, or discolored follicles (Fig. 1). Less frequently there is oviduct impaction, peritonitis (Fig. 2), or ascites. Affected testes may have white foci or nodules.

Young chicks and poults

1. There may be few or no lesions in very young birds that die after a short septicemic course. Occasional dead birds feel wet. Many birds have pasted white feces in the vent area.

2. Classically there are nodules in one or more of the following sites: lungs, liver, gizzard wall, heart (Fig. 3), intestinal or cecal wall, spleen, and peritoneum. Frequently there are petechial hemorrhages or foci of necrosis in the liver. Later there may be swollen joints in occasional birds.

3. When the intestine is opened, white plaques may be found in the intestinal mucosa and cheesy cores of debris may be found in the intestine or ceca. Plaques and cecal cores (Fig. 4) occur more frequently in birds that die later in the course of the outbreak.

4. The spleen frequently is enlarged (Fig. 4 and 5). (This lesion, along with mucosal plaques and cecal cores, also occurs frequently in *Salmonella* infections other than pullorum disease.) The ureters frequently are distended with urates.

DIAGNOSIS

1. In young chicks and poults, typical history, signs, and lesions may suggest pullorum disease. Positive agglutination tests, either plate or tube (Fig. 6), using sera from convalescent surviving birds may strengthen the diagnosis. Chicks hatched by small, noncommercial operators are more likely to be positive for *S.* Pullorum.

2. For a definitive diagnosis, *S.* Pullorum must be isolated and identified. The organism should be typed at a typing center to aid in epidemiologic investigation. Due to state animal health laws, legal complications may occur so identification should be confirmed.

3. The National Poultry Improvement Plan provides details for confirming infection in adult reactor birds. Specified organs are pooled and cultured for *S.* Pullorum.

4. Diseases that must be differentiated from pullorum disease in young birds include:

   A. Chilling. Chilling is often associated with white diarrhea.

   B. Omphalitis (navel infection). Omphalitis occurs in this age group, often with diarrhea.

C. Typhoid, paratyphoid, arizonosis, and colibacillosis. It will be necessary to isolate and identify the etiologic agent to rule out these infections.

CONTROL

1. Prevention is based on establishment and maintenance of pullorum-free breeder and multiplier flocks by serologic testing and other measures. The following tests are used:

    A. The stained antigen, rapid whole blood test is typically performed in flocks in the field. This same antigen can be used for the rapid serum test in the laboratory.

    B. Tube agglutination test is performed on sera and is primarily used to confirm plate test reactions.

    C. An enzyme-linked immunosorbent assay (ELISA) has also been developed for the serologic diagnosis of pullorum disease.

2. Noninfected eggs from tested clean flocks should be hatched in a properly disinfected hatcher and raised on pullorum-free premises, preferably under quarantine.

3. Detailed regulations for control of pullorum disease are given in the National Poultry Improvement Plan. A copy of the plan can be obtained from The National Poultry Improvement Plan, USDA-APHIS-VS, Suite 300, 1506 Klondike Road, Conyers, GA 30094 or consulted on the website of the federal register of the United States: http://69.175.53.20/federal_register/2011/mar/22/2011-6539.pdf.

4. The poultry producer can avoid pullorum disease by purchasing chicks only from those hatcheries that participate in the National Poultry Improvement Plan or a similar eradication program. Exposure of the flock to carriers or a contaminated environment must be avoided.

TREATMENT

Insofar as chemotherapy perpetuates the carrier state, treatment of pullorum-infected birds is indefensible and should not be recommended under any circumstance.

II. FOWL TYPHOID

DEFINITION

Fowl typhoid is an infectious disease, primarily of chickens and turkeys, with many of the clinical and epidemiologic features and lesions that occur with pullorum disease.

In the following material only the differences between fowl typhoid and pullorum disease are emphasized. Most of the facts concerning pullorum disease (see Pullorum Disease) are applicable to fowl typhoid.

OCCURRENCE

Most outbreaks occur in chickens or turkeys but the disease occasionally occurs in other poultry, game birds, and wild birds. In chickens and turkeys most outbreaks occur in recently hatched, young birds, but unlike pullorum disease, the disease often continues for months. Many outbreaks occur in semimature flocks with no history of an earlier onset.

HISTORICAL INFORMATION

A disease that probably was fowl typhoid was recognized in 1888. By the early 1900s many outbreaks, both in the United States and abroad, had been reported. Between 1939 and 1946 there was a marked increase in outbreaks in the United States and fowl typhoid was a major disease of poultry. Application of testing and control measures (now detailed in the National Poultry Improvement Plan) greatly reduced the incidence of both fowl typhoid and pullorum disease. Fowl typhoid is seldom encountered today in the United States but persists as a challenging disease problem in several countries.

ETIOLOGY

The etiologic agent is *Salmonella* Gallinarum. This organism shares many antigens with *Salmonella* Pullorum, the agent that causes pullorum disease, and the two organisms usually cross-agglutinate. As a consequence, birds exposed to or infected with either disease can be identified by the same agglutination test.

EPIDEMIOLOGY

The epizootiology of fowl typhoid is similar to that of pullorum disease. Relatively speaking, transmission of infection through eggshell contamination may be of somewhat greater importance than with pullorum disease. Also, *S.* Gallinarum is more frequently transmitted among growing or mature flocks and the incidence and mortality in older birds is usually higher.

CLINICAL SIGNS

Signs of fowl typhoid and pullorum disease are similar in birds less than approximately 1 month old. Semimature and mature birds with fowl typhoid often have pale head parts (comb, wattles, face), shrunken combs and wattles, and diarrhea. Mortality can be substantial. In one extensive experiment, many broods of birds were hatched from eggs from a typhoid-infected flock of hens. Approximately one third of all hatched birds died with typhoid.

LESIONS

1. Lesions of fowl typhoid and pullorum disease are similar in chicks and young poults (see pullorum disease).

2. Lesions of acute fowl typhoid in older birds include:

A. A bile-stained ("bronzed") enlarged liver with or without small necrotic foci (Fig. 1).

B. Enlargement of the spleen and kidneys.

C. Pallor throughout the cadaver and thin watery blood.

D. Enteritis in the anterior small intestine, often with ulceration.

3. In older birds, chronic fowl typhoid lesions resemble those seen in pullorum disease (see pullorum disease).

DIAGNOSIS

S. Gallinarum should be isolated and identified for diagnosis. It should be carefully differentiated from other salmonella and paracolon organisms.

CONTROL

Control is as for pullorum disease. Fortunately, control of both pullorum disease and fowl typhoid is accomplished by the same program encompassed in the National Poultry Improvement Plan.

III. ARIZONOSIS

DEFINITION

Arizonosis is an egg-transmitted infection, seen primarily in young turkey poults, characterized by variable signs and lesions related to septicemia or to localization of infection in the intestine, peritoneal cavity, eye(s), brain, or other sites.

OCCURRENCE

Most outbreaks occur in turkeys. Although all ages are susceptible, the disease is most common in poults less than 3 weeks old. Chicks, ducklings, canaries, psittacines, and other birds occasionally have been found to be infected. Infection frequently occurs in reptiles, which can serve as reservoirs. Infection in humans has occurred but is not common. The disease probably is worldwide in distribution.

HISTORICAL INFORMATION

1. Arizonosis in chicks was reported in 1936, although it was not clearly differentiated from paratyphoid infection. In 1939 the etiologic agent of arizonosis was definitively characterized and found to cause a fatal septicemia in reptiles in Arizona.

2. In the 1940s to 1960s arizonosis was recognized as an important and widely distributed disease of many turkey flocks.

3. Historically the etiologic agent of arizonosis was considered to be in the genus *Arizona*. Since 1982 the etiologic agent of arizonosis has been characterized as a subspecies of the genus *Salmonella* based on DNA relatedness to this genus.

ETIOLOGY

1. The etiologic agent is *Salmonella arizonae* (syn. *Arizona arizonae* and *Arizona hinshawii*). It is a non-spore forming, Gram-negative, motile bacterium in the family Enterobacteriaceae.

2. *S. arizonae* ferments lactose slowly, usually requiring a few days. Slow-fermenting *S. arizonae* may be mistaken for other *Salmonella* species unless the fermentation tubes are held for a sufficient period.

EPIZOOTIOLOGY

1. *S. arizonae* often localizes in the ovary of carrier birds. When this happens, the organism is included within eggs, infects the developing embryo, and results in infected progeny.

2. Infected adult birds are frequently intestinal carriers and intermittent shedders of *S. arizonae*. Contamination of eggshell surfaces with feces leads to eggshell penetration and infection of progeny.

3. Infected progeny that hatch from infected eggs transmit the organism horizontally to uninfected birds in the hatch and later may become carriers and shedders of the organism.

4. Exposure to the agent can also occur via reptiles, rats, mice, and many other mammals, contaminated hatchers, or fomites. Transmission frequently is via fecal contamination of feed, water, or environment. The organism can persist in a contaminated environment for months.

5. As with many salmonellae, *S. arizonae* has few, if any, species barriers. Interspecies transmission occurs readily and there are many carriers.

CLINICAL SIGNS

In young poults there may be listlessness, diarrhea, pasting of feces in the vent area, huddling near heat sources, ataxia, trembling, torticollis (Fig. 1), excessive mortality (3-5% is most common, although losses up to 50% have been reported), and poor growth. Cloudiness (turbidity) and enlargement of the eye(s) causing blindness may occur in infected poults. Central nervous system signs occur in birds with brain lesions. Signs in young birds closely resemble those seen with paratyphoid. Moderate to marked uneven growth in the flock is seen even after the clinical disease has ended. Affected eyes undergo atrophy and are useful in identifying previously infected flocks. Adult carriers usually show no signs.

LESIONS

1. Typically, lesions of septicemia are observed, including an enlarged, mottled, yellow liver, a retained yolk sac and peritonitis. Occasionally there are cheesy plugs in the intestine or cecum and infected yolk sacs develop into abscesses in poults that survive the initial septicemia.

2. A small but significant number of poults have opacity or turbidity of the eye(s) (ophthalmitis) (Fig. 2). This useful lesion is not pathognomonic because it can also occur with paratyphoid, aspergillosis, or colibacillosis. However, it occurs more frequently with arizonosis.

3. Purulent exudate in the meninges, lateral ventricles of the brain, or in the middle and inner ear is seen in birds with central nervous system signs (Fig. 3).

DIAGNOSIS

The etiologic agent must be isolated and identified. Signs and lesions are inadequate for separating arizonosis from other infection salmonella infections. *S. arizonae* can usually be recovered from liver, spleen, heart blood, unabsorbed yolk, intestine, or other organs. It is readily recovered from infected eyes, ears, and brains. *S. arizonae* persists for several weeks in atrophied eyes, from which the organism can be easily recovered. Also, it may be cultured from nonhatching embryos, eggshells, or organs from infected breeder birds and environmental samples. Enrichment procedures used to isolate other salmonellae are equally effective for detecting *S. arizonae.*

CONTROL

1. If infected breeder flocks can be identified, they should not be used as a source of fertile eggs. Unfortunately, there is no readily available serologic test for identification of infected flocks or individual birds. Such flocks often are identified by culturing the agent from their eggs or progeny. Primary turkey breeder companies in the United States are now free of *S. arizonae* infection, but commercial breeder flocks are still occasionally affected.

2. One-day-old poults are usually inoculated at the hatchery with antibiotics to control mortality from arizonosis. Gentamicin is most commonly used. Strains resistant to gentamicin have been found and have caused high losses in poults. In these cases, the use of injectable tetracyclines or ceftiofur may be helpful.

3. Most of the measures used for prevention of paratyphoid are applicable for control of arizonosis (see under paratyphoid).

TREATMENT

Useful antibiotics and drugs include gentamicin, tetracyclines, and sulfonamides. Treatment does not prevent birds from becoming carriers and shedders of the organism. Experimentally, use of a bacterin in turkey breeder hens has been found to be helpful in reducing shedding and coupled with good management procedures, eventually eliminating the disease from breeder flocks.

IV. PARATYPHOID INFECTION

(Salmonellosis; Paratyphoid)

DEFINITION

Paratyphoid is an acute or chronic disease of poultry, many other birds, and mammals caused by any one of a large group of salmonellae that are not host specific.

OCCURRENCE

Paratyphoid infection occurs in many kinds of birds and mammals; it occurs frequently in poultry. It also occurs in rats, mice, and other rodents, in many reptiles, and in some insects. It is a frequent disease of humans. In most animals the young are more frequently and severely affected. Adults tend to be more resistant but can be infected, especially if stressed prior to exposure. Paratyphoid infection is worldwide in distribution.

These bacterial infections are of much more importance for public health impact than for economic losses in the affected animals. Poultry products have been repeatedly implicated in human outbreaks of salmonellosis and all health management personnel in the poultry industry need to be sensitive to this legitimate consumer concern. Although detailed epidemiology is outside the scope of this book, the technical details in this chapter should provide essential background on the control of avian salmonella infections. The National Poultry Improvement Plan and more recently the Food and Drug Administration are involved in the regulatory aspects of *Salmonella* Enteritidis (SE) infection while NPIP has programming in place to allow poultry breeder farms to attain Salmonella monitored status and FDA a monitoring and control program for SE in egg laying chickens.

ETIOLOGY

1. Paratyphoid salmonella have consisted of salmonella that are motile and not host adapted in contrast with *S.* Pullorum and *S.* Gallinarum, which are nonmotile and highly host adapted.

2. There are more than 2,500 serovars of S. *enterica* but only 10% of these have been isolated from poultry. Distribution of serovars varies geographically and overtime. Frequent isolates in the United States include:

S. Enteritidis	*S.* Typhimurium
S. Heidelberg	*S.* Kentucky
S. Braenderup	*S.* Hadar
S. Muenster	*S.* Senftenberg

3. Most paratyphoid organisms contain endotoxin, which is responsible for their pathogenic effects.

4. Paratyphoid organisms are moderately resistant in their natural environment but are susceptible to most disinfectants and to fumigation with formaldehyde gas.

EPIDEMIOLOGY

1. Paratyphoid organisms often localize in the intestine or gallbladder of carriers. They are intermittently shed in the feces and thus contaminate eggshells, feed and water. Poultry, other birds, reptiles, insects, and various mammals including humans can disseminate salmonella.

2. Infection of young chicks occurs primarily by fecal contamination of eggshells with paratyphoid organisms and subsequent penetration into the eggs. Some chicks are infected at the time of hatch and infection spreads horizontally.

3. Localization of paratyphoid organisms in the ovary with subsequent vertical transmission occurs in some instances (e.g. S. Enteritidis). The frequency of this method of spread is unknown but thought to be transitory or intermittent.

4. Contaminated animal proteins (tankage, meat scraps, etc.) can transmit the agents. These products often are contaminated after processing. Heated, pelleted products seldom contain living salmonella.

5. Interspecies transmission of paratyphoid organisms does occur, often through environmental contamination. Rodents are an important reservoir for paratyphoid organisms. Paratyphoid is an important public health problem and this aspect of the disease is by far the greatest challenge to the poultry industry.

CLINICAL SIGNS

1. Signs usually are seen only in young birds (less than 4 weeks of age). There is somnolence, profuse diarrhea followed by dehydration, pasting or wetting of the vent area, drooping wings, shivering, and huddling near heat sources.

2. There usually is high morbidity and mortality (especially during the first 2 weeks of brooding), although these are variable. The course often is short in individual birds.

3. Increased mortality can be observed in layers naturally infected with S. Enteritidis at the onset of lay.

LESIONS

1. There may be a few or no lesions in birds that die after a short septicemic course, perhaps only a few petechial hemorrhages.

2. There usually is dehydration and marked enteritis, often with focal necrotic lesions in the mucosa of the small intestine. Occasionally there are necrotic foci in the liver. In young birds there often is unabsorbed yolk material in the yolk sac and overt omphalitis. Less frequent lesions include blindness, joint infections, or swollen eyelids, the latter two being common in pigeons.

3. Raised plaques in the intestinal mucosa and cheesy cecal cores are often seen in birds that survive for a few days or longer. These strongly suggest the presence of salmonellosis but are not pathognomonic for any one species of Salmonella.

4. 4. Inflammation of the oviduct and ovaries, with pericarditis, perihepatitis and/ or peritonitis, has been observed in flocks naturally infected with S. Enteritidis (Fig. 1).

DIAGNOSIS

The etiologic agent should be isolated from multiple organs and positively identified. Using polyvalent salmonella antiserum, most labs can identify any isolate as Salmonella. Some, namely S. Enteritidis and S. Typhimurium, have rapid antigen detection kits as well as PCR tests making rapid and accurate diagnosis easier. Typing centers can provide the service of complete species and genetic typing and should be utilized for this specialized work. Selective media often are utilized in isolating Salmonella from the gut.

CONTROL

1. Breeder flocks should be monitored bacteriologically for Salmonella infection in conjunction with efforts to minimize flock exposure.

2. If possible, all birds should be sold after one lay season thus eliminating carriers. While the premise is vacated, it should be thoroughly cleaned and disinfected. Eliminating rodents by trapping them, is an essential step in the eradication of Salmonella from a farm.

3. In flocks provided with nests the nests should be kept clean. Replace nesting material frequently as needed. Maintain a high standard of sanitation in all operations.

4. Gather eggs frequently and store them in a cool place. Separate dirty eggs from clean eggs at the time of gathering. Egg sanitation may be necessary at the poultry farm during the storage that precedes storage and incubation at the hatchery.

5. Fumigate hatching eggs as recommended during incubation (done routinely). Practice scrupulous hatchery sanitation.

6. Raise new broods of birds in the all-in, all-out system. Add no new birds to the started brood. Do not permit contact with wild birds, mammals, rodents, or reptiles. Control insect populations.

7. Provide uncontaminated feed ingredients in the ration. Pelleted feeds are more likely to be free of salmonella as long as there is no cross contamination after the pelleting process.

8. If necessary, specific antigens may be prepared and used in an agglutination test for elimination of carriers. This has been done for S. Typhimurium but not for most other paratyphoid organisms. Once developed, these test antigens can be used on a regular basis at the time of the annual pullorum-fowl typhoid test.

9. One-day-old chicks and poults are often inoculated at the hatchery with antibiotics to control mortality from paratyphoid and other bacterial infections. This practice may help control early mortality but the same positive result can be accomplished by the good breeder flock and hatchery management practices enumerated above.

10. Many commercial vaccines, both live and killed, are available for vaccination of commercial egg layers and broiler breeders, against *S.* Enteritidis and *S.* Typhimurium. Unfortunately, they do not prevent infection but do reduce shedding.

TREATMENT

Treatment with antibiotics does not eliminate *Salmonella* infections and has very minimal value. Use of antimicrobials to eliminate *Salmonella* colonization may imperil their medical usefulness by promoting resistance.

PULLORUM DISEASE

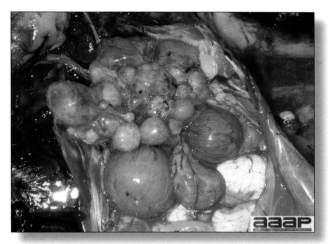

Fig. 1
Abnormal ovary with atrophic and discolored follicles.

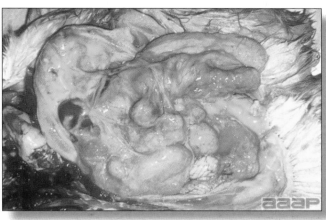

Fig. 2
Severe fibrinosuppurative peritonitis in a hen.

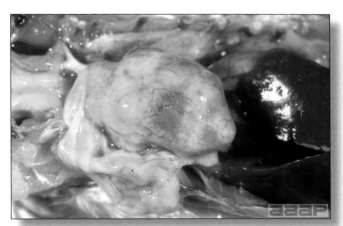

Fig. 3
Misshapened heart in a 6-week-old chicken due to the presence of numerous yellow nodules in the myocardium. Note the thickened pericardium and discolored, mottled liver due to chronic passive congestion.

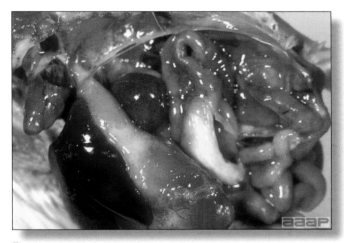

Fig. 4
Splenomegaly and hepatomegaly in a 18-day-old chick. Notice the white cast in the cecum.

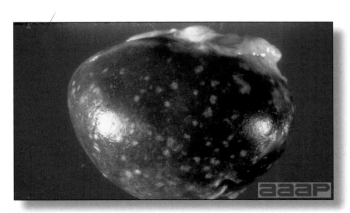

Fig. 5
Enlarged and mottled white spleen in an adult chicken.

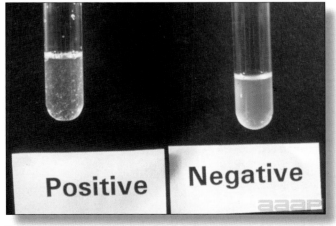

Fig. 6
Tube agglutination test showing Pullorum positive and negative tests. Note floccules in the positive tube.

FOWL TYPHOID

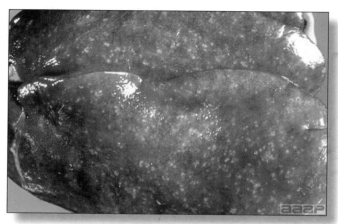

Fig. 1
Bile-stained ("bronzed") enlarged liver with small necrotic foci in a
FP positive adult chicken.

ARIZONOSIS

Fig. 1
Torticollis in a young poult affected with *S. arizonae*.

Fig. 2
Poult with opacity of the eye (ophthalmitis).

Fig. 3
Encephalitis in a young poult.

PARATYPHOID INFECTION

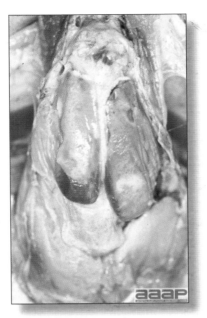

Fig. 1
Pericarditis, perihepatitis and
peritonitis in a naturally SE infected
layer.

SPIROCHETOSIS

DEFINITION
Spirochetosis or nonrelapsing borreliosis is a septicemic disease of most poultry and many other birds. Acute cases are characterized by depression, cyanosis, diarrhea, and leg weakness progressing to paralysis and death.

OCCURRENCE
Spirochetosis occurs naturally in chickens, turkeys, geese, ducks, pheasants, grouse, and canaries. Many other birds can be infected experimentally. In the United States the disease usually has occurred in turkeys, chickens, and pheasants. All age groups are susceptible if not previously exposed. The disease is widely distributed in tropical and temperate regions. In the United States it has been recognized in California, New Mexico, Texas, and Arizona.

HISTORICAL INFORMATION
1. Spirochetosis, one of the major scourges of poultry, was first reported in 1891. Spirochetosis has occurred only a few times in the Southwestern United States. It has been reported in California in 1946 and 1993 as well as in Arizona in 1961. The disease is of major importance in those countries where it is enzootic.

2. There is potential for spread of spirochetosis in the southwestern states because the presence of the tick vector, *Argas persicus*.

ETIOLOGY
1. The etiologic agent is *Borrelia anserina*, a spirochete with 5-8 spirals that is up to 30 microns long.

2. The organism is not very resistant outside the host and must be maintained in some vector between hosts.

EPIDEMIOLOGY
1. *B. anserina* can be transmitted through infectious droppings but usually is transmitted by blood sucking arthropods. *Argas persicus* is the usual vector and mosquitoes of the genus Culex may serve as vectors. Mites may serve as mechanical carriers.

2. *A. persicus* remains infective for up to 430 days after feeding on an infected host. Further, the tick passes the spirochete to its progeny.

3. Infectious vectors and mites transmit the spirochetes to susceptible birds when they feed upon them. Recovered birds clear the infection completely and do not become carriers.

4. Transmission can also occur through ingestion of infected ticks, cannibalism of moribund birds, or scavenging of infected carcasses.

CLINICAL SIGNS
Infected birds are depressed, cyanotic, thirsty, and often have a diarrhea that includes excessive white urates. The birds are weak, squat on the ground, and later may become paralyzed. Morbidity and mortality vary greatly depending on the virulence of the *B. anserina* strain. Morbidity and mortality may approach 100% in highly susceptible flocks.

LESIONS
1. There usually is marked enlargement of the spleen, which is mottled by ecchymotic hemorrhages.

2. The liver frequently is enlarged and may contain small hemorrhages, infarcts, or foci of necrosis. The kidneys may be enlarged and pale. There usually is bile-stained mucoid enteritis. The histopathology has been well described but microscopic lesions are not diagnostic for the disease.

DIAGNOSIS
1. Spirochetosis should be suspected if the tick *A. persicus* is found on typical sick birds. However, nymphs and adult ticks live in the house and feed mostly at night.

2. The spirochetes can be identified in Giemsa-stained blood smears (Fig. 1) or by dark-field or phase-contrast microscopy of blood and other fluids. Spirochetes can be concentrated in the buffy coat of centrifuged blood. This may facilitate identifying birds with low spirochetemia. Spirochetes may not be observed during late stages of the disease.

3. In doubtful cases, the spirochete can be demonstrated by isolating it in six chick embryos inoculated with defibrinated blood from a typical early case. Alternatively, young chicks or poults can be inoculated with serum or tissue suspensions and their blood can be examined daily for spirochetes, which usually appear in 3-5 days.

4. The spirochete can be identified in specially stained tissue sections. Also, the fluorescent antibody test can be used to identify it in tissues or blood. Agar-gel precipitin tests have been used to detect spirochete antibodies and antigens.

CONTROL
1. Spirochetosis can be prevented by controlling or eradicating all the vectors and transmitters of *B. anserina*. It may be difficult to eradicate the fowl tick without destroying infested wooden buildings and all the birds in infected flocks. Isolating the roost by suspending it from wires or placing the supports of the roost in pans filled with oil is helpful in reducing tick feeding.

2. A wide variety of bacterins and vaccines has been prepared abroad but are not available in the United States. They appear to be reasonably effective although they produce a shorter, weaker immunity than is desirable unless revaccination is practiced. Immunity is serotype specific and my not protect against others.

TREATMENT

In countries where spirochetosis is enzootic, numerous drugs and antibiotics, including penicillin, streptomycin, tylosin and tetracyclines have been used successfully for treatment.

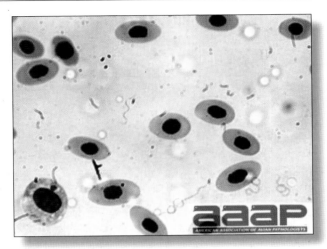

Fig. 1
Spirochetes in Giemsa-stained blood smears.

STAPHYLOCOCCOSIS

DEFINITION
Staphylococcosis is a systemic disease of birds characterized most frequently by purulent arthritis and tenosynovitis.

OCCURRENCE
Staphylococcal infections of poultry occur worldwide and affect all classes of birds. Outbreaks are most important in turkeys and broilers. The organisms are common in the environment and are especially associated with the skin. Most diseases produced by *Staphylococcus* sp. are associated with a break in the skin or beak (trauma, beak trimming, toe trimming, foot pad burns etc.). Avian infections tend to be caused by types occurring in birds rather than human strains. Isolates pathogenic for one class of poultry are usually pathogenic for other classes of birds. Toxigenic strains capable of causing food poisoning can contaminate the skin of processed poultry. The source of these strains at present is in debate. Biotyping indicates processing plant worker origin while plasmid profile indicates poultry origin.

HISTORICAL INFORMATION
Staphylococci were first discovered to be a cause of arthritis in geese in 1892. Since that time they have been identified as the cause of a variety of localized and systemic diseases in many different avian species and in most areas of the world. The disease was more common in turkeys when they were raised on range than it is now.

ETIOLOGY
1. Most staphylococci isolates have been identified as *Staphylococcus aureus*, a Gram-positive coccus occurring in clusters. Pathogenic isolates are usually coagulase positive.

2. Biotyping and phage-typing is frequently used to distinguish strains. Isolates from different geographic areas tend to be different phage types. Particular phage types are often endemic on a particular farm and tend to reappear in successive flocks.

3. Organisms are moderately resistant to common disinfectants. Chlorine-containing disinfectants are efficacious in the absence of organic material.

4. Toxins produced by staphylococci can increase both the virulence and pathogenicity of a particular strain.

LESIONS
Diseases produced by staphylococcus infections include:

1. Omphalitis
Although infections of the yolk sac occur, they are less common than omphalitis caused by other bacteria. Sources of the bacterium include the breeder flock, hatchery environment, and hatchery workers.

2. Gangrenous dermatitis
Affected areas of the skin are dark red, moist, thickened, and clearly demarcated from adjacent normal skin. Usually traumatic lesions such as punctures or scratches are present. The serosanguinous fluid seen in clostridial infections is minimal or absent. Staphylococcal gangrenous dermatitis is typically secondary to immunosuppression caused by infectious bursal disease or chicken infectious anemia virus.

3. Cellulitis
A purulent inflammation is present in subcutaneous tissues. Traumatic lesions may or may not be present. The overlying skin tends to be dry and discolored.

4. Abscesses
There are localized purulent lesions in the skin. The plantar surface of the foot is a common site and results in bumblefoot. Abscesses result from puncture wounds.

5. Septicemia
There is an acute increase in mortality with congestion of the internal organs. It is usually associated with a processing event of the bird such as beak trimming or some other trauma to the skin.

6. Arthritis/periarthritis/synovitis
Any joint, tendon sheath, or synovial bursa can be affected. Arthritis/periarthritis/synovitis is seen clinically as swollen (Fig. 1), hot joints, especially hock joints. It occurs as a sequel to septicemia and can be experimentally reproduced by intravenous injection of pathogenic strains. Initially, affected tissues are acutely inflamed and contain white to yellow soft fibrinopurulent exudate. Later, the exudate becomes caseous. Fibrosis of affected tissues occurs late. Affected birds often have bile stasis of the liver. High numbers of large mononuclear cells are seen in blood smears.

7. Discospondylitis (spondylitis)
The joints of articulating thoracolumbar vertebrae are affected. The process spreads to affect adjacent vertebrae. Lesions may become so extensive that pressure on the spinal cord will develop causing paresis and paralysis.

8. Osteomyelitis
This is a sequel to septicemia. Organisms localize in metaphyseal vessels invading the cartilage of the growth plate of actively growing bones. Initially, pale yellow, friable bone is seen in affected areas adjacent to the growth plate, especially in the proximal tibia and metatarsus. Necrotic areas, abscesses, and sequestra are seen later (Fig. 2).

9. Endocarditis
This is an uncommon sequel to septicemia. There are vegetations on the mitral and/or aortic valves. Emboli from valve lesions cause infarcts in the brain, liver, and spleen.

10. Green liver-osteomyelitis complex in turkey

This is a condition observed at slaughter. Normal-appearing processed turkey carcasses present green discoloration of the liver (Fig. 3) and associated arthritis/synovitis, soft-tissue abscesses, and osteomyelitis of the proximal tibia (Fig. 4). *Staphylococcus aureus* is most commonly isolated from these lesions but other opportunistic bacteria such as *Escherichia coli* have also been isolated.

DIAGNOSIS

1. Gross lesions are suggestive. A rapid, presumptive diagnosis can be made by identifying the typical cocci in smears from lesions.

2. Organisms can be readily cultured and identified from lesions and often from the livers of affected birds.

CONTROL

1. Because staphylococci are ubiquitous in the environment their presence cannot be prevented. When an outbreak is associated with a particular environment, the source should be sought and eliminated.

2. Protect broilers from infectious bursal disease with an appropriate vaccination program.

3. Take measures to reduce the occurrence of traumatic skin lesions and foot pad burns, as well as any enteric disease which would damage the integrity of the intestinal mucosa.

4. The respiratory tract has also been identified as an important portal of entry for pathogenic staphylococci in turkeys. Exposing chickens or turkeys to a live avirulent vaccine, namely strain 115 of *Staphylococcus epidermidis,* by aerosol at 10 days and again at 4-6 weeks substantially reduced the incidence of staphylococcosis and improves overall flock livability.

5. Avoid overly severe feed restriction in breeder replacements which has been associated with an increased incidence of staphylococcosis.

TREATMENT

1. High levels of antibiotics effective against staphylococci may be helpful if given early in the course of the disease.

2. Resistance to antibiotics is common and isolates should be tested for sensitivity.

3. Usually treatment is not cost effective and preventive programs should be relied on.

STAPHYLOCOCCOSIS

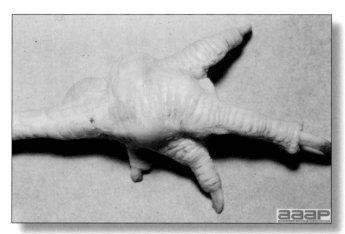

Fig. 1
Swollen and distended metatarsal joint in a broiler chicken.

Fig. 2
Necrotic foci and abscesses in the proximal tibia.

Fig. 3
Green liver discoloration observed at slaughter in a turkey carcass.

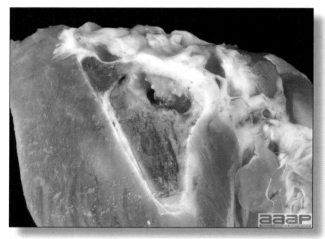

Fig. 4
Osteomyelitis of proximal tibiotarsus of a turkey.

ULCERATIVE ENTERITIS
(Quail Disease)

DEFINITION
Ulcerative enteritis is an acute bacterial infection of upland game birds, turkey poults, and young chickens characterized by ulcerations of the intestinal tract and by focal and/or diffuse hepatic necrosis.

OCCURRENCE
Ulcerative enteritis occurs frequently in young, captive, upland game birds and with increasing frequency in turkey poults and young chickens. Quail are the most susceptible host species. Young birds are affected more frequently than adults although the disease occurs frequently in adult quail. The disease is widespread in the United States and is known to occur in Europe and Asia. In chickens it frequently occurs in association with other diseases, including coccidiosis, chicken infectious anemia, and infectious bursal disease. Recently, the etiology of ulcerative enteritis is thought to be the cause of focal duodenal necrosis (FDN) in egg laying chickens.

HISTORICAL INFORMATION
Ulcerative enteritis was reported in the United States in 1907 and had been observed prior to that time in Great Britain. It was first referred to as quail disease, a name retained for many years despite recognition of the disease in many other birds. The disease was recognized in chickens by 1934 and in domestic turkeys by 1944. The disease has increased in incidence and is now a well-recognized disease. At one time the agent was mistakenly believed to be *Clostridium perfringens*.

ETIOLOGY
The etiologic agent is *Clostridium colinum*, a Gram-positive, anaerobic, spore-forming bacillus. The organism is very resistant. It withstands boiling for 3 minutes or 70 C for 10 minutes. Boiling suspected material is useful in killing other contaminating bacteria during isolation attempts. *C. colinum* can best be isolated from the typically affected fresh liver. The preferred medium is tryptose-phosphate-glucose agar with 8% horse plasma. Cultures are incubated anaerobically.

EPIDEMIOLOGY
1. The etiologic agent is spread primarily through the droppings of acutely affected or recovered carrier birds and spores persist in the soil for years. Interspecies transmission can occur among susceptible birds.

2. Infection can be spread by flies that feed on infectious droppings. The disease is highly contagious, especially among quail.

CLINICAL SIGNS
1. In most species signs are similar to those seen with coccidiosis in chickens. These include listlessness, humped appearance, retracted neck, drooping wings, partially closed eyes, ruffled feathers, diarrhea, anemia, and perhaps bloody feces. In quail, white watery droppings are rather distinctive. In chicken flocks a course of 2 or 3 weeks is common and then the chickens recover slowly.

2. Sudden death may occur without signs being apparent, especially during onset. Birds that die suddenly may be well muscled and fat, especially in quail. Birds with prolonged illness are often emaciated.

3. Mortality may be very high with quail, up to 100% within a few days. Mortality in chickens seldom exceeds 10%. Game birds other than quail usually have a higher mortality than chickens but less than occurs in quail.

LESIONS
1. Lesions are similar in most birds. Most cases presented for necropsy have deep ulcers scattered throughout the intestine, including the ceca, and the ulcers may be numerous enough to coalesce. Ulcers (Fig. 1) may be round or lenticular, the latter shape being more common in the upper intestine. Deep ulcers often can be detected through the serosa of the unopened intestine (Fig. 2) and may penetrate it to induce peritonitis. The intestine may contain blood, thus mimicking coccidiosis. Acute cases have severe enteritis of the small intestine.

2. The affected liver usually contains large, yellow or tan areas or focal yellow lesions, or both. The lesions tend to be colorful and distinctive. The spleen is often enlarged and may be hemorrhagic.

DIAGNOSIS
1. Typical intestinal ulcerations and the distinctive colorful lesions in the liver strongly suggest ulcerative enteritis. Stained impression smears made from the cut surface of the liver may reveal the rod-shaped bacillus with its subterminal spore.

2. For confirmation the etiologic agent should be isolated and identified. The organism must be differentiated carefully from *Clostridium difficile* and *C. perfringens*.

3. Care should be taken to differentiate the disease in chickens from coccidiosis. Coccidiosis often is present in the same bird and assessing the relative importance of the two diseases may be difficult. Both diseases may be contributing to mortality.

CONTROL
1. Raise the flock in facilities and on ground where the disease has never occurred. Do not add birds. Prevent contact with all other species of birds. Raising birds on wire is of value if feasible.

2. Keep old birds and young birds separated. If possible, do not have both age groups on the same premises. Practice careful sanitation including frequent cleaning and disinfection. Promptly remove and destroy all sick birds.

3. Streptomycin, bacitracin, penicillin, lincomycin, and tetracyclines have all been used intermittently in feed or drinking water for prevention of the disease. Rotating the use of these different antibiotics and chemicals will help prevent the emergence of *C. colinum* isolates which are resistant to antimicrobials.

TREATMENT
Most of the antibiotics and chemicals used in feed and water for prevention can be used at higher levels for treatment. These treatments should be administered in the drinking water.

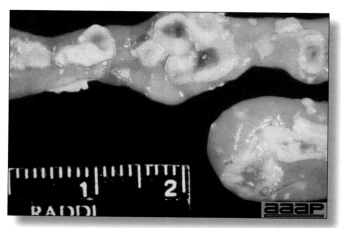

Fig. 1
Deep and coalescing ulcers scattered throughout the intestine.

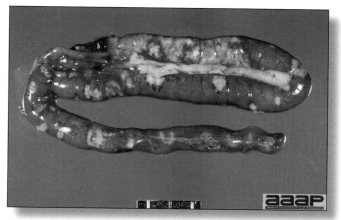

Fig. 2
Deep ulcers detected through the serosa of the unopened intestine.

YERSINIOSIS

DEFINITION
Yersiniosis is a septicemic disease affecting various species of birds caused by the bacterium *Yersinia pseudotuberculosis*. The disease has been reported in turkeys, chickens, ducks, and in Passeriformes (canaries and finches), Psittaciformes (parrots, parakeets), Columbiformes (pigeons and doves), Piciformes (toucans), Cuciliformes (turacos), raptors and other captive and free flying birds. Mammals including humans are also susceptible to *Yersinia pseudotuberculosis.*

OCCURRENCE
Among poultry, sporadic outbreaks have occurred in commercial turkeys.

ETIOLOGY
The etiology of Yersiniosis is *Yersinia pseudotuberculosis.* Other species of Yersinia such as *Y. pestis, Y. enterocolitica, Y. frederiksenii, Y. intermedia* and others have not been associated with disease in birds. Yersinia are Gram-negative rod-shaped bacteria than can be motile or non motile, depending on the incubation temperature. Pathogenic *Y. pseudotuberculosis* carries a virulent plasmid of which six serovars have been identified and serovar 1 is most commonly isolated from birds.

EPIDEMIOLOGY
Yersinia spp. are ubiquitous in the environment and are worldwide in distribution. It has been isolated from many vertebrates and water. It multiplies at low temperatures; therefore, infections are common in the winter and spring. Rodents (rats, mice), hares and rabbits and some wild birds serve as reservoirs. Transmission is probably through contaminated water, feed and environment. Factors such as cold weather, chilling and concurrent diseases can predispose the birds for Yersiniosis.

CLINICAL SIGNS
Clinical signs due to Yersiniosis depend on whether the disease is acute or chronic; the chronic form being more common. In general clinical signs include lethargy, diarrhea, dyspnea and dehydration. In chronic infections symptoms include loss of weight, swollen joints, and paresis.

In one outbreak of 9 and 12 week-old turkeys, anorexia, watery yellow-green droppings, depression and acute lameness were observed with morbidity ranging from 2-15 % with increased mortality due to cannibalism. In acute infections of some species of birds such as wood peckers, toucans and turacos there may not be any clinical signs but birds are found dead.

LESIONS
Gross lesions due to Yersiniosis primarily involve liver and spleen. These organs can be enlarged with either a few pale foci of necrosis or yellow granulomas scattered throughout. Similar foci can also be found in the lungs, heart, kidneys, skeletal muscles and swollen joints. In addition catarrhal enteritis, osteomyelitis and myopathy have been described in turkeys.

Histologically acute lesions generally consist of mild to severe necrosis of the parenchyma accompanied by fibrinosuppurative or fibrinoheterophilic inflammation usually associated with numerous colonies of Gram-negative bacteria of bacilli morphology. In chronic infections these necrotic foci and inflammation will be surrounded by multinucleated giant cells.

DIAGNOSIS
A presumptive diagnosis can be made based on clinical signs, gross and microscopic lesions. Gram stain of smears from lesions can provide a tentative and quick diagnosis. Gross lesions in liver and spleen and other organs have to be differentiated from other bacterial diseases, such as, mycobacteriosis, salmonellosis, etc... *Y. pseudotuberculosis* can be isolated readily from most lesions such as liver, spleen, bone, joints, lungs, intestine and other organs.

CONTROL AND TREATMENT
Y. pseudotuberculosis is ubiquitous in the environment and water; steps should be taken to reduce their numbers. Biosecurity implementation, total confinement of birds, bird and rodent proofing the houses and aviaries are essential for preventing Yersiniosis. Treating chronic infections can be extremely difficult. Prompt diagnosis and use of tetracyclines have been beneficial in reducing mortality in outbreaks of Yersiniosis in turkeys and ducks.

Antimicrobial testing of isolates of *Y. pseudotuberculosis* has shown that antibiotics such as ampicillin, penicillin, enrofloxacin, spectinomycin, tetracyclines, sulfonamides, neomycin, ormetoprim/sulfa, and gentamicin can be effective in treatment. However, sensitivity tests should be performed on each isolate before antibiotics can be used.

FUNGAL DISEASES

Revised by Dr. H.L. Shivaprasad

ASPERGILLOSIS
(Brooder Pneumonia; Mycotic
Pneumonia; Pneumomycosis)

DEFINITION
Aspergillosis is an acute or chronic disease, primarily affecting the respiratory system. Peritoneal, visceral and systemic infections especially involving brain and eyes can also occur. The most common etiology is *Aspergillus fumigatus* but *A. flavus* can be involved. Aspergillosis occurs frequently in turkeys, chickens, and game birds. This condition has also been reported in penguins, raptors, migratory waterfowl, psittacines and zoologic specimens, such as flamingos. All species of birds probably are susceptible. Aspergillosis was first described in a wild duck in 1833 and in turkeys as early as 1898. There are some other species and genera of fungi that may cause similar disease syndromes.

EPIDEMIOLOGY
1. Embryos. *Aspergillus fumigatus* can penetrate egg shells under ideal growth conditions and thus infect the embryos. Such eggs will often appear green when candled (the embryo will be dead). Infected embryos may hatch with well developed lesions.

2. Chicks and poults. If infected eggs break in the hatchery, large numbers of spores are released which contaminate the hatchery environment and air systems can lead to severe outbreaks in very young birds (less than 3 weeks of age). Eggs punctured for *in ovo* injection are particularly susceptible to contamination. Even low-level contamination of hatchers or air systems can result in mortalities of 50% or greater when *in ovo* injection is used. Navel infections can also occur.

3. Adults. Infection usually follows inhalation of large numbers of spores from heavily contaminated feed, litter or environment. Conjunctival infections may occur from heavy exposure to airborne spores following traumatic injuries. It is believed that healthy birds resist infection but that resistance can be overwhelmed by massive exposure combined with depressed host defenses. Debilitated and overcrowded birds are most susceptible. Market age tom turkeys and turkey breeders are commonly affected.

4. *Aspergillus fumigatus* and *A. flavus* are normally present in litter and feed. Enormous numbers of spores can be produced under ideal conditions. Sporulating colonies of *Aspergillus fumigatus* are blue-green and can often be observed grossly.

5. Infections in the brain, posterior chamber of the eye or other visceral tissues result from systemic invasion from the respiratory tract.

6. The difference between localized and disseminated lesions due to aspergillosis has been attributed to the genetic differences between species of Aspergillus strains and their ability to produce elastase.

CLINICAL SIGNS
1. Dyspnea, gasping, cyanosis and accelerated, labored breathing (Fig. 1) frequently are observed. Rales do not usually accompany these respiratory diseases. Other signs include diarrhea, anorexia, somnolence, progressive emaciation, dehydration and increased thirst.

2. Morbidity is variable. Mortality is high in clinically affected birds. Increased mortality will be noted in affected flocks during loadout, hauling and following insemination. Affected birds often die during or just after handling especially if held by their legs.

3. Signs of central nervous system disturbance may occur in a small percentage of the birds if there has been spread to the brain. Signs often include ataxia, falling, pushing over backwards, opisthotonos, paralysis, etc.

4. A gray-white opacity may develop in one or both eyes when there is eye infection. Ocular discharge occurs when the conjunctiva is infected and there can be corneal ulceration. A large mass of exudate typically accumulates in the medial canthus under the third eyelid.

LESIONS
1. Mycelial growth with sporulation may be apparent as fuzzy gray, blue, green or black material (sporulating fungus) or pale yellow plaques on air sac, pleura, pericardium, peritoneum or in the syrinx and main bronchi of the lungs.

2. Pale yellow or gray circumscribed nodules or plaques in the lungs (Fig. 2), air sacs bronchi or trachea (Fig. 3) (usually the syrinx); less often in the brain, eyes, heart, kidneys, liver, or at other sites. In mature birds two patterns of air sac infection are found: disc-like plaques in the recurrent bronchi of the caudal thoracic and/or abdominal air sacs or markedly distended air sacs containing copious fluid and soft fibrinopurulent exudate.

DIAGNOSIS
1. The signs and gross lesions of aspergillosis are very suggestive of the diagnosis which can be confirmed by microscopic demonstration of fungus in fresh preparations made from the lesions or in histologic sections.

2. Microscopic examination reveals septate, branching hyphae within lesions. Hyphae can be seen in fresh preparations cleared with 10% KOH or stained with

lactophenol cotton blue. If fungus is grossly visible in the lesions, the typical fruiting bodies (Fig. 4) and spores can be easily found. In histologic sections, special stains (methenamine-silver, PAS, Gridley) are useful for demonstrating fungi in tissues. Nodules in the lungs usually appear as granulomas containing fungal hyphae.

3. Using sterile technique, the fungus can be cultured by tearing a nodule or plaque open and putting it on fungus media. Aspergillus will usually grow on blood agar in 24-48 hours. Sabouraud's dextrose agar (Fig. 5) is a more selective medium. Since aspergillus spores are common laboratory contaminants, growth of only a few colonies may not be sufficient for a definitive diagnosis. For confirmation, tissue invasion should then be demonstrated.

4. Typical lesions of aspergillosis are unlike those of other avian respiratory diseases except pulmonary granulomas associated with complicated *Mycoplasma gallisepticum* infection. Grossly aspergillus lesions especially in the lungs can resemble lesions caused by *Staphylococcus aureus* in turkey poults. Histopathologic differentiation is usually easy.

5. Another fungus, *Ochroconis* (previously *Dactylaria*) *gallopava*, can cause lesions in the lungs or brain of young chickens and turkey poults. Signs and lesions resemble those caused by aspergillosis. The two fungi can be differentiated by culture. Numerous giant cells are characteristic of microscopic brain lesions caused by *O. Gallopava*. On histopathology, pigmented fungus can be readily recognized but culturing is necessary for positive identification.

CONTROL

1. Collect clean eggs. Disinfect or fumigate eggs before setting. Do not set cracked eggs or eggs with poor shell quality.

2. Thoroughly clean, disinfect and fumigate incubators and hatchers. Inspect air systems and change air filters regularly in hatcheries. Monitor hatchery environment for mold contamination.

3. Use only dry, clean litter and freshly-ground, mold-free feeds. Store feeds and litters properly so as to inhibit growth of mold. Make sure feed bins and feed lines are kept clean, dry and free of mold growth. Do not permit feed to cake in feeders. Avoid wet litter under or around the waterers or feeders. Mold inhibitors may be added to feed to control fungus growth and prevent infection; however, this will add expense.

4. Optimize the ventilation and humidity in the poultry house to reduce air-borne spores. Humidity should be kept in the mid-range, neither too low nor too high. Alternating wet and dry conditions are an ideal situation for *Aspergillus*. The fungus multiplies during the wet period producing abundant spores which then become aerosolized when conditions become dry.

TREATMENT

1. If aspergillosis is diagnosed in a flock, cull clinically affected birds and remove any contaminated feed and litter. Clean and disinfect the house and then spray it with 1:2000 copper sulfate solutions or other fungicide and allow it to dry.

2. Valuable captive birds can be treated with Nystatin or Amphotericin-B or other anti-mycotic agents. Often antibiotics are given simultaneously to prevent secondary bacterial infection. Intravenous fluids may also be required. Ketaconizole, Miconozole and related drugs have been found effective for treating individual birds but are too expensive for commercial flocks.

ASPERGILLOSIS

Fig. 1
Gasping chicks.

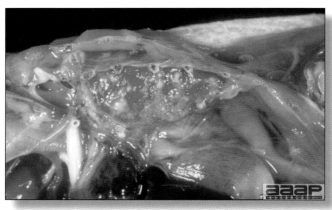

Fig. 2
Bird with yellow to white mycotic nodules in the lung.

Fig. 3
Mycotic nodule in the trachea.

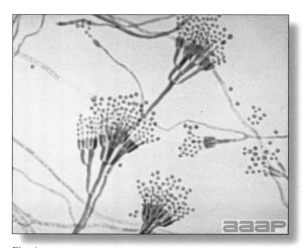

Fig. 4
Aspergillus fruiting body.

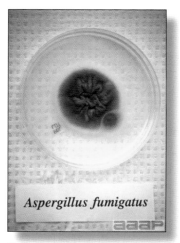

Fig. 5
Aspergillus fumigatus on Sab Dex media.

CANDIDIASIS

(Thrush; moniliasis, crop mycosis, sour crop, muguet, soor, levurosis)

DEFINITION

Candidiasis is a disease of the digestive tract caused by the yeast-like fungus *Candida albicans* and probably other species of *Candida*. The disease generally involves the upper digestive tract and usually occurs as a secondary infection.

EPIDEMIOLOGY

Candida albicans is a common yeast-like fungus that has been recognized as a commensal organism in poultry and mammals for many years. Candidiasis has been reported from a variety of avian species, such as, chickens, turkeys, pigeons, game birds, waterfowl, and geese. In poultry it seldom has been considered a disease of major importance. Young birds tend to be more susceptible than adult birds although all ages can be affected. When birds become debilitated or the normal digestive tract flora is altered, the ingestion of fungus in the feed and water can result in mucosal invasion. The production of a soluble endotoxin may also contribute to pathogenicity. Common predisposing causes include lack of good sanitation, prolonged treatment with antibiotics, heavy parasitism, vitamin deficiency, high carbohydrate diets, and immune suppressing or debilitating infectious diseases.

CLINICAL SIGNS

Signs are non-specific and include, listlessness, inappetence, retarded growth, and ruffled feathers. In advanced cases diarrhea may occur. The signs may be masked by the clinical signs of a primary disease. In advanced cases, the crop may not empty and may become fluid filled. The bird may regurgitate fluid with a sour, fermentative odor, i.e. the name "sour crop".

LESIONS

1. Lesions vary greatly in severity. They are more common in the crop, mouth, pharynx and esophagus, but may involve the proventriculus and, less often, the intestine. Lesions involving the crop are one of the most findings in turkey poults.

2. The affected mucosa is often diffusely or focally thickened (Fig. 1), raised, corrugated and white, looking like terry cloth (Fig. 2). Lesions may also appear as proliferative white to gray pseudomembranous or diphtheritic patches and as shallow ulcers. Necrotic epithelium may slough into the lumen as masses of soft cheesy material.

3. Lesions of a primary predisposing disease may also be present and should be investigated. In particular one should search for evidence of coccidiosis, parasitism or malnutrition especially vitamin A deficiency.

DIAGNOSIS

1. Characteristic gross lesions are generally adequate for diagnosis. Histopathologic examination of the affected mucosa usually will confirm invasion of the tissue by the septate fungal hyphae.

2. *Candida albicans* grows readily on Sabouraud's dextrose agar. However, since *Candida* is commonly present in normal birds, only the demonstration of massive numbers of colonies is of significance.

CONTROL

1. Practice a high standard of sanitation in the poultry operation. Phenolic disinfectants or iodine preparations should be used to sanitize equipment.

2. Prevent other diseases or management practices that might debilitate the birds.

3. Avoid over treatment of birds with antibiotics, drugs, coccidiostats, growth stimulants and other agents that might affect the bacterial flora of the digestive tract.

TREAMENT

1. Copper sulfate at a 1:2000 dilution in drinking water is commonly used both for prevention and treatment but its value is controversial. Nystatin in feed or water has shown efficacy against candidiasis in turkeys.

2. Routine addition of antifungal drugs to rations probably is a waste of money since elimination of contributing factors or other diseases usually will prevent candidiasis. However, if sanitation is at fault and cannot be improved, antifungal drugs may be advisable.

CANDIDIASIS

Fig. 1
Crop mycosis.

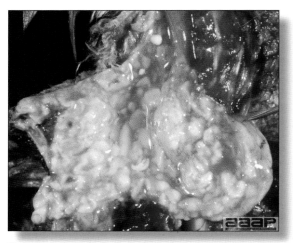

Fig. 2
Severe crop mycosis.

OCHROCONOSIS
(Previously known as DACTYLARIOSIS)

DEFINITION
A neurotropic, mycotic disease of turkey poults and young chickens with many of the clinical and pathologic features of aspergillosis. *Ochroconis gallopava* is a dematiaceous (Phaeohyphomycosis) pigmented septate fungus that has distinctive reddish brown in its walls. It occurs in the soil, saw dust, other litter, decaying vegetation and thermal springs. It is most common as an incidental finding in the granulomas of lungs of turkey poults. Signs of Ochroconosis (incoordination, tremors, torticollis, circling, recumbency) are related to mycotic lesions in the brain (Fig. 1). Lesions also occur with less frequency in the air sacs, liver and eyes (globes). The etiologic agent, *Ochroconis (Dactylaria) gallopava*, grows naturally in old sawdust which often is used as poultry litter.

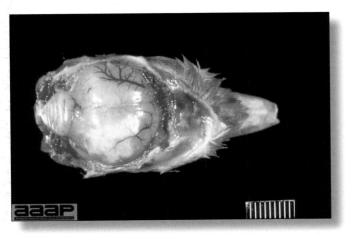

Fig. 1
Mycotic encephalitis.

FAVUS

(Avian ringworm, Avian dermatophytosis)

DEFINITION

Favus is a mycotic infection found primarily in gallinaceous birds. Favus is rare in commercial poultry today, but is occasionally reported in backyard flocks, especially exotic and game chickens. Characteristic lesions include white crusting or powder-like material on the comb and wattles (Fig. 1) that can extend to the feathered portion of the skin to form scutula around the bases of feather follicles. The fungus has predilection for the keratin layer of the epidermis and feather follicles causing hyperkeratosis, hence powder-like material seen grossly. *Microsporum gallinae* is the agent most often isolated, although *M. gypseum* and *Trichophyton simii* have also been isolated. Topical treatment with nystatin has been efficacious on individual birds.

Fig. 1
White crusting of comb (chicken on the right).

MYCOTOXICOSIS

DEFINITION

Mycotoxicosis is a disease caused by a toxic fungal metabolite. Mycotoxicoses may affect both humans and animals. Poultry mycotoxicoses are usually caused by fungi that colonize and invade grains and feeds, but other environmental aspects may be involved.

OCCURRENCE

1. Grains and forages used as foodstuffs support the growth of certain fungi when environmental conditions of temperature and humidity are suitable. Some of these fungi produce metabolites that are toxic to humans and animals and cause disease (mycotoxicosis) by either ingestion or cutaneous exposure.

2. Mycotoxicoses occur throughout animal-rearing regions of the world. Although specific mycotoxins form more frequently in certain geographic locations, interstate and international shipment of grains may result in widespread distribution of a mycotoxin problem.

DIAGNOSIS

1. A definitive diagnosis of mycotoxicosis should involve the isolation, identification, and quantitation of the specific toxin(s). This is usually difficult to accomplish in the modern poultry industry because of the rapid and voluminous use of feed and ingredients.

CONTROL

1. Prevention of mycotoxicoses requires the detection and control of mycotoxin contamination in feed ingredients and the application of feed manufacturing and management practices that prevent mold growth and mycotoxin formation.

2. Feeds and grains can now be screened for several mycotoxins (aflatoxin, T-2 toxin, ochratoxin, zearalenone) using monoclonal antibody detection kits. Many poultry companies already routinely test grain for aflatoxin contamination by a chromatographic procedure (minicolumn technique).

3. Mycotoxins can form in decayed, crusted, built-up feed in feeders, feed mills, and storage bins. This can be prevented by inspection of bins between flocks to certify absence of feed residue and by cleaning bins and feeders when necessary. Use of tandem feed bins on farms allows cleaning between successive feed deliveries.

4. Antifungal agents added to feeds to prevent fungal growth have no effect on toxin already formed, but may be cost-effective management in conjunction with other feed management practices. Several commercial products, most of which contain proprionic acid, should be applied according to manufacturers' instructions.

5. Zeolytes, a class of silica-containing compounds used as anticaking agents in feed formulation, and as aids in the improvement of eggshell quality, show promise as a practical and economical method of reducing the effects of certain mycotoxins. Hydrated sodium calcium aluminosilicate has been shown to bind aflatoxin B1, possibly by sequestration in the digestive tract, and reduce its toxicity to chickens.

TREATMENT

1. Remove the toxic feed and replace it with unadulterated feed.

2. Treat concurrent diseases (parasitic, bacterial) identified in the diagnostic evaluation.

3. Substandard management practices should be immediately corrected as they have increased detrimental effects in a flock stressed by mycotoxins.

4. Vitamins, trace minerals (selenium), and protein requirements are increased by some mycotoxins and can be compensated for by feed formulation and water-based treatment.

I. AFLATOXICOSIS

HISTORICAL INFORMATION

1. During the 1950s, a disease in dogs called hepatitis X occurred in the southeastern United States and was tied to the consumption of moldy dog food. It was later reasoned to have been caused by the same mycotoxin responsible for high mortality in turkeys due to hepatic toxicity (turkey X disease) in England in 1960. Peanut meal imported to England from Brazil was highly contaminated with fungi of the *Aspergillus flavus-Aspergillus parasiticus* group, which produced aflatoxins.

2. The aflatoxin story was historically important because unlike ergotism and alimentary toxic aleukia, which were sporadic and relatively localized phenomena, aflatoxicosis attracted global attention concerning the potential problems of mycotoxins in the food chain, and the ease by which these problems could be widely distributed.

ETIOLOGY

1. Mycotoxins of the aflatoxin group (B1, B2, G1, G2) are the cause of aflatoxicosis. Aflatoxin B1 is the most common in grains and is highly toxic. Aflatoxin forms in peanuts, corn, and cottonseed, and their products, in other grains, and in poultry litter. *A. flavus* is the primary producer of aflatoxin in grains, but not all strains of the fungus are toxigenic.

2. Like other mycotoxins, aflatoxin is produced only when substrate, temperature, and humidity are ideal. Favorable conditions for toxin formation may be localized within a volume of stored or transported grain creating toxic "hot spots". Once formed, the toxin is stable.

3. Grains damaged by insects and drought stress, and broken pieces of grain (screenings) are more likely to support fungal growth and toxin formation.

4. Aflatoxin B1 is a potent, naturally occurring carcinogen and thus has special public health considerations.

CLINICAL SIGNS

Aflatoxicosis in poultry is primarily a disease of the liver (Fig. 1) with important ramifications for other body systems, which may ultimately cause production problems and mortality. Affected birds have reductions in growth, carcass pigmentation, egg production, and immune function, and have increased nutrient requirements for protein, trace elements (selenium), and vitamins. The disease may be fatal.

LESIONS

At necropsy, lesions are minimal with either transient exposure or exposure to a low concentration of toxin. Jaundice, generalized edema and hemorrhages, tan (Fig. 2) or yellow discoloration of the liver, and swelling of the kidneys (Fig. 3) are seen with more severe intoxication. Microscopic changes in the liver occur as necrosis of hepatocytes, lipid accumulation in hepatocytes, bile duct proliferation, and fibrosis. These are common reactions of this organ to toxic insult and although they may be suggestive of aflatoxicosis, are not pathognomonic.

II. CITRININ MYCOTOXICOSIS

ETIOLOGY

Citrinin is a mycotoxin that was first isolated from *Penicillium citrinum* but is also produced by other species of *Penicillium* and by a few species of *Aspergillus*. Citrinin may be a factor in renal disease in food animals in Denmark, but no other documented case studies involving poultry are known.

CLINICAL SIGNS

Experimental citrinin mycotoxicosis in the chicken, turkey, and duckling has shown that chickens are relatively resistant, but all develop clinical illness of marked watery fecal droppings related to increases in water consumption and urine output. Metabolic alterations of electrolytes and acid-base balance occur. Young birds have reduced weight gain.

LESIONS

Citrinin produces marked functional changes in kidneys, however, gross lesions may be slight or overlooked. Swelling of kidneys and microscopic lesions of nephrosis may occur following severe exposure. In these circumstances, lymphoid tissues may be depleted and necrosis occurs in the liver.

III. ERGOTISM

HISTORICAL INFORMATION

1. Ergotism was recognized in central Europe in the Middle Ages and is the oldest known mycotoxicosis. Humans with ergotism (St. Anthony's fire) experienced an initial cold sensation in the hands and feet followed by an intense burning sensation. Gangrene of the extremities developed in both humans and afflicted animals. The disease occurred where bread was made from rye and other grains parasitized by toxigenic strains of the fungus *Claviceps purpurea*. The mold colonizes and replaces kernels of grain to form a hard, dark purple or black mass called an ergot or sclerotium.

2. Although the pharmaceutical properties of the ergot were recognized in China 5,000 years ago, it was not until 1875 that alkaloids present in the sclerotium were recognized as the cause of ergotism.

ETIOLOGY

1. The ergot alkaloids are a large family of compounds, and may cause constriction of blood vessels (vasoconstriction) which results in their pharmacologic and toxicologic effects.

2. *Claviceps* spp. that colonizes wheat, rye, and triticale are the most common causes of ergotism of humans and animals.

CLINICAL SIGNS

In chickens, ergotism causes reductions in growth and egg production, and nervous incoordination.

LESIONS

Lesions include abnormal feather development, necrosis of the beak, comb, and toes, and enteritis.

IV. OCHRATOXICOSIS

HISTORICAL INFORMATION

1. Ochratoxin has been detected in kidneys of chickens with renal lesions in a processing plant in Denmark.

2. Three disease outbreaks in the United States involving 360,000 turkeys were associated with ochratoxin concentrations of up to 16 ppm.

ETIOLOGY

Ochratoxins A, B, and C are usually produced by toxigenic strains of *P. viridicatum* but may be produced by other species of *Penicillium* and by *Aspergillus ochraceus*. Ochratoxin A is the most toxic and is the greatest threat to poultry production.

CLINICAL SIGNS

1. Reductions in feed intake and increases in mortality.

2. Weight loss.

3. Drops in egg production have been reported from Ochratoxin A.

LESIONS
1. Gross and microscopic lesions in the kidneys and liver.

2. Experimental ochratoxicosis in chickens causes a dose-related reduction in weight gain, and gross and microscopic lesions occur in the target organs, liver and kidney. Visceral gout and reductions in plasma carotenoids, immune function, and certain blood coagulation factors also occur.

V. OOSPOREIN MYCOTOXICOSIS

ETIOLOGY
Oosporein is a toxic pigment produced by *Chaetomium* sp. and other fungi and is a contaminant of cereal grains and feedstuff.

CLINICAL SIGNS
Oosporein mycotoxicosis, studied in chickens and turkeys, causes a dose-related decrease in growth and an increase in water consumption. Chickens are more susceptible than turkeys.

LESIONS
Visceral and articular gout as a result of nephrotoxicity.

VI. TRICHOTHECENE MYCOTOXICOSIS

(Fusariotoxicosis)

HISTORICAL INFORMATION
1. A disease called alimentary toxic aleukia occurred in the Russian people in the early 20th century, the 1930s, and especially during the Second World War. Labor shortages during the war necessitated the overwintering of grains (wheat, rye, and millet) in the fields and harvesting was delayed until spring. Bread made from the new grain caused acute gastroenteritis, followed by the formation of ulcers of the face and oral membranes, facial edema and lymph node enlargement, and in the later stages, bone marrow disorders, anemia, and uncontrolled hemorrhages. Morbidity and mortality greater than 50% occurred in some villages. Similar problems occurred in livestock and poultry in the region.

2. The disease is now recognized as a mycotoxicosis caused by colonization of grains by toxigenic species of *Fusarium*. These fungi produce mycotoxins of the trichothecene group, many of which cause caustic injury to mucous membranes and skin, the basis of the facial, oral, and gastrointestinal features of the disease. They also affect rapidly dividing cells (radiomimetic effect) manifested by disorders of the bone marrow (anemia, hemorrhagic disorders) and by abortions.

ETIOLOGY
More than 40 trichothecene mycotoxins are known to exist. T-2 toxin is one of the most toxic to poultry.

CLINICAL SIGNS
1. Chickens with fusariotoxicosis (trichothecene mycotoxicosis) have had reduced growth, abnormal feathering (Fig. 1), severe depression, and bloody diarrhea.

2. In chickens, pigeons, ducks, and geese, the caustic properties of the trichothecenes have been manifested as feed refusal, extensive necrosis of the oral mucosa and areas of the skin in contact with the mold, and symptoms of acute gastrointestinal disease.

3. Experimental fusariotoxicosis, reproduced in chickens with pure T-2 toxin closely resembled the spontaneous disease but lacked the extensive hemorrhages.

LESIONS
1. Trichothecene mycotoxicosis may cause necrosis of the oral mucosa (Fig. 2), reddening of the mucosa of the remainder of the gastrointestinal tract, mottling of the liver, distention of the gallbladder, atrophy of the spleen, and visceral hemorrhages.

VII. ZEARALENONE MYCOTOXICOSIS

HISTORICAL INFORMATION
In experimental studies, chickens have shown relative insensitivity to the effects of zearalenone. Zearalenone mycotoxicosis has been recognized since 1927 as the cause of a syndrome resembling estrogen stimulation in pigs and cattle in the United States and elsewhere.

ETIOLOGY
Zearalenone is a mycotoxin produced by *Fusarium roseum* (*Gibberella zeae*) and other *Fusarium* spp. A period of warm temperature and high humidity followed by low temperature is most conducive to toxin formation on grains

CLINICAL SIGNS
1. Zearalenone-contaminated feed has been associated with high mortality (40%) in a flock of 24,000 broiler breeder chickens. Affected birds had cyanotic combs and wattles and had difficulty walking.

2. Turkeys may develop swelling of the cloaca and reduced fertility.

3. Male geese may have reductions in sperm quantity and viability.

LESIONS
1. Affected chickens have had ascites and cysts both inside and outside of the oviduct. The oviducts were swollen and inflamed, and were obstructed with fibrinous fluid. Some oviducts had ruptured.

AFLATOXICOSIS

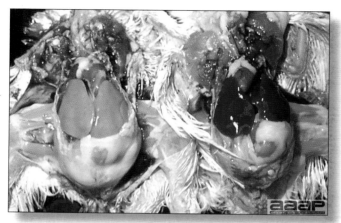

Fig. 1
Tan liver (3ppm aflatoxin in feed) vs. normal liver. (on the right)

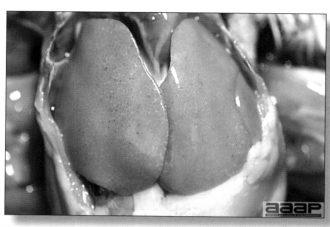

Fig. 2
Tan liver (3ppm aflatoxin in feed).

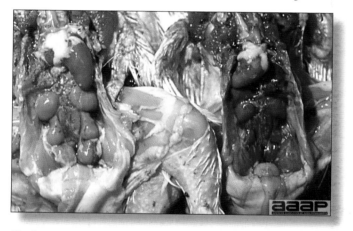

Fig. 3
Swollen kidneys (3ppm aflatoxin in feed) vs. normal kidneys. (on the right)

TRICHOTHECENE MYCOTOXICOSIS

Fig. 1
Abnormal feathering.

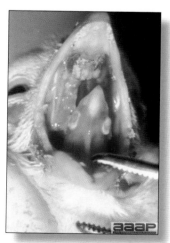

Fig. 2
Oral ulceration & necrosis.

	Aflatoxins	Fumonisins	Oosporein	Tricothecene	Ochratoxins	Citrinin	Ergotism	Zearalenone
Clinical signs	Decreased feed intake Decreased body weight Poor skin Decreased egg production Decreased immunity	Decreased body weight	Dose-related decrease in growth Increased water consumption	Decreased feed intake Reduced growth Severe depression Abnormal feathering Bloody diarrhea	Decreased feed intake, weight loss Increased mortality Increased water consumption Humid litter Decreased egg production	Marked watery fecal droppings Increased water consumption Increased diuresis Reduced weight gain (young birds) Humid litter	Reduced growth Decreased egg production Nervous signs (incoordination)	Reduced fertility
Lesions	Jaundice Generalized edema and hemorrhages, tan or yellow discoloration	Increased liver weight Increased kidney weight	Dehydration Swollen and pale kidneys secondary visceral and	Circumscribed proliferative yellow caseous plaques in oral mucosa Reddening of GI mucosa	Pale and swollen kidneys Secondary visceral gout	Swollen kidneys Degeneration and necrosis of tubular epithelial	Gangrenous-like lesions Necrosis of the beak, comb, toes	Oviduct hypertrophy Cloacal swelling (turkeys)

...of the liver Liver: Periportal necrosis with bile duct proliferation and fibrosis Depletion of lymphoid organs	necrosis with biliary hyperplasia	Mottling of the liver Gallbladder distention Splenic atrophy Visceral hemorrhages	enlarged liver Regression and cellular depletion of lymphoid organs.	proximal and distal tubules.		Reduction in sperm quantity and viability (geese)
Sources						
Peanuts, corn, cottonseed, and their products, In other grains and in poultry litter	Corn and corn based feed	Cereal grains feedstuff				
		Fusarium spp. are important pathogens to plant producing cereal grains, (corn, wheat, barley, oats, rice, rye….)	Widespread natural contaminant of cereal grains (barley, oats, rye, maize)	Often coexists in cereals with ochratoxin A (corn, wheat, barley, oats, rye and rice)	Open inflorescence of graminaceous plants (rye, wheat, triticale barley, oats, sorghum, corn, rice) and several grass species.	Corn, corn products, rice
Treatments						
		None	Vitamin C supplementation might reduce some adverse effects.			

PARASITIC DISEASES

Revised by Dr. Steve H. Fitz-Coy

EXTERNAL PARASITES

I. LICE

DEFINITION
Lice are insect ectoparasites. Lice are generally species specific, meaning for each species of bird or mammal, there are particular species of lice.

ETIOLOGY
In domestic fowl, more than 40 species of lice have been reported. Some of the most important chicken lice include the Body Louse (*Menacanthus* stramineus), Head Louse (*Culclotogaster heterographa*), Shaft Louse (*Menopon gallinae*), Wing Louse (*Lipeurus caponis*), Fluff Louse (*Gonicocotes gallinae*) and the Brown Chicken Louse (*Goniodes dissimilis*). Also important are the Large Turkey Louse (*Chelopistes meleagridis*), and the Slender Pigeon Louse (*Columbicola columabae*). Birds may be parasitized simultaneously by more than one species.

EPIDEMIOLOGY
As is evident from the common names, certain lice prefer different regions of the bird. They feed on scales of the skin and feathers. The life cycle is approximately 3 weeks. The entire life cycle is on the host. Lice will die in 5 to 6 days if separated from their host. Spread from bird to bird is dependent on close contact. Lice problems tend to be worse in the autumn and winter months.

CLINICAL SIGNS
Lice probably are not highly pathogenic for adult birds but the discomfort caused by the biting louse can have tremendous effects on flock performance. Heavy infestation of young birds is especially harmful due to the disruption of sleep.

LESIONS
Careful examination of the vent area, the underside of the wings, the head (crest and beard) and legs will reveal these pests. Most bird lice are straw-colored and vary in size from 1-6 mm, but some may reach 10 mm. Louse eggs often can be found attached to feathers in clumps called "nits" (Fig. 1).

DIAGNOSIS
Diagnosis is based on gross observation of skin and feathers.

CONTROL
Pesticides treatments are highly regulated chemicals and all have hazardous potential for animal and human health tissue residues, and environmental contamination. A current listing of approved pesticides should be obtained from local agricultural authorities or university specialists. Examples of classes of products approved by the EPA for control of mites and lice include: organophosphates, carbamates, pyrethrins, and pyrethroids. Treatment is simplified by the fact that the louse is only found on the bird and not in the environment. In general, the efficacy of the treatment is dependent on the application of the chemical. The agent must penetrate to the skin in order to kill the lice. Also the entire bird must be treated as the lice are very mobile and will move away from the treated areas. Lice eggs are not affected by insecticide treatment, therefore a minimum of two treatments are required. Treatment should be performed on 7 to 10 day intervals. Egg-laden feathers should be removed from the premises. Routine examination for infestation should be performed for resident flocks on a bi-weekly or monthly basis.

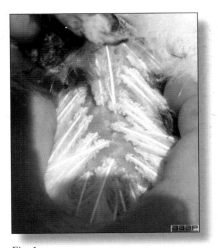

Fig. 1
Clumps of louse eggs attached to feathers.

II. MITES

DEFINITION

Mites are very small, barely discernible without magnification. They are not host specific and will infest any avian species. Mites feed on blood, feathers, skin or scales. Some mites do not spend their entire life on the bird; only visiting to feed, therefore, the bird and the environment must be treated to affect control. The most common mites of poultry species are listed below.

A. CHICKEN MITE – RED MITE
(*Dermanyssus gallinae*)

Common red mites are up to 0.7 mm long x 0.4 mm and appear black or gray or red if engorged with blood. Red mites feed mostly at night and may not be found on the hosts during the day. Inspection at night is usually necessary to confirm infestation. Their life cycle can be completed in as few as 7 days. During the day, they may be found in colonies in the cracks or joints of roosts or nests. They can survive for over 30 weeks without food. This makes treatment of the facility imperative in control of this pest. Anemia and mortality can result from heavy infestations, especially in young birds. They have been reported to transmit the agents of fowl cholera and spirochetosis. They frequently parasitize caged layers. Treat both birds and facilities. Repeat treatment in one week.

B. NORTHERN FOWL MITE
(*Ornithonyssus sylviarum*)

These mites are common bloodsuckers of a wide variety of poultry and other birds. They are known to or suspected of harboring agents that cause fowl pox, Newcastle disease, ornithosis and certain encephalitides. These mites stay on the birds continuously looking like moving red to black specks. Heavy infestations appear as blackened feathers, often near the vent with scabbed and cracked skin (Fig. 1). Mite egg sacs are in white or off-white clusters located under the wings, above and below the vent, in the beard and crest and on the feathers of the proximal thigh. After handling or at necropsy they may transfer to the hands and arms of the handler or diagnostician. Northern fowl mites parasitize caged layers, especially during the wintertime, and often are seen crawling on eggs. All birds in the flock should be treated twice in a 5-7 day interval.

C. DEPLUMING MITES (*Knemidokoptes gallinae*)

A variety of feather mites live on the feathers or in the quills of domestic and wild birds. Feather mites tend to be somewhat host specific. They cause breakage or complete loss of feathers. The depluming mite of chickens, pheasants and pigeons (*Knemidokoptes gallinae*) burrows into the basal shafts of the feathers producing intense irritation that will cause the host to pull out its body feathers. Loss of feathers can lead to inability to control body temperature and may increase susceptibility to other diseases. These mites are difficult to treat as the feather shaft protects them from chemical agents. Isolate affected birds. Treat as for *Dermanyssus gallinae* (red mites).

D. SCALY LEG MITE. (*Knemidokoptes mutans*)

This mite represents a group of closely related mites that live primarily within unfeathered skin, often on the shanks and feet. It cannot be seen without magnification. The affected skin becomes thickened and hyperkeratotic with white, powdery, exfoliating crusts (Fig. 2). Some mites in this group attack the beak of birds. Without treatment, the affected bird will eventually become crippled. Long term treatment of individual birds is necessary for recovery. Spread through a flock is usually slow.

CONTROL

Treatment must be tailored to the species of mite affecting the bird, i.e., accurate identification is important. In many cases, the bird and the environment must be treated simultaneously. Insecticides for mites can be applied as powders, dusts or sprays. It is important that the insecticide penetrates to the skin. For mites that live off the host (e.g. *Dermanyssus gallinae*), facilities can be treated by spraying a wet-able powder or liquid spray that will penetrate small cracks and crevices. Floors and bedding should also be treated. Pesticides are highly regulated chemicals and have hazardous potential for animal and human health and environmental contamination. A current listing of approved pesticides should be obtained from local agricultural authorities or university specialists. A notable exception is the Scaly leg mite (*Knemidocoptes mutans*) which can only be treated by direct application of an oil based product, such as petroleum jelly or cooking oil and kerosene (50:50) on a daily basis for at least 2 weeks or until the appearance of the legs returns to normal. Gentle washing of affected areas with soapy water to loosen and mildly hyperkeratinized areas is also helpful.

MITES

Fig. 1
Northern fowl mites infestation.

Fig. 2
Severe hyperkeratosis of the feet of a silkie chicken.

III. MISCELLANEOUS PESTS

A. BEDUGS (*Cimex lectularius*)

Bedbugs attack mammals and birds, including poultry and pigeons. Adult parasites are up to 5.0 mm long and have 8 abdominal segments. Bedbugs usually feed at night and may be observed on parasitized birds. Bedbugs can survive in houses for a year in the absence of poultry. Birds parasitized by bedbugs soon become unthrifty and anemic.

B. CHIGGERS (*Neoschongastia americana*)

Larval mites of *Neoschongastia americana* are a serious pest of turkeys and birds in southern states. In turkeys the larvae attach to the skin and cause localized skin lesions that lead to market downgrading. The lesions resemble pimples and may be very numerous.

C. DARKLING BEETLE AND LESSER MEALWORM. (*Alphatobius diaperinus*)

This beetle (Fig. 1 and 2) and its larvae (Fig. 3) often are present in poultry house litter. They may act as disease vectors for Marek's disease virus and Infectious Bursal disease virus. Botulism toxin has been found in these beetles. They probably serve as vectors for tapeworms. Their primary importance to the commercial poultry industry is economic due to the severe damage they can cause to the insulation and wall materials of modern poultry houses.

D. STICKTIGHT FLEAS
(*Echidnophaga gallinaceae*)

In poultry these parasites usually are tightly attached in clusters to the skin of the head. They irritate the skin, cause anemia, lower egg production and may kill young birds. The adult fleas are about 1.5 mm long and reddish brown.

E. PIGEON FLIES (*Pseudolynchia canariensis*)

Dark brown, bloodsucking flies, about 6.0 mm long, that often parasitize pigeons, especially nestlings. The flies cause anemia and dermatitis. These flies also transmit *Hemoproteus columbae*, the cause of a malaria-like disease of pigeons.

F. BLACKFLIES (*Simuliidae*)

Grey-black, thick, humpbacked insects up to 5.0 mm long. Swarms of the flies attack mammals and birds, including poultry. Blackflies are the vectors for protozoan (leukocytozoonosis) and filarial worms (*Ornithofilaria fallisensis)* in ducks. These biting insects require a blood meal for the maturation of eggs. The blood sucking blackflies can produce severe anemia and may kill young birds.

G. MOSQUITOS

Multiple genera and species of mosquitoes feed on birds, including poultry. Mosquitos are most important as vectors of viral diseases such as Pox and Equine Encephalitis. Generally a blood meal is necessary for egg production by the female mosquito. They also transmit protozoa that cause avian malaria-like syndromes. Mosquitos' bites cause irritation which may affect performance.

H. FOWL TICKS (*Argas persicus*)

Soft ticks including the "fowl tick" parasitize a wide range of poultry, wild birds and, occasionally, mammals. Some of the ticks not only cause anemia, skin blemishes or tick paralysis but also transmit *Borrelia anserina*, the agent that causes spirochetosis. Fowl ticks occur more frequently in the southwestern states, including California. The ticks spend relatively little time on their hosts and are easily overlooked.

MISCELLANEOUS PESTS

Fig. 1
Darkling beetles.

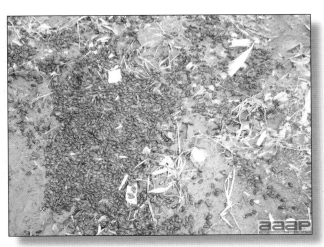

Fig. 2
Darkling beetles.

Fig. 3
Darkling beetles larvae.

INTERNAL PARASITES

I. NEMATODES, CESTODES AND TREMATODES ("WORMS" AND FLUKES)

DEFINITION

The most important internal parasites of poultry belong to the taxonomic group Nematodes. They have spindle shaped bodies with tapered ends and are also known as roundworms. Eggs are shed in the host droppings. Infection is established when a bird ingests an embryonated egg from the environment (direct life-cycle) or when a bird consumes an intermediate invertebrate host that is infected with the parasite (indirect life-cycle). Other internal parasites of diagnostic significance are Cestodes (tapeworms) and Trematodes (flukes).

A. ASCARIDS (Large Intestinal Roundworms)

One of the most common parasitic roundworms of poultry (*Ascaridia galli*) occurs in chickens and turkeys. Adult worms are about one and a half to three inches long and about the size of an ordinary pencil lead (Fig. 1). Heavily parasitized birds may be droopy, emaciated and show signs of diarrhea. Feed efficiency is usually impaired in severe cases.

The life cycle of this worm is simple with no intermediate host, females lay thick heavy-shelled eggs in the intestine that pass in the feces. Over two to three weeks embryo develops into the infective stage in the egg; the embryonated eggs may stay viable for long periods. Birds become infected by eating eggs that have reached the infective stage. Normal cleaning and disinfecting agents do not kill the eggs.

B. CECAL WORMS

These worms (*Heterakis gallinae*) are found in the ceca of chickens, turkeys and other birds (Fig. 1 and 2). The worms themselves are not considered a major threat, but they are highly considered a major carrier/vector for the agent that causes blackhead (*Histomonas meleagridis).*

The life cycle is similar to that of the large roundworm. Eggs (Fig. 3) are produced in the ceca and pass in the feces and become infective in approximately two to three weeks.

C. CAPILLARIA (Capillary or Thread Worms)

There are many Capillaria species that affect birds; but in commercial poultry the commonly encountered are *Capillaria annulata* and *Capillaria contorta* (Fig. 1). These occur in the crop and esophagus of the hosts (Fig. 2). These may cause thickening and inflammation of the mucosa, and occasionally severe losses are sustained in turkeys and game birds. Severe infestation may lead to mortality.

The life cycle is direct, adult worms are embedded in the lining of the intestine. Eggs (Fig. 3) are laid and passed in the droppings, embryonation occur in six to eight days.

D. TAPEWORMS

Tapeworms or cestodes are flattened, ribbon-shaped worms composed of numerous segments or division (Fig. 1). Tapeworms vary in size from very small to several inches in length. Several species of tapeworms affect birds but the most commonly found in poultry are *Raillietina cesticillus* (Fig. 2) and *Choanotenia infundibulum*.

Tapeworms use intermediate hosts for their lifecycles, birds become infected by eating the intermediate hosts. These hosts include snails, slugs, beetles, ants, grasshoppers, earthworms, houseflies and others. The intermediate host becomes infected by eating the eggs (Fig. 3) of tapeworms that are passed in the bird feces.

The pathology or damage tapeworms produce in poultry is questionable; however, it is believed that occlusion of the intestines is a fairly common finding.

EPIDEMIOLOGY

Animals in modern commercial poultry systems have a lower incidence and worm burden by the less access to many parasites and intermediate hosts; however, the incidence in backyard and free range flocks can be higher with a significant worm burden. Also, clinical disease in all-in-all-out production systems for commercial broilers and turkeys is rare. The short life span of the commercial broiler or turkey also circumvents severe parasitism. In meat birds, ascaridiasis is the most common parasitism observed. Commercial caged layers rarely have parasites without intermediate hosts (e.g.tapeworms) because of their lack of contact with the soil. Commercial turkeys and broilers reared on built-up litter, breeder flocks, non-caged commercial laying hen flocks, backyard flocks, game birds and pet or zoo animals are likely to exhibit higher rates of parasitism.

CONTROL

Control measures that interrupt the life-cycle are effective for most nematodes with direct cycles of infection. For parasites with indirect life-cycles (some nematodes, cestodes and trematodes), control is often aimed at elimination of the intermediate host such as beetles or other insects, snails or slugs, or preventing access of poultry to the intermediate host.

Piperazine is the only FDA approved treatment for internal parasites in meat and egg producing fowl. It is effective against ascarids in areas where resistance has not developed. It is applied in the drinking water. FDA regulations now allow the off-label use of drugs approved for other food animals for treatment of poultry. Fenbendazole has been used as a feed or water additive has been successfully used against *Capillaria,* and *Heterakis* infections. Thiabendazole, mebendazole, cambendazole, levamisole and tetramisole have been used against *Syngamus* and other nematodes such as *Trichostrongylus.* Pyrantel tartrate and citarin have also been effective against some nematode infections. Butynorate is approved for treatment of some cestodes of chickens.

SEE TABLE OF COMMON WORMS
AND FLUKES ON PAGE 177

II. BLOOD-BORNE PROTOZOAL PARASITES

INTRODUCTION
Avian species act as host for a number of blood-borne protozoal parasites. Although these infections are for the most part inapparent and undiagnosed, they can offer a diagnostic challenge to the uninformed. Diagnosis is generally by microscopic evaluation of blood smears or histologic sections. The parasites can be differentiated by their tissue distribution, size and physical characteristics.

PREVENTION AND TREATMENT
These diseases are rarely treated. Few effective treatments are known. Clopidol (Coyden) has been administered in the feed to turkeys for treatment of Leukocytozoonosis at 0.0125 to 0.025% for 14-16 weeks with apparent success. Prevention is attempted by vector control. In enzootic areas, screening of housed flocks and programs to control the black flies and midges have failed to control the disease. Most avoid raising poultry where conditions are ideal for propagation of vector species. Elimination of carriers by yearly disposal of old poultry may be helpful if contact with local, wild birds that are carriers is avoided.

A. ATOXOPLASMA
A coccidial protozoan that primarily parasitize captive and possibly wild passerine birds; these organisms have had a controversial taxonomic association for many years with other known members of the suborder Eimeriorina such as Toxoplasma, Lankesterella, or Isospora. The parasite is spread by fecal-oral route; the lifecycle consists of an intestinal and an extraintestinal forms. The oocysts are shed in the droppings. Differentiation of the oocysts from Isospora based on microscopic evaluations may be challenging; both organisms, the sporulated oocysts have two sporocysts and four sporozoites. Acute deaths may occur in young birds, older birds may be chronically infected and show no signs. Hepatosplenomegaly is common. Tumor-like lesions may be present in the liver and spleen of chronically infected birds. The parasite can be visualized microscopically as a clear notch in the nucleus of mononuclear cells and sometimes in erythrocytes. Cytology may be useful for the extraintestinal stages and histopathology for the intestinal forms.

B. HAEMOPROTEUS
DESCRIPTION
Haemoproteus belongs to the family Plasmodiidae and shares similarities with Plasmodium and Leucocytozoon. Over 120 species have been reported from birds, mostly in wild waterfowl, raptors, passerines, and others. Infection is host-specific. Species occurring in domestic poultry and pet birds include *Haemoproteus meleagridis* which has been diagnosed in domestic and wild turkeys, *H. columbae* and *H. saccharovi* in pigeons and doves and *H. nettionis* in waterfowl.

EPIZOOTIOLOGY
Biting flies and midges act as vectors. Sporogony occurs in the insect host and enters the avian host through fly bites. Schizonts (meronts) commonly infect pulmonary vascular endothelium or other visceral endothelial cells. Merozoites invade erythrocytes and mature. In some Haemoproteus species a second cycle of schizogony occurs involving the development of megaloschizonts in cardiac and skeletal muscles before merozoites invade red blood cells.

CLINICAL SIGNS, LESIONS & DIAGNOSIS
Most infections are inapparent and remain undiagnosed. Clinical signs have been reported in quail, turkeys, pigeons, and some psittacines. Generally, only a small proportion of birds infected exhibit clinical symptoms. Turkeys experimentally infected with *H. meleagridis* have developed severe lameness, diarrhea, depression, emaciation and anorexia. Anemia and hepatomegaly have also been reported. Myopathy associated with megaloschizonts has been reported in wild turkeys where skeletal muscles contained fusiform cysts oriented parallel to the muscle fibers. Enlarged gizzards have been rarely reported in pigeons. Lameness, dyspnea, sudden death associated with edematous lungs and visceral organ enlargement has been rarely reported in Muscovy ducks infected with *H. nettionis.* Gametocytes and pigment granules can be visualized adjacent to the nuclei of red blood cells in blood smears stained with Giemsa or Wright's stain (Fig. 1). Schizonts can be seen in vascular endothelial cells of the lung and visceral organs.

C. LEUKOCYTOZOA
DESCRIPTION
Leucocytozoonosis is an acute or chronic protozoal disease of birds, including turkeys, ducks, geese, guinea fowl and chickens. The disease was first observed in wild bird species and reported in turkeys in 1895. Relatively few reports of acute outbreaks of leucocytozoonosis in domestic poultry have been made during the last decade. The disease is generally clinically inapparent. Most acute outbreaks are in young poultry whereas the chronic form of the disease usually occurs in older birds (breeding stock). Black flies (*Simuliidae*) and culicoid midges serve

as intermediate hosts. The disease is most common in poultry housed near slow streams, shallow lakes or near marshy areas. Most outbreaks occur during the warmer seasons of the year when black flies are numerous. In the United States leucocytozoonosis occurs more frequently in southeastern states and in the upper Midwest.

EPIDEMIOLOGY
Birds appear to be the only hosts of most *Leucocytozoons*. The etiologic agents are *Leucocytozoons* but their classification may be inaccurate. Commonly applied names include: *L. smithi* (turkeys), *L. simondi* (ducks), *L. neavei* (guinea fowl) and *L. andrewsi* (chickens). Wild or domestic birds that survive the disease are inapparent carriers of leucocytozoons during the winter. During the warm seasons black flies (*Simuliidae*) and midges (*Culicoides*) feed on the carriers and become infected by the leucocytozoons. The parasites undergo sporogeny in the insects and pass to glands in their oral cavity. The insects then act as vectors and transmit leucocytozoons to young susceptible birds on which they feed. Birds that survive the disease become carriers in turn.

CLINICAL SIGNS, LESIONS & DIAGNOSIS
There is usually a sudden onset and numerous birds soon show signs. There is depression, anorexia, thirst, loss of equilibrium, weakness, and anemia. There may be rapid labored breathing. The course often is short and affected birds die or improve within a few days. Mortality varies but often is high. In birds that live a few days there may be splenomegaly (Fig. 1), hepatomegaly and evidence of anemia. The most pronounced lesion often is splenic enlargement. Microscopically, gametes usually are seen within enlarged, distorted erythrocytes, leucocytes or both types of cells on blood smears stained with Wright or Giemsa stains (Fig. 2 and 3). Histologic sections of the liver and brain often reveal megaloschizonts or schizonts.

D. PLASMODIUM
Plasmodium infections have been rarely reported in domestic pigeons, canaries, turkeys, penguins, falcons, bald eagles and cliff swallows. Plasmodia species are often not host-specific. The parasite causes a disease similar to malaria in man and is spread through the injection of infected blood by mosquito vectors. Parasites are found in red blood cells (Fig. 1).

E. TRYPANOSOMES
Motile protozoa found in the plasma of numerous species of wild and domestic birds. Pathogenic significance in domestic poultry appears to be minimal or nil. A wide range of insect vectors includes mosquitoes, Culicoides (midges), Hippoboscids (black flies), mites and Simulids.

NEMATODES, CESTODES AND TREMATODES ("WORMS" AND FLUKES)

ASCARIDS

Fig. 1
Ascaridia.

CECAL WORMS

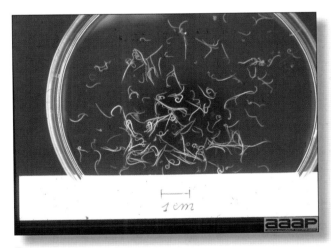

Fig. 1
Heterakis worms.

Fig. 2
Heterakis in caecum.

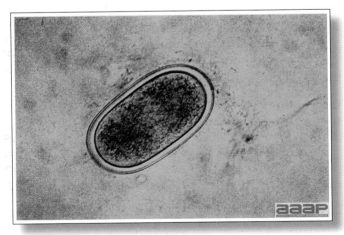

Fig. 3
Heterakis egg.

NEMATODES, CESTODES AND TREMATODES ("WORMS" AND FLUKES)

CAPILLARIA

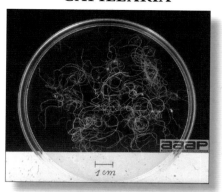

Fig. 1
Capillaria worms.

Fig. 2
Capillaria in crop.

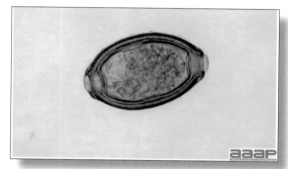

Fig. 3
Capillaria egg.

TAPE WORMS

Fig. 1
Severe tapeworm infestation.

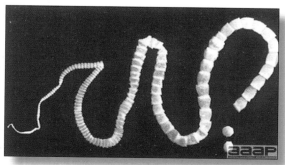

Fig. 2
Adult tapeworm *Raillietina*, 12-13 cm.

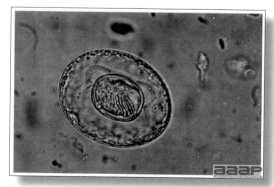

Fig. 3
Tapeworm egg (*Raillietina* spp).

BLOOD-BORNE PROTOZOAL PARASITES

HAEMOPROTEUS

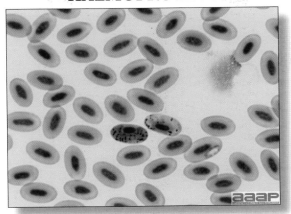

Fig. 1
Haemoproteus: Gametocytes and pigment granules
adjacent to the nuclei of erythrocytes.

LEUKOCYTOZOA

Fig. 1
Splenomegaly.

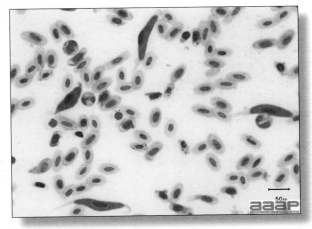

Fig. 2
Numerous leucocytozoons in blood smear.

PLASMODIUM

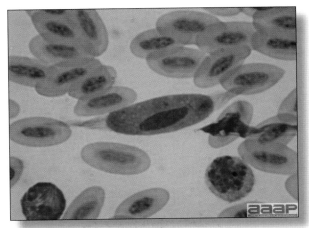

Fig. 3
Leucocytozoon (gamete) in distorted, elongated leucocyte.

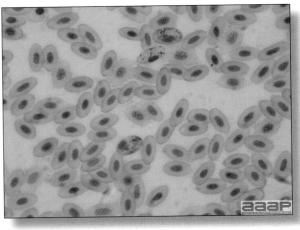

Fig. 1
Plasmodium in erythrocytes.

III. PROTOZOAL INFECTIONS OF THE DIGESTIVE TRACT

A. COCCIDIOSIS

DEFINITION

Avian coccidiosis is a common protozoal disease of poultry and many other birds characterized by diarrhea and enteritis. Coccidiosis in poultry affects the intestinal tract, except for renal coccidiosis in geese.

OCCURRENCE

Coccidiosis is found in all segments of the poultry industry and has a world-wide distribution. The development of intensive confinement production systems has increased the economic significance of this disease. Subclinical disease has been recognized as having important impact on performance in commercial meat-bird production and negative impacts on flock uniformity of layer and breeder pullets. The development of effective anticoccidial drugs and vaccines has made prevention of clinical disease and negative effects on performance more manageable. Uncontrolled coccidiosis may cause of loss of production and even mortality. Coccidiosis can be one of the predisposing factor for necrotic enteritis caused by *Clostridia perfringens*. In gamebird production, coccidia control is important in preventing outbreaks of ulcerative enteritis or "quail disease", caused by *Clostridia colinum*.

ETIOLOGY

1. Coccidiosis in chickens and turkeys is caused by the protozoal species of *Eimeria*. There are nine described species of *Eimeria* in chickens and seven in turkeys, but not all are severe pathogens. Coccidia are host specific; hence do not pass among the various classes of poultry, with the exception of *E. dispersa* that affect turkeys, quail and other gallinaceous birds.

2. Coccidia have a direct but complex life cycle. Infection is by the fecal-oral route. Ingestion of infected feed, water, litter and soil results in infection. Sporulated (infective) coccidial oocyst is ingested, sporozoites are released to initiate a series of asexual replications followed by a sexual cycle that lead to development of thousands of new oocysts in the intestine or ceca. Unsporulated oocysts are shed in the feces. These oocysts sporulate within 24 hr and then are infectious for other chickens or turkeys. A single oocyst may give rise to more than 100,000 progeny.

3. Coccidia produce lesions in the gut by destruction of the epithelial cells in which they develop and multiply, and by trauma to the intestinal mucosa and submucosa. Intestinal damage is directly proportional to the number of sporulated oocysts and the species ingested by a susceptible host.

4. Speciation of coccidia by microscopic features of oocysts (size, shape, color, length and width), the region of the gut parasitized, the nature of lesions produced, pre-patent periods, sporulation times, etc. Identification frequently can be made with reasonable accuracy by using several of these features by someone skilled at the art. Recently, molecular tools have developed and demonstrated to be effective in the identification of several *Eimeria* species. Precise identification is of some value in selecting the most effective anticoccidial(s) for control.

5. With controlled exposure via the use of live coccidia vaccines, the birds develop long term immunity, often without clinical signs of infection. Current vaccines are given either at pre-hatch or post-hatch; immunity depends upon self exposure to selected *Eimeria* for the stimulation and development of immunity. Poultry maintain their immunity to a species of coccidia by repeated re-exposure. The host remains susceptible to coccidial species not yet encountered. Birds may be infected simultaneously with more than one species.

6. Oocysts are maintained in the litter of the poultry house or can be easily transported to poultry farms on boots, shoes, clothing, crates, vehicle wheels, by other animals and insects. People are important vectors of coccidia. Wet litter and warm temperatures facilitate sporulation and precipitate outbreaks of coccidiosis.

COCCIDIA OF CHICKENS

Nine species of *Eimeria* have been described in chickens: *E. acervulina*, *E. necatrix*, *E. maxima*, *E. brunetti*, *E. tenella*. *E. mitis*, *E. mivati*, *E. praecox* and *E. hagani*. Clinical disease is determined by the species of the infecting coccidia. Less pathogenic species produce few or no lesions. The more pathogenic species often cause diarrhea which may be mucoid or bloody. Dehydration often accompanies the diarrhea. Diarrhea and dehydration are soon followed by ruffled feathers, anemia, listlessness, weakness, retraction of the head and neck and somnolence. Growth rate is often adversely affected. In laying hens coccidiosis is usually manifested by a drop in egg production. Depigmentation of the skin may be apparent in well established cases. Morbidity and mortality within a flock may vary greatly, but both can be very high.

A. *Eimeria acervulina.*
E. acervulina is a moderately severe pathogen causing enteritis in the anterior one third of the intestinal tract (Fig. 1). The enteritis can be mild to severe and cause thickening of the mucosa. May affect skin pigmentation due to malabsorption of carotenoids and reduce feed conversion. Transverse white to gray striations are visible in the mucosa (Fig. 2). Oocysts in mucosal scrapings are moderate in size and egg shaped (18.3 x 14.6 microns). This type of coccidiosis occurs rather frequently in older birds. This location is also favored by other less pathogenic species i.e. frequently multiple species will be present, obscuring the diagnosis. This species is one of the most prevalent in commercial poultry operations.

B. *Eimeria necatrix.*
Eimeria necatrix causes severe enteritis characterized by congestion, hemorrhage, necrosis and blood in the middle small intestine with bloody feces (Fig. 3). The intestine often is markedly dilated, inflamed and thickened. White to yellow foci and petechial hemorrhages may be seen through the serosa of the unopened gut (Fig. 4); these lesions are the development of the large schizonts predominantly in the mid small intestines (Fig. 5). This species is often mistaken or confused with *E. maxima*; the lesions with *E. necatrix* have the appearance of salt ands pepper (dark red). Oocysts develop only in the ceca, and the oocysts may not be numerous; moreover, mortality may precede the appearance of oocysts in the feces. Often causes high mortality. Often causes disease in commercial broiler breeders or layer pullets.

C. *Eimeria maxima.*
Eimeria maxima is moderately pathogenic and may cause moderately high mortality. It causes mild to severe enteritis sometimes with thickening of the intestinal wall and marked dilatation of the the middle small intestine, these resemble *E. necatrix*, but the lesions of *E. maxima* are bright red (Fig. 6). Intestinal content may be bloody. Very large oocysts (30.5 x 20.7 microns), often with a golden color and large gamonts are diagnostic for this species. Subclinical infections may impede absorption and result in poor skin pigmentation. This species is very prevalent in commercial poultry operations.

D. *Eimeria brunetti.*
Eimeria brunetti causes enteritis in the lower small intestine, rectum and proximal cecum. In severe cases, a fibrinous or fibrinonecrotic mass of debris may cover the affected mucosa or produce caseous cores in the ileum and rectum (Fig. 7). These oocysts are fairly large (24.6 x 18.8 microns), each with a polar granule. *E. brunetti* is a moderately severe pathogen that can produce moderate mortality, loss of weight gain, and poor feed conversion.

E. *Eimeria tenella.*
Eimeria tenella is highly pathogenic, causes a marked typhlitis (Fig. 8) with occasional involvement of the adjacent areas of the intestine. Blood is often apparent in the ceca (Fig. 9) and feces in the early stages of the infections; later, cheesy cecal cores may be found. Large clusters of schizonts may be seen in microscopic scrapings of the ceca. *E. tenella* can cause high morbidity, mortality and reduced weight gain in commercial broilers or layer pullets. This species is commonly found in commercial poultry.

F. *E. mitis.*
There are no clinical lesions with this species, the lower small intestine which may appear pale and flaccid. Pathogenic effects on weight gain and a cessation in egg production of Single Comb White Leghorn hens have been demonstrated.

G. *E. mivati.*
Causes reduced weight gain and mortality. This species is moderately pathogenic causing bloody mucoid enteritis, in severe cases lesions may extend throughout the entire small intestine (Fig. 10). White spots may be seen scattered throughout the serosa and are visible in the mucosa. Oocysts in mucosal scrapings are relatively small and broadly ovoid in shape (15.6 x 13.4 microns). This location is also favored by other species such as *E. acervulina*, thereby obscuring the diagnosis.

H. *E. praecox.*
Causes watery intestinal contents with mucus and mucoid casts in the duodenum. There may be reduced weight gain, loss of pigmentation, dehydration and poor feed conversion.

I. *E. hagani.*
Reportedly causes watery intestinal contents and catarrhal inflammation. This species is relatively rare in commercial broiler chickens.

COCCIDIA OF TURKEYS

Seven species of *Eimeria* have been described in turkeys in the USA. The four pathogenic species of *Eimeria* in turkeys are: *E. adenoeides*, *E. meleagrimitis*, *E. gallapovonis*, and *E. dispersa*. Nonpathogenic species include: *E. innocua*, *E. meleagridis*, and *E. subrotunda*. Coccidiosis in turkeys resembles the disease in chickens; the diarrhea may be watery, mucoid and bloody and mortality may occur.

A. *Eimeria meleagrimitis*.

Eimeria meleagrimitis causes spotty congestion and petechiae from duodenum to ileum, dilation of jejunum, and mucosal casts (Fig. 11) in the anterior two thirds of the intestine. Lesions are most severe in the jejunum and from there extend anteriorly and posteriorly. Oocysts in mucosal scrapings are small in size (19.2 x 16.3 microns) and ovoid. This species is considered a moderate pathogen. Mortality, morbidity, weight loss, dehydration and general unthriftiness may occur in young poults as a result of infection. Non-pathogens *E. innocua* and *E. subrotunda* may also be found in this region. This species is one of the commonly found *Eimeria* in litter from commercial turkey facilities. This species is fairly prevalent in commercial turkey operations.

B. *Eimeria dispersa*.

Eimeria dispersa produces a cream-colored serosal surface, dilation of intestine and yellowish mucoid feces in the middle one third of the intestine. Lesions are principally located in the midgut region, but some infection may extend from the duodenum to the cecal necks. Oocysts are large (26.1 x 21.0 microns) and broadly ovoid. The oocyst wall is distinctively contoured and lacks the double wall common to other species. The prepatent period is the longest of the turkey coccidia, 120 hours. This species is considered mildly pathogenic but can cause reduction of weight gain and diarrhea in young poults. This species also parasitized other avian hosts (quail and other phesants). This species is fairly prevalent in commercial turkey operations.

C. *Eimeria gallapovonis*.

Eimeria gallapovonis causes edema, ulceration of mucosal ileum, yellow exudate, and flecks of blood in feces. Lesions are principally located in the posterior one third of the intestine (ileum and large intestine) and ceca. Oocysts are relatively large (27.1 x 17.2 microns), elongated and ellipsoidal. The prepatent period is 105 hours. This species can cause high mortality in young poults; however, the prevalence is low.

D. *Eimeria adenoides*.

Eimeria adenoides affects the posterior one third of the intestine and is responsible for liquid feces with mucus and flecks of blood. There is edema and swelling of the cecal and/or intestinal wall. Cecal contents often contain hardened mucosal debris appearing as loose whitish cecal cores (Fig. 12). Lesions are principally located in the ceca but generally extend to the lower small intestine and cloaca. Oocysts are ellipsoidal and very elongated (25.6 x 16.6 microns). This species is one of the most pathogenic of the turkey coccidia. Infections in young poults may produce high mortality and infection in older turkeys can cause considerable weight loss. *E. meleagridis* may also be found in this area. This species is relatively common in litter samples from commercial operations.

DIAGNOSIS

1. Gross necropsy should be performed on fresh (< 1 hour) dead birds typical of the flock. Post mortem changes can quickly obscure gross lesions.

2. Diagnosis is made primarily on the basis of clinical signs and the appearance and location of gross intestinal lesions. Large numbers of oocysts may be present in mucosal scrapings from affected birds. Fresh mount wet smears and or histologic examination of the intestine can confirm the presence of asexual or sexual stages of coccidia (sporozoites, merozoites, schizonts). A history indicating recent flock exposure to a large source of sporulated oocysts may be helpful.

3. Subclinical infection is common. The presence of a few oocysts in the feces does not justify a diagnosis of clinical disease.

4. Coccidiosis frequently occurs in association with other avian diseases, such as necrotic enteritis, ulcerative enteritis, salmonellosis and histomoniasis. Immunosuppressive diseases may increase the severity and incidence of clinical disease.

CONTROL

1. Anticoccidial compounds in feeds are the most common method of control. However, coccidia may become resistant to the anticoccidials, therefore rotation of types of products may be used to prolong efficacy. No anticoccidial is highly effective against all species of coccidia although some are effective against multiple species. Several anticoccidials are approved for prevention of coccidiosis, although not all are commercially available such as Amprolium, Monensin, Clopidol, Nicarbazin, Robenidine, Decoquinate, Lasalocid, Halofuginone, Narasin, Diclazuril and Semduramycin. Not all products might be able to be used for all poultry species; turkeys are extremely sensitive to Salinomycin and Narasin. Care should be taken in choosing the product to be used.

2. Immunization. Commercial coccidiosis vaccines are available. Planned exposures of young chicks or poults to small numbers of oocysts by coarse spray at the hatchery or in feed, water or gel blocks or *in ovo* at 18 to 19 days incubation have been used successfully. The number of oocysts of each *Eimeria* species provided in the vaccine is critical to initiating immunity

without causing clinical disease. Some vaccines contain drug-sensitive strains of *Eimeria,* facilitating the establishment of drug-sensitive populations and extending the usefulness of anticoccidials. Some vaccines may contain attenuated and or precocious lines of *Eimeria.*

3. Natural Exposure. If chickens are exposed to modest numbers of oocysts in their environment, they develop immunity to the species of coccidia represented. Exposure must be moderate or clinical signs will appear. Exposure can be limited if dry litter conditions are maintained. Wet litter (including wet areas around waterers) is especially to be avoided. This practice is rarely used in large commercial operations.

TREATMENT
Prevention is emphasized. However, chemical agents widely used for treatment include amprolium, sulfadimethoxine, sulfaquinoxaline, sulfamethazine. Sulfas should not be used in layers. Required withdrawal times are usually required prior to marketing. Increasing vitamins A and K in feed or water may reduce mortality and hasten recovery, respectively.

B. HEXAMITIASIS/SPIRONUCLEOSIS

DEFINITION
Hexamitiasis is a protozoal disease characterized by catarrhal enteritis and by foamy or watery diarrhea.

The etiologic agent in turkey poults and most other susceptible birds is *Spironucleus meleagridis* (previously known as *Hexamita meleagridis*). The etiologic agent in pigeons is *Spironucleus columbae.*

HISTORICAL INFORMATION
1. For many years hexamitiasis was confused with trichomoniasis. In 1938 Hinshaw and others first clearly identified hexamitiasis and its etiologic agent. Rather extensive losses were attributed to hexamitiasis during the early development of the turkey industry, in range flocks.

EPIDEMIOLOGY
1. In turkeys, Hexamitiasis usually is seen in 1 to 9-week-old poults. Hexamitiasis also occurs in gamebirds (pheasants, quail, chukar partridge, etc.), peafowl and ducks. Pigeons have their own distinct form of hexamitiasis.

2. The parasite is shed in feces which contaminate feed, water and range. Recovered birds are often inapparent carriers. Susceptible birds get the organism by ingestion. Interspecies transmission, e.g., from game birds to turkey poults, occurs readily. This method of transmission is more likely in poults on range.

3. When consecutive broods of poults are raised in the same facilities, especially if sanitation is poor, there appears to be an increase in the number of *Spironucleus* in each brood and a corresponding increase in signs of the disease in the later broods.

4. The disease has been reported in many different countries and states. It usually occurs during the warmer months of the year and on facilities that maintain a poor standard of sanitation. Heximitiasis is currently rare in commercial production, but may be seen in backyard or ornamental flocks.

CLINICAL SIGNS
1. Initially the affected birds are nervous and very active. They chirp excessively, shiver, crowd around any heat source and have subnormal temperatures. There is watery or foamy diarrhea and the birds dehydrate rapidly.

2. Later the birds are more depressed, stand with their heads retracted, feathers ruffled and wings drooping. Terminally the birds go into a coma, struggle and die. Terminal signs appear to be related to hypoglycemia.

3. Morbidity is high. Mortality varies with age and the quality of husbandry provided. Mortality may be very high (75-90%) in young birds that are poorly housed and which receive no treatment.

DIAGNOSIS & LESIONS
1. The cadaver is dehydrated. The intestine is flabby, may have areas of bulbous dilatation and contains excessive mucus and gas. The proximal one-half of the intestine is inflamed. Cecal tonsils may be congested.

2. *Spironucleus meleagridis* is found in the crypts of Lieberkuhn, especially in the duodenum and upper jejunum of infected birds, including older carriers. The protozoan is roughly 3 by 9 microns, has 8 flagella and two nuclei that resemble eyes. It moves with a rapid, darting motion.

3. Duodenal scrapings should be examined from a freshly killed, infected bird using reduced light or phase contrast microscopy. Many *S. meleagridis* in the upper intestine suggest hexamitiasis. If only recently dead birds are available, warm saline added to the scrapings may revive enough of the *Spironucleus* for identification. Dead *S. meleagridis* are difficult to identify so it is preferable to submit live birds for necropsy.

4. The history, signs and lesions are suggestive of the disease but must be differentiated from coronaviral enteritis, paratyphoid infection, trichomoniasis and blackhead.

CONTROL
1. Hexamitiasis seldom is reported now, or may be undiagnosed. Improved sanitary practices and management on turkey farms may have reduced

the incidence or antihistomonal chemicals routinely given for blackhead control may also be controlling hexamitiasis.

2. Short periods of depopulation combined with thorough cleaning and disinfection of buildings will greatly reduce the population of *S. meleagridis*.

3. Clean and disinfect feeders and waterers and keep them on large wire platforms so concentrations of droppings are not available to the birds.

4. Do not hold possible carriers from old flocks. Raise only birds of the same age group together.

5. Prevent contact of poults with captive or wild gamebirds. Avoid using gamebird ranges for poults.

TREATMENT
No current medication is approved for food animals in the United States. Antiprotozoal drugs such as Metronidazole may be considered in non-food animals. Supportive treatment such as increasing the brooding house temperature to a point where the birds appear comfortable may be beneficial for young birds.

C. HISTOMONIASIS
(Blackhead; Enterohepatitis)

DEFINITION
Histomoniasis is a protozoal disease caused by *Histomonas meleagridis* affecting a wide range of birds including turkeys, chickens, peafowl, grouse, quail, other gallinaceous birds and ducks. The disease is characterized by necrotizing lesions involving the ceca and liver.

OCCURRENCE
Histomoniasis occurs in commercial and non-commercial turkeys, especially young turkeys and if left untreated, most may die. Blackhead also occurs in chickens, but to occur at a lower prevalence and severity. Young birds are more frequently and severely affected.

HISTORICAL INFORMATION
1. Histomoniasis once limited the expansion of the turkey industry. Prior to development of safe antihistomonal drugs, it could be controlled only by cumbersome and relatively ineffective measures designed to prevent exposure of turkeys to the embryonated ova of cecal worms (*Heterakis gallinarum*).

2. Significant growth of the turkey industry occurred after safe antihistomonal drugs were developed. The disease is now uncommon and these drugs are no longer used routinely. It occurs sporadically when turkeys are raised where chickens were previously located. The disease is still common in chickens but its effect on production is mild and rarely recognized. Histomoniasis remains an important cause of death

among other galliformes including peafowl, pheasant and quail.

3. Recently, blackhead has become a big concern for commercial pullet flocks and many of these outbreaks are reoccurring.

ETIOLOGY
1. The etiologic agent is the protozoan *Histomonas meleagridis*, assisted by secondary bacteria. In the experimental absence of bacteria, the histomonad appears not to be pathogenic. *H. meleagridis* is a flagellate in the lumen of the cecum but assumes an ameboid form in tissue.

2. A larger histomonad distinguished by its 4 flagella, *H. wenrichi*, also occurs in the cecum but is not pathogenic.

EPIDEMIOLOGY
Transmission of *H. meleagridis* to susceptible birds is possible via three routes:

1. Ingestion of fresh feces. This route probably is relatively unimportant except for spread within a flock.

2. Ingestion of embryonated cecal worm ova containing the protozoan. Within these resistant ova the histomonad can survive for years. *H. meleagridis* is liberated in the intestine when ingested ova hatch and then invades the cecal wall and initiates the disease.

3. Ingestion of earthworms containing cecal worm larvae within their tissues. Earthworms serve as transport hosts for the cecal worm and the cecal worm acts as a transport host for the histomonad. Infection results after the cecal worm larvae are liberated during digestion.

DIAGNOSIS
1. Diagnosis can be made on the basis of clinical signs and characteristic lesion. Typical well-developed lesions are pathognomonic and consist of typhlitis and characteristic hepatic lesions (Fig. 1).

2. In turkeys histomoniasis appears 7-12 days after exposure. Initially there is listlessness, moderate anorexia, drooping wings and yellow ("sulfur colored") feces. Head parts may be cyanotic ("blackhead") although they often are not. In chickens with histomoniasis there may be some blood in the feces.

3. Later the affected turkey is depressed and stands with its wings drooping, eyes closed, head drawn close to the body. Emaciation is common in chronic cases, usually in older birds. In young turkeys morbidity and mortality are high, up to 100%. Older birds tend to be more resistant.

4. Gross lesions. There is a bilateral enlargement of the ceca with thickening of the cecal walls (Fig. 2). The mucosa usually is ulcerated. The ceca often contain caseous cores which are yellow, gray or green and may be laminated. In chronic cases the cores may

have been expelled. Peritonitis occurs when the cecal wall becomes perforated.

5. Liver contains irregularly-round, depressed, target-like lesions that vary in (Fig. 3). They often are yellow to gray but may be green or red. They vary greatly in diameter but often are 1-2 cm and may coalesce to produce larger lesions.

6. Lesions may not be entirely typical in birds under treatment, less susceptible avian species or young turkeys in the early stages of the disease. In most infected flocks typical lesions usually can be found if an adequate number of birds are examined. In quail, cecal lesions may not occur even though mortality is high.

7. Microscopically, histomonads can be found in the inflamed cecal walls and necrotic foci which develop in the liver (Fig. 4). In birds killed for necropsy the agent sometimes can be identified in smears from the ceca or in scrapings from the margin of hepatic lesions.

CONTROL

1. Histomoniasis usually can be prevented by adding antihistomonal drugs to the ration in proper dosage. No current preventive medication is approved in the U.S. Histostat (nitarsone) is still used in poultry outside the U.S. In quail, a cholinesterase inhibiting carbamate (Sevin) increases susceptibility to histomoniasis.

2. Control by the use of antihistomonal drugs may fail unless reasonably good sanitation is practiced.

3. Control by the use of attenuated *H. meleagridis* organisms; recently experimental demonstrations showed good promise of immunizing poults against severe challenges.

4. Other measures that assist in control follow:

   A. Do not keep chickens and turkeys (or other susceptible birds) on the same farm.

   B. Do not use chicken ranges for turkeys or other susceptible birds unless those ranges have been free of chickens for at least 4 years.

   C. *H. meleagridis* is quickly destroyed by disinfectants and drying unless protected within earthworms or within the cecal worm ova. Avoid exposure to vectors. If possible, raise susceptible birds on sandy, dry, loose soil. Prevent access to earthworms after rains. In range birds, rotate ranges periodically if possible. Some operators with small lots replace the top few inches of soil every few years using power equipment or plow the lots to reduce the number of cecal worm ova and other pathogens.

   D. Reduce access of birds to their own droppings or to feed and water contaminated with droppings. Place feeders and waterers on large wire platforms or keep them outside of the lot but accessible through a wire fence.

TREATMENT
There is currently no approved medication for treatment of histomoniasis in food animals. Small groups of birds not being raised for consumption can be effectively treated individually with metronidazole at a dose of 30 mg/kg orally SID for 5 days. Antihelmintic treatment may help suppress the population of cecal worms.

D. TRICHOMONIASIS
(Canker in pigeons and doves; Frounce in falcons)

DEFINITION
Trichomoniasis is caused by *Trichomonas gallinae*, a flagellated protozoan. Pathogenicity varies greatly by strain. The disease is characterized by raised caseous lesions in the upper digestive tract, but may extend to other tissues. Pigeons, doves, turkeys, chickens and raptors are commonly affected.

HISTORICAL INFORMATION
Trichomoniasis was once recognized as an important disease of turkeys and chickens, especially of ranged turkeys, but is seldom reported now. Conversely, trichomoniasis in pigeons and doves continues to be a common and significant disease. Trichomoniasis may have played a role in the extinction of the carrier pigeon. Trichomoniasis can be a consequential disease in raptors.

EPIDEMIOLOGY
1. The organism is fragile in the environment and transmission occurs only through contact with infected oral secretions or through water contaminated by oral secretions of carriers. Pigeons are believed to be the natural hosts and primary carriers. The prevalence of the infection is near 100% for adult pigeons. Carrier birds show no signs or lesions.

2. Pigeons and doves transmit trichomonads to their young during feeding of regurgitated partially digested crop content (pigeon milk). Transmission in raptors (hawks, owls, eagles, etc.) and their young, is through ingestion of infected prey. Turkeys and chickens probably contract the disease after consuming stagnant surface water containing *T. gallinae*. Other disease may predispose turkeys to clinical trichomoniasis.

3. In established flocks or lofts, trichomoniasis may only be noted as a clinical disease after introduction of a more virulent strain of *T. gallinae* by new birds or exposure to wildlife carriers.

DIAGNOSIS
1. Typical signs and lesions are very suggestive of the diagnosis. In pigeons, doves and raptors, yellow

plaques or raised cheesy masses involve the upper digestive tract. Masses often are large and conical or pyramidal and can be surprisingly invasive in soft tissue (Fig. 1). Lesions are usually most extensive in the mouth, pharynx or esophagus (Fig. 2) but may occur at other sites including the crop, proventriculus or sinuses. In raptors, lesions may also occur in the liver and are accompanied by peritonitis.

2. Infected squab (baby pigeons) become depressed and die at 7-10 days of age. Lesions of the oral cavity are most common but may also occur in the nasal turbinates and brain. The infection may become systemic, with lesions in the liver and other visceral organs.

3. Adults pigeons, doves and raptors often have difficulty in closing their mouth because of lesions in the oral cavity. They drool and make repeated swallowing movements. Watery eyes may be apparent in occasional birds with lesions in the sinuses or periorbital area. Rare cases with penetrating cranial lesions may show signs of central nervous disturbances, including loss of balance.

4. Lesions are similar in turkeys but frequently are found only in the crop and upper or lower esophagus. Occasionally the proventriculus contains lesions. Infected turkeys often have a gaunt appearance with a hollowed area over the crop. Swallowing movements often are apparent and infected birds may have an unpleasant odor ("sour crop"). Morbidity and mortality in affected birds varies but can be quite high. Demonstration of trichomonads in the oral fluids may not be significant in the absence of lesions since many normal birds have some trichomonads. Small plaques in the mucosa should not be confused with pox or candidiasis.

CONTROL

1. Eliminate any known infected birds and all suspected carriers. If possible, depopulate at regular intervals and thoroughly clean and disinfect the premises. Add no birds to an established flock since they may be carriers of a more virulent strain. Permit no contact among pigeons, doves and susceptible poultry.

2. Provide a source of clean, fresh water, preferably running water being replaced constantly. Eliminate all sources of stagnant water. Disinfect watering containers and water lines regularly (e.g. chlorine).

3. Avoid feeding infected pigeons and doves to captive raptors.

TREATMENT

There is currently no approved medication for treatment of trichomoniasis in food animals. Birds not being raised for food can be effectively treated individually with metronidazole (Flagyl) at a dose of 30 mg/kg orally SID for 5 days, or with Enheptin.

IV. OTHER PROTOZOAL INFECTIONS

A. CRYPTOSPORIDIOSIS

DEFINITION

Cryptosporidiosis is a protozoal disease, characterized in avians by acute or chronic disease of the respiratory or digestive tracts. Two avian species of *Cryptosporidium* are recognized, based upon tissue (site) specificity and organism morphology. *Cryptosporidium meleagridis* infects only the small intestine, while Cryptosporidium *baileyi* infects the digestive tract (especially the bursa of Fabricius and cloaca) and can also infect the respiratory tract. The organism has a coccidian life cycle that is completed in an intracellular, but extracytoplasmic location on the microvillous border of epithelial cells.

HISTORICAL INFORMATION

1. Dr. Earnest Tyzzer first described infections by *Cryptosporidium muris* (gastric mucosa) and *C. parvum* (small intestine) in laboratory mice (1907).

2. Tyzzer also was the first to describe and report (1929) infections by *Cryptosporidium* in the cecum of chickens, but no signs of disease were noted. The first report of avian morbidity and mortality caused by *Cryptosporidium* (1955) involved the distal small intestine of turkeys. The first report of respiratory cryptosporidiosis in any species also involved turkeys (1978).

3. Cryptosporidiosis was recognized in humans in 1976, and was soon identified as a life-threatening enteric disease associated with acquired immune deficiency syndrome (AIDS) and other immunosuppressive disorders.

EPIDEMIOLOGY

The disease is spread through ingestion or inhalation of oocysts shed in feces or in respiratory secretions. Oocyst will remain viable in the environment for long periods. Infections by *C. meleagridis* and *C. baileyi* have been reported in chickens, turkeys, quail, pheasants, peafowl, psittacines, finches, and waterfowl. Avian species of *Cryptosporidium* spp. do not appear to infect mammals, including man, i.e. no known public health threat exists. Other species of *Cryptosporidia* have documented zoonotic potential, passing between man and other mammalian species, reptiles, amphibians, and fish. Cryptosporidiosis can be a severe and life-threatening disease in an immunosuppressed host, but immunosuppression is not a requirement for either infection or disease. Some infections are asymptomatic.

CLINICAL SIGNS

1. Clinical signs of cryptosporidiosis are not specific for the disease and other pathogens are frequently involved. Involvement of the small intestine causes

diarrhea that may be fatal in young turkeys, quail, and psittacines. Mortality in young quail may exceed 95%. Digestive tract infections in broilers may cause reduced carcass pigmentation.

2. Respiratory cryptosporidiosis produces signs related to the site of infection: naso-ocular discharges, swollen sinuses, coughing, sneezing, dyspnea, and rales. Fatalities may occur with deep lung and air sac infections. Concurrent infections by other respiratory pathogens are the rule rather than the exception.

LESIONS

1. The small intestine becomes dilated with fluid. Ceca may be distended with foamy, fluid contents. Histopathology includes shortening of villi with necrosis and loss of enterocytes from the villous tips, and the presence of numerous organisms in the brush border of the mucosal epithelium (Fig. 1). Organisms may also be found in the cecal tonsil mucosa, and in the mucosa of the bursa of Fabricius and cloaca. The bursal mucosal epithelium may be hyperplastic with heterophilic infiltration.

2. Respiratory cryptosporidiosis causes gray or white mucoid exudate on the mucosal surface infected: conjunctiva, sinuses or nasal turbinates, trachea, bronchi, or air sacs. Marked enlargement of the infraorbital sinuses may occur in turkeys. Infected lungs may appear gray and firm.

DIAGNOSIS

1. Most cases of cryptosporidiosis are recognized during histologic examination of either biopsy specimens or tissues collected at necropsy. The organisms are basophilic with H&E stain, are 2 to 4 microns in diameter, and are intimately associated with mucosal brush borders.

2. Cytology can be used to confirm the diagnosis. Touch impressions of mucosal surfaces can be examined by phase-contrast or interference phase-contrast microscopy. They can also be air dried and stained with red carbon fuschin or with Giemsa stain. Differentiation from yeasts may be necessary with Giemsa stain.

3. Flotation techniques for oocyst identification from feces include Sheather's sucrose solution, zinc sulfate (33% to saturated), and sodium chloride (36% to saturated). Oocysts are 4 microns in diameter, with a thick cell wall, prominent residual body, smaller globular dense bodies, and curved sporozoites.

4. Concurrent respiratory tract pathogens identified with respiratory cryptosporidiosis include *Escherichia coli*, *Pasteurella multocida*, Newcastle disease virus, adenovirus, infectious bronchitis virus, and reovirus.

5. Most broiler chickens with bursal cryptosporidiosis have had infectious bursal disease, but it is not a pathogenic requirement. Among other avian hosts, diseases that occur with cryptosporidiosis include salmonellosis, candidiasis, turkey viral hepatitis, intestinal reovirus infections, and other parasites.

CONTROL

1. Reduction or elimination of oocysts from the environment is the primary means of controlling the spread of *Cryptosporidium*.

2. Stringent cleaning followed by disinfection with either steam or autoclaving is necessary for oocyst destruction to break the cycle of infection. Most disinfectants will not inactivate oocysts. Either formol saline (10%) or 5% ammonia will inactivate oocysts after a minimum of 18 hours contact. Undiluted commercial bleach is also effective.

TREATMENT

1. Drugs used for coccidiosis and other protozoan diseases have been ineffective for cryptosporidiosis in animals. Therapeutic agents are minimally effective for reducing the severity of clinical disease and for interfering with replication of the organism.

2. Supportive therapy for individual animals is a consideration, but with multiple animals, continued shedding of infective organisms may infect others in a flock, cage, or aviary. Treatment and control of concurrent or secondary infections reduces mortality.

B. SARCOSPORIDIA

DEFINITION

A parasitic infection caused by species of the genus Sarcocystis first described by Lankester in 1882. Sarcocystis has a world-wide distribution but is rarely reported in domestic fowl. The incidence is high in wild waterfowl and some passerines, such as grackles. Not an economically important disease in domestic poultry, but does occur extensively in wild ducks and game birds. Although not a public health hazard, the parasite is killed by cooking or freezing. Most affected carcasses are discarded for esthetic reasons.

ETIOLOGY

1. At least five species of Sarcocystis have been identified. *S. horwathi* is regarded as the etiologic agent in chickens (also call *S. gallinarum*). *S. anatina* and *S. rileyi* (*Balbiani rileyi*) are found in ducks.

2. All species of Sarcocystis have similar developmental stages and an obligatory two host life cycle. Sarcocysts in cardiac, smooth, or skeletal muscle tissues are eaten by a definitive host, releasing cystozoites which penetrate the intestinal wall and develop in the subepithelial tissue. Oocysts are produced and shed in the feces fully sporulated. When sporocysts are ingested by the intermediate host they go through a series of asexual reproductive cycles in endothelial cells of various organs. Merozoites are eventually released, invade muscle tissue and eventually develop

into the macroscopically visible sarcocysts (1.0-6.5 x 0.48-1.0 mm), each containing numerous banana-shaped cystozoites (bradyzoites ;2-3 x 8-15 microns).

DIAGNOSIS

Usually based on gross lesions, i.e. the presence of large, pale sarcocysts in the muscle tissue (Fig. 1), arranged in parallel with the muscle fibers. Diagnosis can be confirmed histologically.

COCCIDIA OF CHICKENS

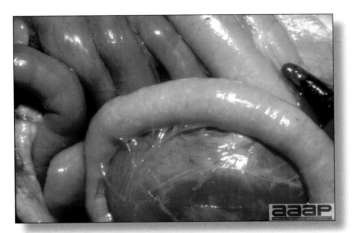

Fig. 1
Eimeria acervulina.

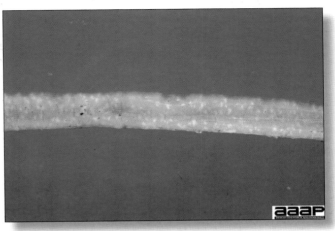

Fig. 2
Eimeria acervulina; transverse white to gray striations are visible in the mucosa of the jejunum.

Fig. 3
Severe *E. necatrix* infection. Infections are usually midintestinal and characterized by hemorrhage and ballooning.

Fig. 4
Eimeria necatrix: ballooning and distention of the intestines.

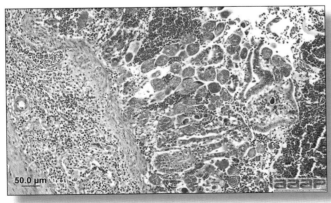

Fig. 5
Eimeria necatrix: presence of large schizonts predominantly in the mid small intestines.

Fig. 6
Moderate *E. maxima* infection. Serosal surface are speckled with numerous red petechiae and intestine contains orange mucus.

COCCIDIA OF CHICKENS

Fig. 7
E. brunetti.

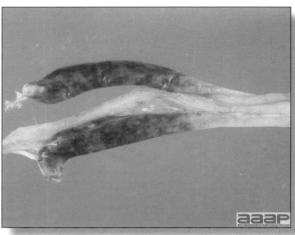

Fig. 8
Eimeria tenella: blood can be seen through the cecal wall.

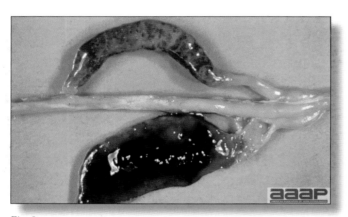

Fig. 9
Severe *E. tenella* infection. The cecal wall is distended with blood.

Fig. 10
Eimeria mivati.

COCCIDIA OF TURKEYS

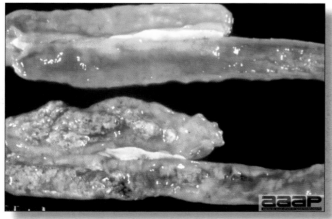

Fig. 11
Eimeria meleagrimitis.

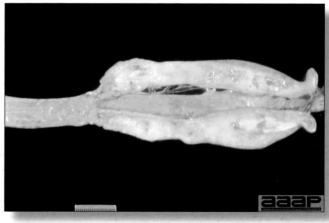

Fig. 12
Eimeria adenoides.

HISTOMONIASIS

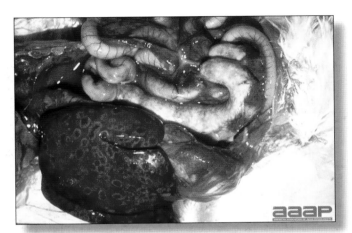

Fig. 1
Hepatitis and typhlitis.

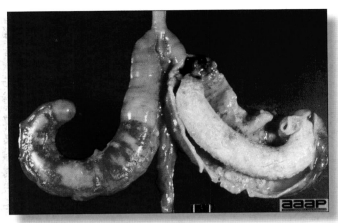

Fig. 2
Cecal cores.

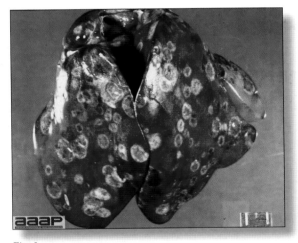

Fig. 3
Hepatitis: typical irregularly-round, depressed, target-like lesions.

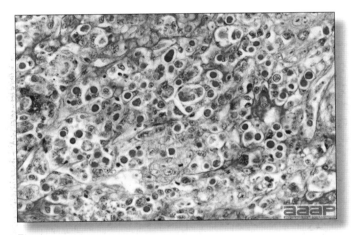

Fig. 4
Histomonads in liver at histology.

TRICHOMONIASIS

Fig. 1
Large masses invading the soft tissues of the upper digestive tract.

Fig. 2
Extensive lesions in the mouth, pharynx, esophagus and crop.

CRYPTOSPORIDIOSIS

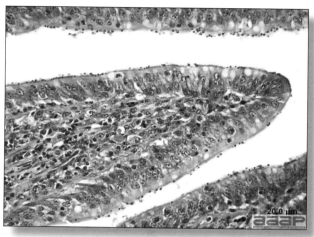

Fig. 1
Presence of numerous organisms in the brush border of the
mucosal epithelium at histology.

SARCOSPORIDIA

Fig. 1
Presence of large, pale sarcocysts in the muscle tissue.

COMMON INTERNAL PARASITES OF POULTRY

Parasite	Common name	Description	Lifecycle	Site of Infection	Lesions/Clinical signs	Comments
Oxyspirura sp.	Eyeworm	Nematode up to 2.0 cm	Indirect or direct. Cockroaches are intermediate hosts for some species.	Conjunctival sac, often beneath the nictating membrane, or in nasolacrimal duct	Conjunctivitis, opthalmitis, protrusion of nictitating membrane, adherence of eyelids	Occurs in all types of poultry and wild birds
Syngamus trachea	Gapeworms	Red nematodes up to 2.0cm long, forked appearance, male and female in permanent copulation	Indirect or direct earthworms, slugs, snails are intermediate hosts for many species	Trachea and possibly large bronchi.	Gasping, gaping, dyspnea and head shaking	Occurs in all species of poultry. Generally difficult to treat. *Cyathostoma bronchialis* may cause gasping in domestic geese.
Capillaria sp. C. annulata C. contorta or Gongylonema ingluvicola	Cropworms	Thin, threadlike nematodes up to 60 mm long	Direct	Crop or crop and esophagus	Infected tissues become thickened and inflamed. Causes malnutrition, emaciation and severe anemia	Occurs in all species of poultry. Best identified in mucosal scrapings, difficult to see grossly
Ascaridia galli	Roundworms	Large, thick, yellow-white nematodes	Direct. Eggs ingested by insects remain infective	Lumen of small intestine	Weight loss, potential intestinal blockage and blood loss	Affects chicken, turkey, dove, duck and goose. Fowl > 3 mo. develop immune resistance. Similar to *A. dissimilis* reported in turkeys and *A. columbae* in doves, pigeons
Capillaria sp.	Intestinal worms	Hairlike nematodes 6 to 25 mm long	Direct or indirect with earthworm intermediate host	Mucosa of small intestine and ceca	Huddling, emaciation, diarrhea, hemorrhagic enteritis, or death. Thickened upper intestine with catarrhal exudate	Occur in many poultry species

Parasite	Common name	Description	Lifecycle	Site of Infection	Lesions/Clinical signs	Comments
Dispharynx sp. *Tetrameres* sp. *Cyrnea* sp.	Proventricular worms	Grossly visible, 3 to 18 mm long nematodes	Indirect. Grasshoppers, cockroaches, sowbugs and pillbugs are intermediate hosts	Mucosa and glands of proventriculus	Diarrhea, emaciation and anemia. Mucosal ulceration, necrosis, hemorrhage and swelling	Occur in poultry and other birds
Cheilospirura sp. *Amidostomum* sp.	Gizzard worms	Small, broad nematodes up to 25 mm long	Indirect. Grasshoppers, beetles, weevils and sandhoppers are intermediate hosts	Under the gizzard lining	Muscular wall of gizzard may be sacculated or ruptured. Mucosa ulcerated, necrotic or sloughed.	Numerous poultry species.
Heterakis gallinarum	Cecal worms	Small white nematode up to 15 mm long	Direct. Eggs may be ingested by earthworms where they hatch and live for months	Most numerous in the tips of the ceca	Marked inflammation and thickening, nodules in cecal walls. May see hepatic granulomas	Affects numerous poultry species. Acts as carrier for *Histomonas meleagridis* (blackhead)
Cestodes *Raillietina* sp. *Choantaenia* sp., *Davainea* sp., *Hymenolepis* sp. *Amoebotaenia* sp.	Tapeworms	Flattened, ribbon-shaped, segmented. Usually grossly visible	Indirect. Many invertebrate intermediate hosts; flies, snails, beetles, earthworms, crustaceans	Intestine	Small lesions at point of attachment. Enteritis proportional to degree of infection	Usually infect a definitive host, but are not host specific.
Trematodes	Skin flukes	Hemispherical, flattened shape, up to 5.5 mm long	Requires a molluscan intermediate host and may use a second intermediate host	Skin, usually near vent	Forms cutaneous cysts, usually with 2 flukes in vent area	Not host specific. Affect poultry and wild birds. Other flukes may invade oviduct, digestive organs, kidney, circulatory organs and eye

NUTRITIONAL DISEASES

Revised by Dr. H.L. Shivaprasad

Introduction:

Nutrients including amino acids, carbohydrates, fats, vitamins, inorganic chemical elements, energy, water and oxygen are essential for normal growth and development, livability, work and reproduction. These nutrients should be in proper concentration and balance to be effective. Nutrient requirements have been well established for growing chicks and poults, as well as for laying –type hens and for broiler and turkey breeders.

Various factors can influence nutritional diseases in birds in general. It is important to understand these factors as they are necessary for taking corrective steps. These include human errors such as omission of an ingredient or two or groups of vitamins (water soluble and fat soluble), improper mixing, improper storage, miscomputations in feed formulations and misfeeding to wrong species or sex or age of the birds. Other important factors that can influence nutritional diseases include poor nutritive value of an ingredient, nutrient and mineral interactions, poor shelf life and insufficient feed availability. Poor health of the birds due to bacteria, viruses, parasites and other causes can result in anorexia, dysphagia, maldigestion, malabsorption, decreased storage or utilization, increased excretion or secretion and increased requirements that can also cause malnutrition.

Malnutrition in poultry can result in a generalized or a specific disease. Specific conditions such as encephalomalacia or rickets are easy to diagnose and treat. But generalized or subclinical disease or signs such as loss of weight or failure to thrive due to marginal deficiencies of nutrients are very difficult to diagnose and treat. The latter are probably more common in the field than realized. Malnutrition can also suppress immune system, decrease reproductive performance, decrease weight gain, cause feather problems and decrease response to therapeutic agents.

Diagnosis of malnutrition in poultry can generally be based on history such as recent feed change, clinical signs, gross and microscopic lesions, analysis of feed if available, liver and serum, serum chemistry, radiography for the birds and estimating peroxide levels in the feed which indicates rancidity. Feeding trials with suspected feed can also be performed but are time consuming, expensive and often disappointing. Treatment is another option to diagnosing diseases; for example, curled-toe paralysis due to riboflavin (vitamin B2) deficiency where the birds will respond to supplementation of riboflavin or multiple B vitamins if the condition is diagnosed and treated promptly.

VITAMIN DEFICIENCIES

Chickens are particularly susceptible to vitamin deficiencies because they get little or no benefit from microbial synthesis of vitamins in their digestive tract. These birds also have a fast growth rate and are raised in modern management conditions hence submitted to various stresses. Their feed can be shipped or stored for various periods of time, at different temperatures, submitted to oxidative stress, these accounting for possible potency loss. For the above reasons, increased dietary vitamin levels are required in commercial diets in order to optimize growth and performances, when compared to the minimal requirements of the NRC established to prevent clinical signs in ideal conditions.

If birds received vitamins below requirement levels for any length of time, classical deficiency pathologies will develop at various speeds, depending on the age, the quantity of vitamin passed on from the breeder hen, and the vitamin storage level. For example, clinical signs will appear rapidly in young chicks, with the young embryo being the most sensitive model. Problems with water soluble vitamins such as B are soon observed because they are not stored to any extent, even excreted via the urine if in excess, while fat soluble vitamin deficiencies can take longer to develop because of adipose tissue and liver storage in older birds.

Since it is today a common practice to include vitamins and minerals in feed composite premixes, we are less likely to observe individual classical vitamin deficiencies, such as the ones described below, unless one has been omitted from a premix. It is therefore not unusual to see a situation where the entire premix has been inadvertently excluded with as a consequence, the development of a complex array of clinical signs.

BIOTIN DEFICIENCY

DEFINITION
Biotin is common in poultry feedstuffs, yet recent evidence concludes some of this biotin may be biologically unavailable. Turkeys seem more sensitive to biotin deficiency than chickens.

ASSOCIATED DISORDERS
Biotin acts as a coenzyme in carboxylases which are involved in lipid and carbohydrate metabolism. First signs of biotin deficiency are related to reduced cell proliferation.

1. Dermatitis. Biotin deficiency causes an exudative dermatitis around the mouth and eyes (Fig. 1) and on the feet and legs (Fig. 2), which must be differentiated from pantothenic acid deficiency. In turkeys dermatitis and cracking of foot pads occurs (foot pad dermatitis), soon followed by chondrodystrophy.

2. Chondrodystrophy (old name = perosis). A growth-plate disorder, where long bones are shorter than normal, is a sign of biotin deficiency in growing chickens and turkeys. Biotin may also play a role in varus leg deformities because it affects bone remodeling.

3. Fatty liver and kidney. Fatty liver and kidney syndrome in broiler chickens is a biotin-responsive disease. Chicks exhibit depressed growth, hypoglycemia, increased plasma-free fatty acids, and increased ratio of C 16:1 to C 18:0 fatty acids, in liver and adipose tissue. Necropsy reveals pale liver and kidney with accumulation of fat. It has been suggested that biotin deficiency may impair gluconeogenesis by decreasing the biotin-containing enzyme pyruvic carboxylase and, therefore, increasing the conversion of pyruvate to fatty acids.

4. Fatty liver syndrome. Biotin may also be a complicating factor in fatty liver syndrome in laying hens.

5. Embryonic abnormalities and reduced hatchability. Biotin is essential for embryonic development. Embryos from deficient hens display parrot beak, chondrodystrophy, micromelia and syndactyly. Two peaks of embryonic mortality occur: one during the 1st week and another during the last 3 days of incubation.

TREATMENT
Water-soluble vitamins are readily available and may be administered as needed. To avoid problems, most starter and grower rations are supplemented with 0.1 to 0.3 mg/kg of biotin.

Fig. 1
Exudative dermatitis around the mouth.

Fig. 2
Exudative dermatitis on feet.

RIBOFLAVIN DEFICIENCY
(Vitamin B2 Deficiency; Curled Toe Paralysis)

DEFINITION
Riboflavin supplementation is generally required in rations for growing chicks, poults, and ducklings because few poultry feedstuffs contain large quantities.

CLINICAL SIGNS
Riboflavin is essential for growth and tissue repair in all animals. Many tissues may then be affected by riboflavin deficiency, the most severely being the epithelium and myelin sheaths of various nerves.

Chicks
The characteristic clinical sign is "curled toe" paralysis caused by lesions to the sciatic nerve; however, if the deficiency is absolute or very severe, the chick will die before curled toe paralysis develops. In mild cases, chicks tend to rest on their hocks and the toes are only slightly curled (Fig. 1) or not at all. In moderate cases, there is marked weakness of the legs and a distinct curling of the toes on one or both feet. In severe cases, the toes are completely curled inward or under (Fig. 2), the legs are extremely weak, and birds walk on their hocks with the aid of their wings ("wing-walking"). At rest, the wings may droop. Leg muscles are atrophied and the skin is dry and harsh. Other signs include stunting, diarrhea after 8-10 days, and high mortality at about 3 weeks. Feather growth in chicks is not generally impaired.

Poults
A dermatitis resulting in encrustations on eyelids and corners of the mouth will develop in approximately 8 days. The vent becomes encrusted, inflamed, and excoriated. Other clinical signs are similar to the chick. Growth is retarded or completely ceases by about day 17. Mortality occurs at about day 21.

Ducklings
Ducklings and goslings usually have diarrhea, stunting, and a bowing of the legs in conjunction with chondrodysplasia.

Laying hens
Riboflavin deficiency will cause decreased egg production and decreased hatchability that is roughly proportional to the degree of deficiency, with an increase in hepatic size and fat content.

Embryos
Embryonic mortality peaks at 4, 14, and 20 days of incubation are typical of riboflavin deficiency, with peaks more prominent early as deficiency becomes severe. In severe cases, embryonic death due to circulatory failure occurs at 4 days. Embryos with moderate inadequacies die at 14 days incubation, with the appearance of shortened limbs, mandible malformations, possible edema, and defective down. But one study failed to reproduce typical 'clubbed down' condition experimentally in the progeny of broiler breeders fed riboflavin deficient diet. Clubbed down is caused by the failure of the down feathers to rupture the sheaths and may be seen in the neck and vent areas of late-stage embryos or hatched chicks. In marginal deficiencies mortality will be delayed until pipping with dwarfism and clubbed down the major signs.

LESIONS
In young poultry, riboflavin deficiency produces specific changes in the main peripheral nerve trunks. There may be marked swelling and softening of sciatic and brachial nerves and loss of cross striations (Fig. 3). Myelin degeneration, Schwann cell proliferation, and axis cylinder fragmentation have been observed (Fig. 4). Congestion and premature atrophy of the thymic lobes may also be observed.

TREATMENT
Marked and dramatic improvement and alleviation of clinical signs can be expected if treatment occurs early in the course of the disease. Water-soluble vitamins are readily available and can be administered in the water if needed. Irreversible damage will occur over time.

RIBOFLAVIN DEFICIENCY

Fig. 1
Curled toes in a riboflavin deficient poult.

Fig. 2
Severe case of riboflavin deficiency.

Fig. 3
Swollen sciatic nerve with loss of cross striations.

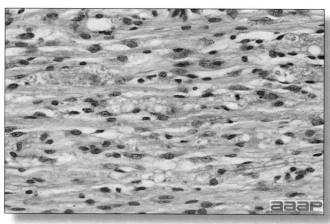

Fig. 4
Myelin degeneration, Schwann cell proliferation, and axis cylinder fragmentation in sciatic nerve.

VITAMIN A DEFICIENCY

OCCURRENCE
Most outbreaks of vitamin A deficiency occur in young birds, usually 1-7 week-old chicks or poults. Other outbreaks occur in pullets or hens. Because rations compounded by owners of small, backyard flocks are more likely to be deficient, most outbreaks are seen in those flocks. But vitamin A deficiency can also occur in commercially raised flocks.

ETIOLOGY
1. Most poultry rations contain some alfalfa meal or new yellow corn, both excellent sources of provitamin A carotenoids which are easily converted into vitamin A by enzymes in the intestinal mucosa. If rations do not contain alfalfa meal and if stored (depleted) corn is utilized, the ration may be low in vitamin A unless a vitamin A supplement is added. Vitamin A is naturally derived from fish oils.

2. Birds hatched from layers low in vitamin A have very low vitamin A reserves. If they are placed on deficient rations after hatching, they soon will be deficient in vitamin A.

3. Birrds raised on range get large amounts of vitamin A from green plants. During confinement, this source of vitamin A is not available and deficiency may develop unless the formulated ration includes other sources.

4. Since vitamin A is a fat soluble vitamin and if the fat used in the diet is rancid, there can be a deleterious effect on its availability.

CLINICAL SIGNS
Vitamin A is involved in numerous processes: it plays a role in vision, maintains mucous membrane integrity and cerebrospinal fluid pressure, is required for normal growth and reproduction. It acts as an antioxidant, is photoprotective and an anticarcinogen.

Recently hatched birds (chicks and poults)

1. Signs appear in 1-7 weeks according to the amount of vitamin A stored in the egg and present in the feed. First there is anorexia, growth retardation, then drowsiness, mild ataxia and increased mortality.

2. Birds usually die before the development of eye lesions. However, in birds surviving over 1 week, eyelids become inflamed and perhaps adhered with a cheesy like material present in nostrils and eyes (Fig. 1).

Adult birds
1. Depending on liver storage levels, severe deficiency of vitamin A over a period of 2-5 months is necessary before signs develop. Unthriftiness, decreased egg production, decreased hatchability and embryonic mortality are observed.

2. Scattered birds in the flock have inflammation of the eyes or sinuses and the eyes and sinuses may be swollen. Mucoid or caseous exudate accumulates in the conjunctival sac and may be voluminous. There is nasal and ocular discharge. Owners often report "the birds have a cold".

LESIONS
1. In young birds the eyelids are inflamed, often adhered, by sticky exudate. There may be excessive urates in the ureters, in collecting tubules of the kidneys, and in the bursa of Fabricius.

2. In layers, 1-3-mm white pustule-like lesions are present in the mucosa of the mouth, pharynx, esophagus (Fig. 2), and sometimes in the crop (Fig. 3). Mucoid exudate often is present in nasal passages. Conjunctival sacs or sinuses contain mucoid or caseous exudate and may be greatly distended. There may be a delicate pseudomembrane lining the trachea.

3. Microscopically the original epithelium is replaced by a keratinizing epithelium. There is squamous cell metaplasia of the secretory and glandular epithelium (Fig. 4) of the upper respiratory and digestive tracts, which blocks the mucous gland ducts resulting in glands becoming distended with necrotic material.

DIAGNOSIS
1. A careful study of the formula used in compounding the ration may reveal the likelihood of deficiency of vitamin A. One should consider not only the ingredients used but the quality of those ingredients. Analysis of the ration is expensive and time consuming and may be misleading unless the sample is truly representative.

2. Signs and lesions often are suggestive of vitamin A deficiency. Microscopic demonstration of squamous cell metaplasia is nasal passages may assist in diagnosis.

3. Low vitamin A levels in the liver are indicative of vitamin A deficiency.

4. Swollen infraorbital sinuses and exudate in the conjunctival sacs occur with other diseases of poultry. Differential diagnosis should consider infectious coryza of chickens, chronic fowl cholera, infectious sinusitis of turkeys, and influenza of turkeys, ducks, geese, and quail.

CONTROL
1. Prevention is easily accomplished by feeding a ration with adequate vitamin A (broiler chickens and turkeys: 7 000 to 12 500 U.I/kg, with the highest recommended levels in breeders: 10 000 to 14 000 U.I./kg).

2. Avoid long storage of prepared feeds or ingredients for those feeds. Buy or prepare feeds only in relatively small quantities.

3. Add chemical antioxidants to feeds at the time of preparation to protect vitamin A content or add stable forms of vitamin A.

TREATMENT

Treat affected flocks by adding a water-dispersible vitamin A supplement to the drinking water. Alternatively, add a stabilized vitamin A supplement to the ration at 2-4 times normal levels for about 2 weeks. Then feed a balanced ration at normal levels.

Fig. 1
Caseous exudate present under the eyelids.

Fig. 2
Distended, impacted mucosal glands resembling pustules in the oesophagus.

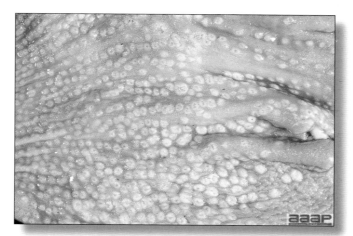

Fig. 3
Distended and impacted mucous glands forming small white nodules in crop.

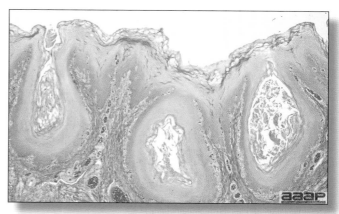

Fig. 4
Squamous metaplasia.

VITAMIN E DEFICIENCY

DEFINITION
Three distinct disorders (syndromes) related to or caused by vitamin E deficiency have been recognized in poultry. Each disorder usually occurs alone, although there are occasional overlaps. The three disorders are:

1. Encephalomalacia (crazy chick disease).

2. Exudative diathesis.

3. Muscular dystrophy.

Although each of these syndromes is associated to some degree with vitamin E deficiency, each can be prevented by dietary changes unrelated to the vitamin E content of the ration. There is some interaction with synthetic antioxidants, selenium and sulfur-containing amino acids, especially in preventing exudative diathesis and muscular dystrophy.

OCCURRENCE
Vitamin E deficiencies usually are seen in young chicks or turkey poults but also occur in ducklings and, perhaps, in other poultry. Deficiencies usually occur in birds raised in confinement i.e., birds compelled to eat only what is offered to them. Most outbreaks occur in birds fed rations that are high in polyunsaturated fats (e.g., cod liver oil, soy bean oil), that oxidize and become rancid. Vitamin E is very unstable with oxidative destruction enhanced by minerals and polyunsaturated fats in diet.

ETIOLOGY
1. Vitamin E and the selenium-containing enzyme glutathione peroxidase prevent cell membrane destruction caused by peroxides and other powerful oxidants produced as metabolic by-products.

2. There is evidence that vitamin E, selenium, and sulfur-containing amino acids perform separate functions but still act together to prevent the accumulation of harmful peroxides in tissue. Peroxides are derived, in part, from polyunsaturated acids in feeds.

3. 3. The following facts are of interest in considering etiology:

   A. Encephalomalacia can be prevented by adding synthetic antioxidants to the feed.

   B. Exudative diathesis can be prevented by adding selenium to the feed.

   C. Muscular dystrophy can be prevented by adding cysteine, a sulfur-containing amino acid, to the feed.

CLINICAL SIGNS
Vitamin E is involved in several metabolic functions but mostly play a role of natural antioxidant.

Encephalomalacia
Signs are those associated with lesions of the central nervous system and include ataxia, loss of balance, falling over backwards while flapping the wings, sudden prostration on the side with legs outstretched, toes flexed, and head retracted (Fig. 1). Birds that show clinical signs often continue to eat. The deficiency usually occurs between the 15th and 30th day of life; however, it may occur as early as the 7th and as late as the 56th day.

Exudative diathesis
There is a severe edema caused by increased capillary permeability. This edema is located along the ventrum of the thorax, the abdomen, and perhaps under the mandible. Birds with extensive edema may have difficulty in walking and may stand with their legs far apart because of accumulation of subcutaneous fluid ventral to the abdomen.

Muscular dystrophy
Signs are usually inapparent but there may be locomotor problems.

LESIONS
Encephalomalacia
The swollen cerebellum often contains yellow or congested, hemorrhagic, or necrotic areas visible on the surface (Fig. 2). Swollen and hemorrhagic cerebella are quite striking in turkey poults suffering from encephalomalacia. Lesions occur less frequently on the cerebrum. Lesions are accentuated by formalin fixation for a few hours. In turkeys, poliomalacia of the lumbar spinal cord can be found microscopically.

Exudative diathesis
There is green-blue blood-stained viscous edema in the skin and subcutis of the ventrum (Fig. 3). Muscular dystrophy occasionally is apparent in breast or leg muscles of the same birds. Distention of the pericardium with fluid has been the cause of sudden deaths in birds (Fig. 4).

Muscular dystrophy
In chicks white to yellow degenerative muscle fibers give a streaked appearance to skeletal muscles of the breast or legs (Fig. 5). In poults the musculature of the gizzard may contain gray areas of muscle degeneration (Fig. 6).

DIAGNOSIS
1. The diagnosis can usually be made on the basis of typical signs and gross lesions.

2. Examination and analysis of the ration may indicate rancidity or likelihood of deficiency of vitamin E and/or selenium. Feed analysis for vitamin E activity can be time-consuming and expensive, therefore care should

be taken to submit truly representative samples. Liver can also be analyzed for vitamin E and selenium. Storage temperature and duration are very important in evaluating the quality of the vitamin E ingredient.

3. Gross and microscopic examination of typical lesions is of considerable value in confirming suspected vitamin E deficiency, especially with encephalomalacia or muscular dystrophy.

CONTROL

1. Mix new batches of feed at frequent intervals. Use only high quality ingredients. Avoid storage of mixed feeds for periods longer than 4 weeks. If prolonged storage is necessary, add chemical antioxidants.

2. Use only stabilized fats in the feed.

3. Store feeds in a cool, dry place to reduce vitamin and other quality losses.

4. Avoid improperly compounded do-it-yourself -type rations. Most well-known, commercially prepared feeds are superior in quality to unplanned, self-mixed feeds.

TREATMENT

1. Recommended vitamin E levels are 30 to 150 mg/kg in the diet. Be sure an antioxidant (0,25kg of BHT or santoquin per 1000kg of feed) is in the feed if storage is long or environmental temperatures high. However the newest forms of vitamins are enveloped hence more resistant to heat treatments, humidity and storage. A dose of 0.3 ppm of selenium is recommended in the broiler chicken and turkey diets. Zero to 3 week-old chicks and 0 to 6 week-old turkeys should receive half of this selenium in an organic form which is more readily available to the bird.

2. Oral administration of a single 300 IU of vitamin E per bird will often cure exudative diathesis or muscular dystrophy. Birds with encephalomalacia do not usually respond well to treatment.

VITAMIN E DEFICIENCY

Fig. 1
Chicks with paresis/paralysis.

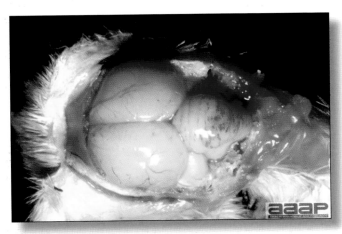

Fig. 2
Hemorrhagic cerebellum.

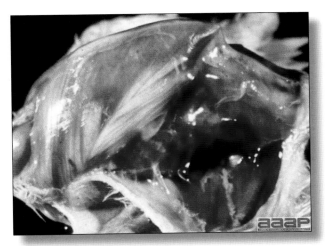

Fig. 3
Exudative diathesis: edema of subcutaneous tissues.

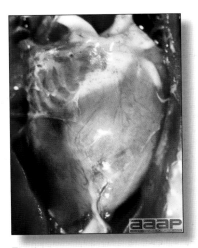

Fig. 4
Pericardium distended with fluid.

Fig. 5
Nutritional myopathy with degeneration of the muscle fibers.

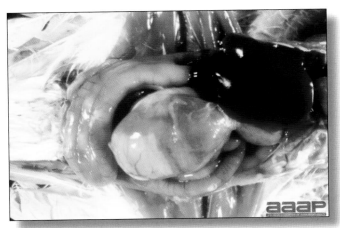

Fig. 6
Gizzard muscle degeneration.

Other nutritional diseases

RICKETS

DEFINITION

In poultry a deficiency of vitamin D_3, phosphorus or a wide imbalance in the calcium:phosphorus ratio of the diet can cause rickets. The term osteomalacia has been used to denote similar condition in laying chickens.

OCCURRENCE

Deficiencies of vitamin D_3 and phosphorus are encountered most frequently in young chicks or poults a few weeks old. Calcium deficiencies usually affect birds of the same age or adult layers. Rickets occurs more frequently in small flocks of poultry raised on carelessly formulated rations. Most commercial feeds are carefully compounded and are adequate in required nutrients. The incidence of rickets is increased in chickens with infectious stunting or malabsorption syndrome.

ETIOLOGY

1. There are numerous forms of vitamin D, however only cholecalciferol or vitamin D_3 acts as the nutritional precursor of 1,25 dihydroxycholecalciferol, the hormone that stimulates active transport of calcium and phosphorus across the intestinal epithelium, bone and shell formation. Because of this role in calcium transport, as the ratio of calcium to phosphorus becomes wider or narrower, vitamin D_3 requirements increase.

2. Although rickets can occur because of phosphorus deficiency, most outbreaks are the result of inadequate vitamin D_3. Sometimes vitamin D_2 is erroneously fed to poultry instead of D_3.

3. Young, rapidly growing chickens or turkeys experiencing malabsorption or any intestinal condition impairing nutrient absorption, may develop rickets even with adequate dietary levels of phosphorus and vitamin D_3. Failure to absorb and/or utilize nutrients in the ration is considered to be secondary to other causes.

4. Calcium deficient laying hens may suffer from cage layer fatigue (up to 30 weeks of age) or bone breakage (old hens).

CLINICAL SIGNS

1. In young growing flocks, affected birds develop a lame, stiff-legged gait. There is retardation of growth. There may be enlargement of the ends of long bones, especially noticeable in the hocks. Birds often rest in a squatting position.

2. Egg layers can experience increased number of thin-shelled and soft-shelled eggs followed soon by decreased egg production.

3. Laying hens suffering from cage layer fatigue lie on their side, with their legs extended or they crouch in the corner of their cage.

LESIONS

1. In young birds bones, beaks and claws are soft and rubbery and the epiphyses of long bones often are enlarged. There is a characteristic beading of the ribs, most noticeable at their junction with the spinal column. Ribs are thickened and tend to bend so that the thorax is flattened laterally. The beak becomes soft and rubbery and can be bent or flexed easily (Fig. 1). Parathyroids often are markedly enlarged.

2. Feathering is usually poor and an abnormal black banding of feathers has been observed in colored breeds such as red or buff chickens.

3. In laying hens bones are soft and easily broken, ribs may become beaded (Fig. 2) and parathyroids are enlarged (Fig. 3).

DIAGNOSIS

1. In young birds, their age, signs, and lesions are all useful in diagnosis. Softening of the beak and beading of the ribs are almost pathognomonic.

2. Careful calculation of the calcium:phosphorus ratio, and vitamin D_3 levels of the ration may reveal that it is deficient or imbalanced. Chemical analysis of the ration for minerals and D_3 can be done but is expensive and time consuming. A single sample may not be typical of the ration. Extensive sampling may be advisable for forensic reasons.

CONTROL

1. Feed a balanced ration with adequate calcium, phosphorus, and vitamin D_3 levels. Rations should be carefully compounded to fit the age, purpose, and production of the flock. Note that poultry requires vitamin D_3 a form differing from vitamin D_2, which often is fed to other types of livestock. Rovimix Hy-D, a commercial form of 1,25 dihydroxycholecalciferol can also be given to the birds alone or in combination with vitamin D_3.

2. The following calcium:phosphorus ratios are recommended:

   Broiler chickens: 1,35 to 1,5 :1,
   Turkeys (0 to 6 weeks): 1,5 :1
   Turkeys (end of growth): 1,35 to 1,5 :1
   Commercial layer: 5,8 to 7:1 (varies according the egg production period)
   Breeders: 4,5 to 5,5 :1

TREATMENT

1. Adjust the ration to fit the age and production level of the flock. If the ration has been deficient in vitamin D_3 give three times the usual amount for a period of 2-3 weeks. Then go back to a balanced ration with the usual recommended level. Liquid vitamin D_3 will also treat the birds.

2. Calcium-deficient, paralyzed, or down layers can be given 1g of calcium carbonate in a gelatin capsule daily for a few days. If the affected layers are caged layers, they should be removed from their cage and confined on the floor until fully recovered.

3. If housed birds are deficient in vitamin D_3, it may be useful to turn them out on range or otherwise expose them to sunlight.

4. Removing hens affected with cage layer fatigue from cages during the early stages of lameness may result in recovery.

Fig. 1
Rubbery beak.

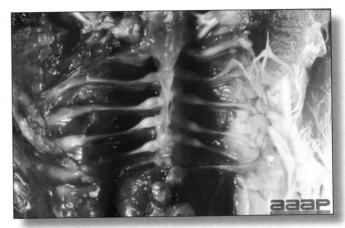

Fig. 2
Beaded ribs.

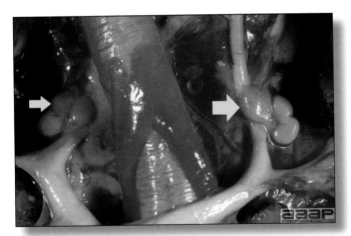

Fig. 3
Enlarged parathyroids (arrows).

FATTY LIVER-HEMORRHAGIC SYNDROME
(FLHS; Fatty Liver Syndrome)

OCCURRENCE
Fatty liver-hemorrhagic syndrome (FLHS) is a sporadic disease with worldwide distribution that occurs primarily in caged layers. Outbreaks are most common in high-producing flocks during hot weather.

HISTORICAL INFORMATION
Fatty liver syndrome was first reported in 1956 and was soon observed by many other diagnosticians. The appearance of the syndrome coincided with the practice of confining layers to cages. There has been much speculation as to the cause of the syndrome. In 1972, the syndrome was reproduced experimentally by force-feeding hens. The lesions closely resembled those of the natural disease. This appears to be a multifactorial problem.

ETIOLOGY
1. Excessive consumption of high-energy diets combined with restricted activity is believed to result in excessive fat deposition in the liver.
2. Contributing factors may include a genetic component.
3. The syndrome may be caused by a deficiency of lipotrophic agents, which are necessary for mobilization of fat from the liver.
4. Aflatoxin in laying hen diets has been shown to increase fat content (dry weight basis) approximately 20% over controls and may play a contributing role.
5. FLHS and caged layer fatigue are often diagnosed simultaneously.

CLINICAL SIGNS
Outbreaks of FLHS are often associated with a sudden drop in egg production (from 78-85% to 45-55%). The flock overall may be obese (body weights 25-30% above normal). Some birds may have pale combs and wattles covered with flaking epidermis. Mortality increases moderately with occasional hens in full production dying suddenly and unexpectedly. Often hens are found dead with pale heads. Mortality rarely exceeds 5%.

LESIONS
Dead birds have large blood clots in the abdomen, often enveloping the liver, as a result of subcapsular hepatic hemorrhage and rupture of the parenchyma (Fig. 1). Subcapsular hematocysts or hematomas may be visible within the parenchyma (Fig. 2). Liver is generally enlarged, pale, yellow and friable. Fat content in livers generally exceeds 40% dry weight and may reach 70%, hence the yellow coloration. Clinically healthy birds in the same flock may also have hematomas in the liver, either dark red (recent) or green to brown (older). Large amounts of fat are present within the abdominal cavity and surrounding the viscera.

TREATMENT
There has been no clear elucidation of dietary causes of FLHS other than excessive caloric intake. Reducing obesity of laying hens is the only successful preventive measure to date. However, further loss of production may result from diet changes during the laying cycle. Lipotropic agents such as vitamin E, vitamin B12, biotin, methionine and choline have been widely used with variable results. Management practices that reduce heat stress and minimize mold growth in feed may also be helpful. Results of feeding particular nutrients or formulations of nutrients to treat FLHS are inconsistent.

Fig. 1
Subcapsular hepatic hemorrhage with rupture of the parenchyma.

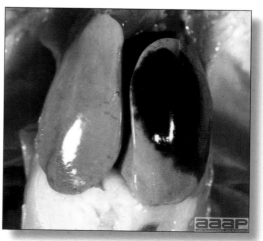

Fig. 2
Enlarged, pale, yellow liver with subcapsular hematocyst.

MISCELLANEOUS DISEASES

Revised by Dr. H.L. Shivaprasad

CARDIOVASCULAR DISEASES OF CHICKENS

I. ASCITES OR PULMONARY HYPERTENSION SYNDROME

DEFINITION
Ascites secondary to pulmonary hypertension syndrome (PHS) is one of the most important causes of mortality in broiler chicken flocks. It is associated with rapid growth and a high metabolic rate.

OCCURRENCE
Ascites occurs worldwide in rapidly growing broiler chicken flocks.

HISTORICAL INFORMATION
Ascites was first reported in 1968 in broiler chickens raised at a high altitude. However, the incidence of ascites caused by PHS, where broilers are grown at a low altitude, has increased over the past several years and coincides with genetic and nutritional improvements that resulted in better growth rate and feed conversion.

ETIOPATHOGENESIS
Four pathophysiological mechanisms are recognized to cause ascites: increased hydrostatic vascular pressure, decreased oncotic pressure, increased capillary permeability, and impaired lymphatic drainage. Although numerous chemical toxicities have been reported to cause ascites in broiler chickens through one of these mechanisms, the most common form of ascites in fast-growing broiler chickens is caused by increased hydrostatic vascular pressure.

Rapid growth, elevated metabolic rate, and therefore a high oxygen demand impose an increased workload on the heart. This, combined with the insufficient pulmonary capillary capacity of the modern broiler chicken, aggravates the pulmonary hypertension and further precipitates right ventricular hypertrophy.

Hypertrophy is soon followed by dilation, right ventricular failure, passive congestion, and then ascites. This process is accelerated in birds because of an anatomical particularity. The right atrioventricular valve is a muscular flap, an extension of the right ventricular wall. Any hypertrophy of the latter affects the valve and its apposition against the septum, facilitating venous regurgitation, passive congestion, and ascites.

CLINICAL SIGNS
Clinically affected broiler chickens are smaller than normal and depressed with ruffled feathers. Severely affected birds show abdominal distension with reluctance to move, respiratory distress, and cyanosis (Fig. 1).

LESIONS
1. Hypertrophy and dilation of the right ventricle (Fig. 2) with or without accumulation of straw-colored ascitic fluid in the peritoneal cavities, (Fig. 3), and a generalized passive congestion are characteristic of ascites secondary to PHS (Fig. 4).
2. Hydropericardium, protein clots in the ascitic fluid, and a fibrotic liver (Fig. 5) may be present in chickens with chronic PHS.
3. Microscopic lesions show generalized passive congestion.

DIAGNOSIS
Macroscopic lesions are diagnostic.

If mortality in a flock is abnormally high, look for causes decreasing oxygen availability to the broiler chicken (poor ventilation, high altitude, concomitant respiratory pathology, etc.), or increasing oxygen needs (rapid growth, cold rearing temperature stimulating the metabolic rate).

Other pathological mechanisms can be involved in the development of ascites, and toxicities due to sodium, phenolic compounds, coal-tar derivatives, and dioxin, among others, might also be considered.

CONTROL
Lowering the oxygen requirement by slowing the metabolic rate will reduce, and if severe enough, prevent ascites. A variety of feed restriction and light programs have been used or recommended. The goal is to find a program that will maintain feed efficiency while reducing metabolic rate without increasing days to market.

TREATMENT
There is no treatment.

II. SUDDEN DEATH SYNDROME OF CHICKENS

DEFINITION
Apparently healthy fast-growing broiler chickens, mainly males, die suddenly after a short terminal wing-beating convulsion. Dead birds are found lying on their back (Fig. 1). This is a common cause of "normal mortality" in a flock.

OCCURRENCE
This condition occurs from 1- 8 weeks of age in most intensive broiler-growing areas of the world. The incidence

in a flock varies from 0.5% to more than 4% in some cases. Sixty to 80% of the affected birds are males.

HISTORICAL INFORMATION
This syndrome has been recognized for 30 years and has been described as acute death syndrome, heart attack, flip-over, dead in good body condition, and lung edema.

ETIOLOGY
The cause is unknown but this condition affects highly performing broiler chickens. It is suggested that death is the result of ventricular fibrillation secondary to a possible imbalance of metabolites or electrolytes. It is classified as a metabolic disease and the incidence appears to be affected by genetic, environmental, and nutritional factors.

CLINICAL SIGNS
There are no premonitory signs. Large healthy broiler chickens will start to convulse and wing flap, and rapidly die lying on their back.

LESIONS
Birds are in good body condition with a full digestive tract (Fig. 2). There is red and white mottling of the breast muscle (Fig. 3), the ventricles of the heart are contracted, and the auricles dilated with blood. Lungs might be congested secondary to postmortem blood pooling. There are no specific histopathologic lesions.

DIAGNOSIS
Dead birds appear healthy and there are no lesions except the findings described above.

CONTROL
Various feed and light regimens have been tried with little success in decreasing the incidence of sudden death without decreasing feed conversion.

TREATMENT
There is no treatment.

III. ROUND HEART DISEASE OF CHICKENS

This myocardial degeneration used to affect mature chickens (> 4 months of age) but has not been diagnosed in commercial poultry flocks for years. Birds die with a bilateral ventricular hypertrophy and dilation. Histopathology reveals myocardial fatty infiltration. The etiology is unknown.

ASCITES OR PULMONARY HYPERTENSION SYNDROME

Fig. 1
Broiler that died of ascites following right ventricular failure. Note the cyanosis and enlarged abdomen.

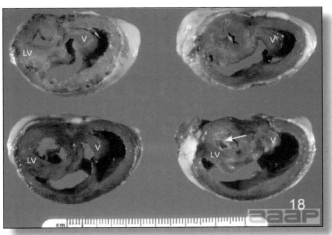

Fig. 2
Cross-section of fixed hearts showing marked right ventricular hypertrophy and dilation and hypertrophy of the right AV valve (V).

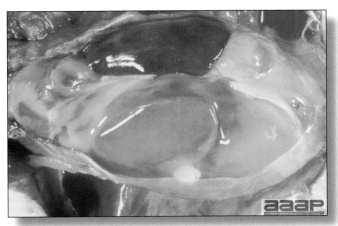

Fig. 3
Accumulation of straw-colored ascitic fluid in the peritoneal cavities.

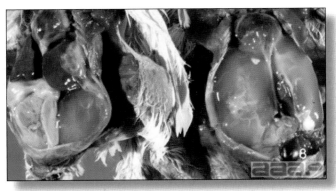

Fig. 4
Broiler carcasses showing generalized passive congestion, enlarged heart and large quantity of yellow clotted protein (fibrin) and fluid that can accumulate in the peritoneal cavities.

Fig. 5
Hydropericardium, protein clots in the ascitic fluid, and a fibrotic liver may be present in chickens with chronic PHS.

ASCITES OR PULMONARY HYPERTENSION SYNDROME

Fig. 1
Dead broiler lying on its back, a regular finding in a broiler barn.

Fig. 2
Chicken carcasses in good body condition with a full digestive tract (arrows). The ventricles of the heart are contracted, and the auricles dilated with blood.

Fig. 3
Broiler chicken showing a full crop and red and white mottling of the breast muscle.

CARDIOVASCULAR DISEASES OF TURKEYS

I. AORTIC RUPTURE OR DISSECTING ANEURYSM

Aortic rupture is an occasional cause of mortality in 12-16-week-old heavy turkeys, characterized by massive internal hemorrhage from a ruptured lower or, less commonly, upper aorta. The cause is unknown but some contributing factors such as the relatively high blood pressure of turkeys, their natural susceptibility to atherosclerosis, and the absence of an intramural *vasa vasorum* (intrinsic vascularization of the artery) in the lower aorta, might all play a role in the pathogenesis. Copper and zinc levels of the liver are normal in affected birds. Genetics might play a role in the incidence as the progeny of certain breeder lines seem to have a higher incidence. The carcass is typically in good body condition but pale. Blood may be seen in the mouth or nostrils. A large amount of clotted blood is found in the body cavity, surrounding the kidneys or filling the entire body cavity (Fig. 1) if rupture occurs in the lower aorta, or surrounding the heart if it occurs in the upper aorta. Careful examination will reveal a longitudinal tear in the wall of the aorta (Fig. 2). Microscopically there is subintimal fibrosis and decreased elastic tissue in the tunica media. Management procedures to decrease the incidence of this condition consist of avoiding excitement in the birds.

Similar to aortic rupture, coronary artery aneurysm and rupture can also occur in rapidly growing male turkeys. Necropsy of such birds reveals hemopericardium and hemorrhage in the coronary groove. The cause and pathogenesis of this condition is unknown but is probably similar to aortic rupture

II. DILATED CARDIOMYOPATHY OF TURKEYS (ROUND HEART DISEASE)

This condition causes mortality in turkey poults most commonly between 1 and 4 weeks of age and is a common occasional finding in commercial turkey flock. Affected poults are found dead with a severe bilateral dilated cardiomyopathy (Fig. 1 and 2) often accompanied by secondary ascites and hydropericardium (Fig. 3), and congestion of other organs. If the poult survives with this cardiac disorder, growth will stop and the bird will soon show ruffled feathers, unwillingness to move, respiratory distress, and death. Livers will often show chronic passive congestion and centrolobular hepatocyte degeneration with cytoplasmic vacuolation of hepatocytes. Microscopic changes in the myocardium are usually nonspecific. The etiology is unknown, but several factors such as genetic factors, early viral myocarditis and hypoxic conditions during incubation have been suggested. Because of its similarities to human dilated cardiomyopathy, turkey is being used as an animal model.

III. SUDDEN DEATH SYNDROME OF TURKEYS (PERIRENAL HEMORRHAGE)

DEFINITION
Sudden death syndrome of turkeys (SDS) or perirenal hemorrhage syndrome causes death in heavy-turkey flocks, particularly during grow-out period. Turkeys in good body condition, mainly males, die suddenly with postmortem lesions of acute generalized passive congestion.

OCCURRENCE
SDS is the main cause of mortality in fast-growing turkeys 8-15 weeks of age, but has been reported in turkeys older than 20 weeks of age. This syndrome is uncommon in female turkeys.

HISTORICAL INFORMATION
The condition was first reported in 1973 under the name sporadic renal hemorrhage. The disease has also been named perirenal hemorrhage syndrome, acute hypertensive angiopathy, or sudden death with perirenal hemorrhage. These confusing terms that describe lesions but give no indication of etiology and pathogenesis likely refer to the same condition.

ETIOLOGY
Through intense genetic selection and high-energy diets, the industry has developed a rapidly growing, heavily muscled turkey. SDS of turkey occurs during a period of fast growth and often follows exposure to stress or increased activity level in the flock.

Experimental studies have demonstrated the inability of the cardiovascular system of the domestic turkey to meet metabolic needs generated by exercise; within minutes, turkeys develop hypotension combined with severe lactic acidosis. Turkeys dying of SDS also show greater ventricular weights and cardiac changes described as concentric left ventricular hypertrophy.

It has been hypothesized that a certain percentage of the turkey population has concentric left ventricular hypertrophy which reduces myocardial blood flow and impairs coronary vascular reserve. Exercise or stress could therefore prompt acute myocardial ischemia triggering ventricular arrhythmias and terminal ventricular fibrillation. Ventricular arrhythmias could also be precipitated by the severe lactic acidosis developing during exercise subsequent to tissue hypoxia. Thus, an inadequate cardiovascular response of the turkey to stress or exercise may create hemodynamic instability leading to sudden death.

CLINICAL SIGNS
There are no clinical signs, except violent agonal wing flapping preceding death.

LESIONS
Turkeys dying of SDS are in good body condition with the digestive tract filled with ingesta, demonstrating the suddenness of death. Lesions are indicative of an acute generalized passive congestion with subcutaneous varicoses, pulmonary congestion and edema, perirenal hemorrhage (Fig. 1), a swollen severely congested spleen, and congestion of other organs.

Perirenal hemorrhage has been reported to occur in other conditions and is not pathognomonic of the so-called SDS of turkeys. Birds possess a renal portal system with a superficial peritubular capillary plexus at the periphery of the renal lobule. Local passive congestion would therefore result in pooling of blood in the perirenal area and possible diapedesis, explaining the hemorrhages observed at the surface of the kidneys.

DIAGNOSIS
Lesions are diagnostic. Aortic aneurysm affects the same-age turkey, but in aortic aneurysm, birds are pale and free blood is present in the body cavity of the turkey.

CONTROL
Avoid excitement in the flock.

TREATMENT
There is no treatment.

AORTIC RUPTURE

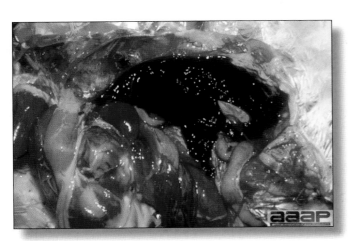

Fig. 1
Aortic rupture: massive internal hemorrhage.

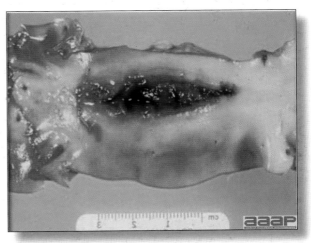

Fig. 2
Ruptured aorta.

DILATED CARDIOMYOPATHY OF TURKEYS

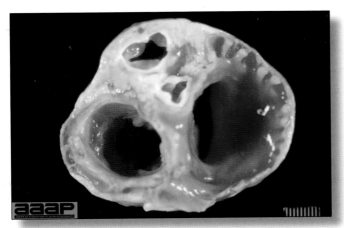

Fig. 1
Severe bilateral dilated cardiomyopathy: cross-section of the heart.

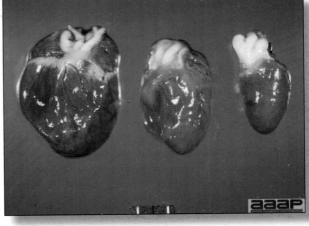

Fig. 2
Various degrees of dilated cardiomyopathies (Normal heart on the right).

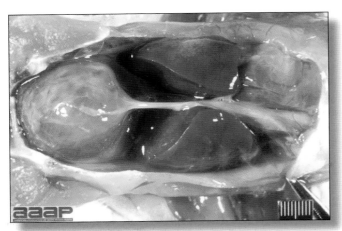

Fig. 3
Severe bilateral dilated cardiomyopathy with hydropericardium and secondary ascites.

SUDDEN DEATH SYNDROME OF TURKEYS

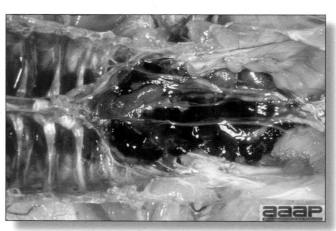

Fig. 1
Perirenal hemorrhage.

DIGESTIVE DISORDERS

I. PENDULOUS CROP

Pendulous crop is a condition sporadically observed in broiler chicken and turkey flocks. Birds have a greatly distended crop (Fig. 1) filled with feed and malodorant material. Crop mucosa can be ulcerated or infected by *Candida albicans*. Affected birds keep eating but since feed transit is affected, they soon lose weight, eventually become emaciated and die. Affected carcasses reaching market age are condemned at slaughter. Increased water intake during sudden hot weather has been proposed as a possible cause since birds will over drink and eat at night when the temperature cools down. This appears to "over stretch" the muscular wall of crop and even sometimes the proventriculus, leading to permanent distention. Hereditary predisposition has also been suggested in turkeys.

Fig. 1
Broiler chicken with a greatly distended crop.

II. PROVENTRICULAR DILATION

Proventricular dilation has been reported in birds fed a finely ground diet. The gizzard of birds fed such a diet does not need to contract much, hence its poor muscular development with secondary proventricular dilation. The proventriculus is enlarged with thin walls and no clear demarcation between the gizzard and proventriculus (Fig. 1). Excessive histamine amounts will also cause proventricular dilation and flaccidity along with gizzard erosions.

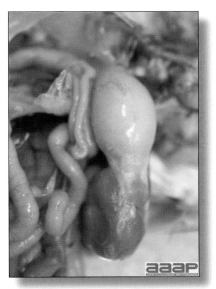
Fig. 1
Enlarged proventriculus.

DIGESTIVE DISORDERS OF CHICKENS

I. DYSBACTERIOSIS

DEFINITION
Terminology used in Europe to describe an intestinal microflora imbalance and overgrowth characterized by enteritis and mild diarrhea.

OCCURRENCE
Dysbacteriosis is commonly observed after 21 days of age in European commercial broiler chicken flocks but can occur as early as 15 days of age.

HISTORICAL INFORMATION
There has been an increase in the number of broiler chicken flocks affected with dysbacteriosis with the ban of growth promoters in Europe in 1999.

ETIOLOGY
Overgrowth of an abnormal bacterial duodenal population has been demonstrated in birds affected with dysbacteriosis. *Clostridium spp.* has been shown to contribute to this overgrowth. The absence of antimicrobial growth promotors, animal protein and animal fat appear to predispose birds to the disease. Other predisposing factors may include non-specific stress, mycotoxins and systemic disease.

CLINICAL SIGNS
Dysbacteriosis is characterized by normal water consumption, humid litter, poorly formed and wet feces and a reduction in feed intake.

LESIONS
Thinning and ballooning of the small intestines accompanied by viscous or watery intestinal contents (Fig. 1).

DIAGNOSIS
History of diarrhea, wet droppings. Elimination of any other causes of diarrhea and wet litter. Empirical therapeutic response to antimicrobial effective against *Clostridium perfringens* or other enteric pathogens might be a diagnostic indicator for both necrotic enteritis and dysbacteriosis.

CONTROL AND TREATMENT
Monitoring litter quality with a litter box and an underlying absorbent paper might help in assessing any changes in fecal water content and alert to early signs of diarrhea. Antibiotics might be required if there is associated mortality or subsequent necrotic enteritis. Competitive exclusion products might help.

II. POLYCYSTIC ENTERITIS OF BROILER CHICKENS or RUNTING-STUNTING SYNDROME OF BROILER CHICKENS

DEFINITION
Polycystic enteritis (PE) is most common in the Southeastern United States but is seen sporadically on the West coast as well as other parts of the United States. It is a disease of economic concern in other parts of the world most notably in Australia and Europe. It is characterized by large numbers of chicks with marked growth depression, watery diarrhea, and cystic enteritis. The condition is named after the characteristic severe microscopic cystic dilation of intestinal crypts.

OCCURRENCE
PE may appear in chicks as early as 6-7 days of age, but the usual peak of the problem occurs at around 10-12 days of age, mostly during winter and spring. Farms that have short downtimes between flocks appear to be at higher risk for the disease. Turkeys are not known to be affected.

HISTORICAL INFORMATION
Runting Stunting Syndrome (RSS) has been recognized in chickens since the late 1970s. This condition occurs sporadically, usually with increasing severity over a year period within a given farm, then declines afterwards. During 2003-2005 a new clinical and pathological presentation appeared and caused economically significant problems in the Southeastern United States, and some countries in Asia, Middle East, and Latin America. In contrast to RSS, persistent problems with PE have been noted on specific "problem" farms as successive flocks are affected. Many research institutions are actively studying this condition, further characterizing enteric viruses, developing diagnostic tests and searching for potential vaccine candidates.

ETIOLOGY
The disease has been reproduced by placing broiler chicks on contaminated litter obtained from previously affected farms, and by gavaging birds with intestinal contents from affected chickens. These resulted in severe weight depression. Multiple viruses have been identified from chicks with PE including rotavirus, reoviruses and astroviruses. Bacteria do not appear to be involved in the disease as primary agents. Vertical transmission is considered a possibility and is being investigated.

CLINICAL SIGNS
Affected flocks show large numbers of depressed chicks huddling around feeders and drinkers within hours after placement. Litter quickly becomes damp. Feed consumption decreases, there is loss of flock uniformity and many chicks will show severe growth depression (5 up to 20%) (Fig. 1). If allowed to remain in the flock, stunted chickens do not recover. This will translate in

increased need for culling, reduced livability, increased feed conversion, and days to market.

LESIONS
At necropsy, affected chicks have small livers with enlarged gallbladder, pale, dilated thin-walled intestines with watery contents and undigested food (Fig. 2 and 3). Histologically, intestinal lesions consist of numerous large cysts involving intestinal crypts with degenerating or necrotic cells and mucin inside the lumen of these cysts (Fig. 4). As the condition progresses, intestinal villi become shortened and clubbed.

DIAGNOSIS
History, clinical signs, and microscopic intestinal lesions are suggestive of the disease.

CONTROL
Built-up litter and short downtime may contribute to PE. Proper brooding temperature minimizes early poor uniformity and delayed growth. Heat treatment of affected houses during downtime is likely to mitigate the condition.

TREATMENT
There is no specific treatment. Good husbandry and symptomatic support of an affected flock will lessen economic losses. Severely stunted chicks will not recover and should be culled.

III. TRANSMISSIBLE VIRAL PROVENTRICULITIS

DEFINITION
Transmissible viral proventriculitis (TVP) is a transmissible proventricular inflammation of viral etiology found in commercially raised broiler chickens and associated with increased proventricular fragility, impaired feed digestion, poor growth performances, and increased contamination and decreased efficiency at processing.

HISTORICAL INFORMATION
Within the past few years, commercial broiler chickens from Southeastern United States, the West coast and probably others regions of the States have sporadically been affected with this disease.

ETIOLOGY
TVP has experimentally been reproduced with homogenates from proventricular tissue of affected birds and a virus that is consistent with a novel Birnavirus (different from Infectious Bursal Disease Virus). Presence of the virus in proventricular lesions, in natural and experimentally infected birds, indicates it is the cause of the disease. Chicks can be experimentally infected by oral or intracoelomic inoculation, but the natural route of infection is unknown.

CLINICAL SIGNS
Affected birds are pale and significantly smaller than uninfected flock mates. They show poor growth rate, increased feed conversion, and the passage of undigested or poorly digested feed in the feces.

LESIONS
At necropsy affected broilers show proventricular enlargement, especially the isthmus between the proventriculus and ventriculus (Fig. 1), with mottled thickened, firm walls. Attenuation of mucosal papilla where ducts from the glands open into the lumen may be seen. The mucosa appears roughened. Dilated, cystic glands are not indicative of TVP since this is a postmortem change occurring rapidly following death. Four lesions characterize the microscopic changes in the proventriculi (Fig. 2): 1) necrosis of the glandular epithelium, 2) lymphocytic infiltration in the interstitium of proventricular glands and mucosa, 3) hyperplasia of ductal epithelium, and 4) replacement of lost glandular epithelium by ductal epithelium. Epithelial cell nuclei are swollen, pale, and often have prominent nucleoli. Lesions do not occur in other tissues.

DIAGNOSIS
TVP is difficult to identify on the basis of gross lesions. In contrast, microscopic lesions are sufficiently characteristic to provide a diagnosis. Confirmation requires demonstration of the novel Birnavirus by PCR. Correlation between histopathology and virus presence is very high.

CONTROL, PREVENTION AND TREATMENT
There are no specific treatment, prevention, or control measures for TVP other than biosecurity measures effective against infectious agents.

IV. FOCAL DUODENAL NECROSIS
Focal Duodenal Necrosis (FDN) is a pathology that has been observed in laying hens in United States since 1996. This condition is characterized by lower egg production goals, and/or lower egg weight standards. Lesions are located in the duodenal loop where single to multiple dark grey focal areas can be seen through the serosa (Fig. 1). These areas correspond to areas of necrosis or ulceration of the intestinal epithelium (Fig. 2). Histologically, focal necrosis of the duodenal epithelium, with numerous gram-positive bacteria covering the villi tips, is observed (Fig. 3).

Although the presence of *Clostridium colinum* has been identified by molecular methods, other bacteria have also been found, no causes and effects have been established and there remains much uncertainty about the cause, nature and risk factors for the condition. FDN can be treated and prevented with the addition of antimicrobials commonly used to prevent necrotic enteritis in broiler chickens, such as zinc bacitracin, in the feed.

DYSBACTERIOSIS

Fig. 1
Thinning of the small intestines accompanied by mucoid intestinal contents.

POLYCYSTIC ENTERITIS OF BROILER CHICKENS

Fig. 1
Loss of flock uniformity with many chicks affected with severe growth depression (in the middle).

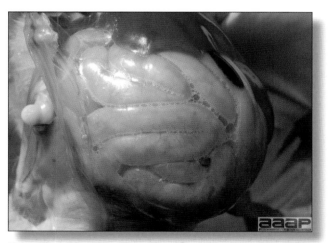

Fig. 2
At necropsy, affected chicks have small livers with enlarged gallbladder, pale, dilated thin-walled intestines with watery contents and undigested food.

Fig. 3
Chick showing pale, dilated thin-walled intestines.

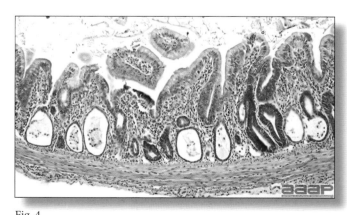

Fig. 4
At microscopy, numerous large cysts involving intestinal crypts with degenerating or necrotic cells and mucin inside the lumen of these cysts are observed.

TRANSMISSIBLE VIRAL PROVENTRICULITIS

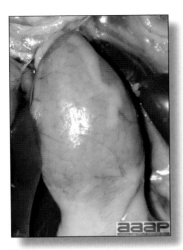

Fig. 1
28 day-old broiler with proventricular enlargement (especially the isthmus between the proventriculus and gizzard).

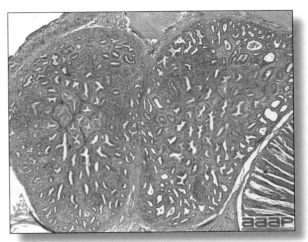

Fig. 2
At microscopy of the proventriculus there is necrosis of the glandular epithelium, lymphocytic infiltration in the interstitium of proventricular glands and mucosa, hyperplasia of ductal epithelium, and replacement of lost glandular epithelium by ductal epithelium.

FOCAL DUODENAL NECROSIS

Fig. 1
Multiple grey areas can be seen through the serosal surface of the duodenal loop.

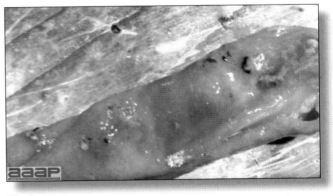

Fig. 2
Numerous circular to coalescing areas of necrosis, at various stages of inflammation, can be observed on the duodenal mucosa.

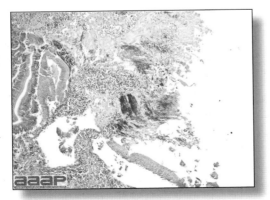

Fig. 3
Focal necrosis of the duodenal epithelium, with numerous Gram-positive bacteria covering the villi tips.

DIGESTIVE DISORDERS OF TURKEYS

I. POULT ENTERITIS COMPLEX
(PEC)

Poult enteritis complex is a terminology used to describe various infectious intestinal diseases of young turkeys. It includes numerous conditions such as turkey coronavirus (TCV), poult malabsorption or runting-stunting syndrome, and poult enteritis mortality syndrome (PEMS). These diseases all have the following common features: less than six-week-old turkeys develop diarrhea, soon followed by growth retardation and secondary nutritional deficiencies. However, while TCV is well characterized, poult malabsorption or runting-stunting syndrome and PEMS remain poorly defined in terms of etiology. Basic pathogenesis involves intestinal mucosal injury by one or more viruses, and possible secondary opportunistic infection by bacteria.

A. TURKEY CORONAVIRUS
(TCV). See page 70.

B. POULT MALABSORPTION / RUNTING-STUNTING SYNDROME

DEFINITION
Poult malabsorption / runting-stunting syndrome is an intestinal disease condition of young turkeys (Fig. 1 and 2) characterized by malabsorption/maldigestion of nutrients that may result in stunting, secondary nutritional diseases such as rickets or encephalomalacia, and secondary infections such as cryptosporidia or bacterial enteritis.

OCCURRENCE
Poult malabsorption / runting-stunting syndrome generally occurs in turkeys between 7 and 28 days of age. Nutrient absorption and/or digestion are inhibited, causing decreased growth rate, stunting, poor feathering, skeletal problems, and an uneven flock. Lack of uniformity and skeletal lesions may persist throughout the grow-out period.

ETIOLOGY
Poult malabsorption/runting-stunting syndrome is a multifactorial disease of unclear etiology. Viruses regularly isolated from intestinal tracts of affected poults include astrovirus, enterovirus, parvovirus and rotavirus. However, the detection of a viral agent from a diseased host does not in itself constitute a cause and effect relationship for that disease. In addition, *Cryptosporidia, Cochlosoma,* and coccidia have been identified and will increase severity and disease duration. *Salmonella* species and Gram-positive filamentous bacteria are also commonly isolated from affected birds. Dietary factors such as high protein levels in starter feeds, poor quality fats and fish meal, and

mycotoxins have also been implicated in increasing the severity of the disease.

C. POULT ENTERITIS MORTALITY SYNDROME (PEMS)

DEFINITION
Two clinical forms of PEMS have been identified; an acute form with a sharp peak of mortality (mortality is greater or equal to 9% between 7 and 28 days of age, and daily mortality on three consecutive days is greater than 1%), and a less severe form (mortality exceeds 2% between 7 and 28 days but daily mortality does not reach 1% during three consecutive days), which has been referred to as Excess Mortality of Turkeys (EMT).

Sick poults show diarrhea, dehydration, anorexia, growth depression, immunosuppression, and mortality, but also a variety of physiological abnormalities, including reduced body temperature, reduced energy metabolism and hypothyroidism.

OCCURRENCE
The disease occurs only in turkeys, when they are 7 to 28 days of age. There appears to be an age susceptibility; the younger the flock, the more severe the clinical expression. In the field, hens economically are more affected than toms. The acute form of PEMS has mostly been observed in the southeastern United States and presents a seasonal pattern i.e., from late spring to early fall. While this form was prevalent in the late 90's, it is now an uncommon occurrence. However, EMT is still regularly reported.

ETIOLOGY
The etiology of PEMS is still unknown. The disease can be experimentally reproduced by either contact exposure or oral inoculation of healthy poults with intestinal contents of infected poults, and it is believed that transmission is strictly horizontal. Several agents, including turkey coronavirus, rotavirus, astrovirus, reovirus, Group 1 aviadenovirus, torovirus and unidentified small round viruses, have been isolated or identified from PEMS cases. However, most have been found capable of reproducing the disease alone or has been consistently associated with the disease. In addition to viruses, certain atypical *Escherichia coli* strains including attaching and effacing *E. coli* as well as other bacteria, and various protozoa have also been associated with PEMS.

CLINICAL SIGNS
Affected poults are initially hyperactive and vocal but within twenty-four hours they become depressed, anorexic, and huddle together near heat sources. Feed and water consumption drop, while diarrhea develops. Litter quality rapidly deteriorates from abundant and watery droppings. Marked lack of uniformity can be observed few days after

the onset of the disease (Fig. 1). Clinical signs will wane within seven to ten days, but unevenness will worsen and remain for the duration of the life of the flock (Fig. 2).

LESIONS

Lesions are characteristics of an acute severe diarrheal disease. Carcasses are dirty and exhibit signs of dehydration (Fig. 3) and emaciation. The digestive tract can be empty with occasional presence of some litter material in the gizzard while small intestines are thin-walled and dilated with fluid and gas (Fig. 4 and 5). Ceca can also be distended (Fig. 6) with frothy contents. Lymphoid organs are atrophied in more severely affected birds (Fig. 7 and 8).

DIAGNOSIS

Diagnosis of PEC requires flock records comparison for analysis of growth and brooding performance, clinical evaluation, collection of diagnostic samples such as sera, fecal droppings, water and feed samples, necropsy, and isolation and identification of enteric pathogens.

CONTROL

Biosecurity is of primary importance to control PEC. Biosecurity procedures include management of dead bird disposal, litter management, movement of used litter, controlling traffic patterns of people and vehicles, rodent control and water sanitation. Affected farms should be placed under quarantine and premises should be thoroughly cleaned, disinfected and fumigated. All-in/all-out production or separate brooding and finishing units are helpful. No vaccines are available.

TREATMENT

Supportive care for affected flocks includes raising house temperatures slowly until poults appear comfortable. Water-soluble vitamins and/or electrolytes should be added to the drinking water. Vitamin E added to the feed at twice the recommended level has been shown to be helpful. Antibiotics have been used with mixed success. They should be directed toward Gram-positive bacteria since those with Gram-negative activity may further upset normal intestinal flora.

Any action that will increase feed intake, such as walking frequently through the flock, remixing feed, top dressing feed with rolled oats, whole grains, etc., should have a positive effect of PEC. On farms considered at high risk of experiencing PEC, it is recommended to avoid placing birds from young breeders because their progeny is smaller and would be more susceptible to the disease.

POULT MALABSORPTION / RUNTING-STUNTING SYNDROME

Fig. 1
Dilated and thinned-walled intestines
with gaseous and watery contents.

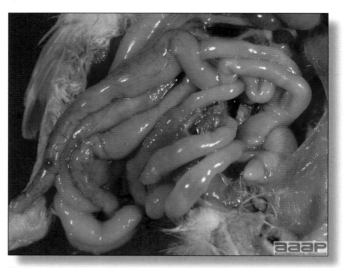

Fig. 2
Dilated and thinned-walled intestines with gaseous and watery
contents.

POULT ENTERITIS MORTALITY SYNDROME

Fig. 1
Marked lack of uniformity can be observed few days after the onset of the disease.

Fig. 2
Uneven turkey flock.

Fig. 3
Dehydrated poult (on the left) vs a normal poult (on the right). Note the darker shank characteristic of dehydration and the vent soiled with diarrhea of the PEMS affected poult (left).

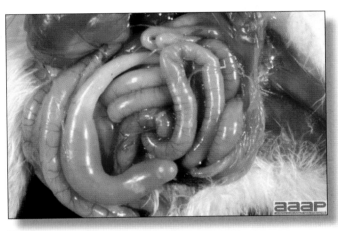

Fig. 4
Poults with small intestines that are thin-walled and dilated with fluid and gas.

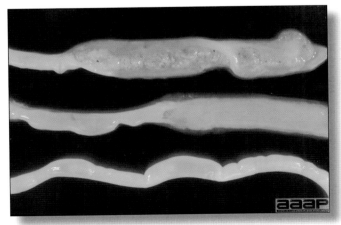

Fig. 5
Small intestines are thin-walled and filled with fluid, gas and/or poorly digested, liquid intestinal content.

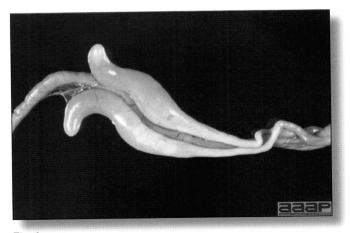

Fig. 6
Distended ceca.

POULT ENTERITIS MORTALITY SYNDROME (PEMS)

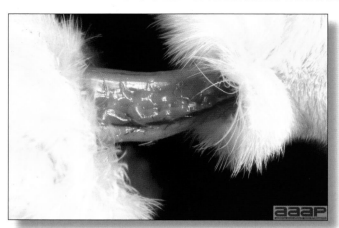

Fig. 7
Normal thymus in a poult. Note the thymic lobes running alongside the jugular vein.

Fig. 8
Atrophied thymus in a PEMS affected poult.

MUSCULOSKELETAL DISORDERS

It is generally agreed that skeletal disorders are a major source of economic loss in poultry. The specific cause of most leg problems is difficult to determine; they are often considered to have a genetic basis but may be influenced by or due to environmental or nutritional factors. With the advance of nutrition and the use of computer-generated diets, deficiencies responsible for arthroskeletal conditions are quite uncommon. However, most of the actual problems are associated with rapid growth and the incidence can be reduced by restricting growth rate.

I. ANGULAR BONE DEFORMITY
(Valgus-Varus Deformity of the Intertarsal Joint)

Angular bone deformity or valgus and varus deformity of the intertarsal joint is the most common form of long bone distortion found in broiler chickens and turkeys. There is lateral (Fig. 1) or medial (Fig. 2) angulation of the shaft of the distal tibiotarsal bone resulting in deviation of the lower part of the leg and frequent bending of the proximal shaft of the tarsometatarsus. Flattening of the tibial condyles and displacement of the gastrocnemius tendon may also occur. With severe angulation, the birds will walk on the hock joint with bruising of the area, ulceration of the overlying skin, and sometimes secondary infection. This condition results in significant trimming at processing.

II. CHONDRODYSTROPHY

Chondrodystrophy is a generalized disorder of the growth plate of long bones that impairs growth, while mineralization and appositional growth remain normal. It occurs in young growing poultry. In the past this condition was often described as perosis. Any condition, whether of genetic, nutritional, or environmental origin, resulting in a failure of physeal chondrocytes to proliferate can be called a chondrodystrophy. Chondrodystrophy results in shortened long bones, enlargement of hock joints, and often secondary valgus or varus deformity (Fig. 1) and subluxation of the gastrocnemius tendon.

III. CONTACT DERMATITIS OF FOOT PADS
(PODODERMATITIS OR BUMBLEFOOT)

This condition is characterized by a local injury to integument of the avian foot, usually the digital or plantar metatarsal pads, which lead to scab formation and inflammation of the subcutaneous tissues (Fig. 1). Common sequelae include tendonitis, septic arthritis, and osteomyelitis. Trauma, poor litter condition, increased ammonia and devitalization of the weight-bearing plantar structures are generally suggested to initiate the disease. Severe foot lesions result in lameness, reluctance to move, body weight depression and might lead to sternal bursitis (breast burn or breast blister), and a cause of carcass downgrading at slaughter.

IV. DEEP PECTORAL MYOPATHY

This condition, also named green muscle disease, is an exertional myopathy involving the supracoracoideus (deep pectoral) muscle (Fig. 1 and 2) of heavy meat-type birds. Vigorous wing beating increases subfascial pressure in this muscle and results in ischemic necrosis. The lesion is unilateral or bilateral and the macroscopic appearance varies according to the age of the condition. Earlier lesions consist of edema followed by hemorrhage and necrosis. In chronic cases, the necrotic muscle has contracted due to fibrosis and is uniformly green. The defect is then visibly apparent at the abattoir as a depression in the breast over the affected muscle and causes downgrading. Turkey leg edema is another condition occurring secondary to exertional myopathy, for example during transportation to slaughter.

V. FEMORAL HEAD NECROSIS
(FHN)

Femoral head necrosis (FHN) is a poorly defined and often inappropriately used term for numerous lesions of meat-type birds. It has been used for and/or confused with the following conditions (VI, VII, and VIII).

VI. IATROGENIC TRAUMA TO THE FEMORAL HEAD

During growth the femoral head is cartilaginous, and separation of the proximal femoral epiphysis from the femur on disarticulation of the coxofemoral joint is common during routine necropsy. The *teres* ligament and joint capsule frequently pull the articular cartilage and, occasionally, the femoral epiphysis detaches from the femoral shaft and remains in the acetabulum. This epiphyseal separation exposes dark, rough and pitted physes. This is not a lesion but an artifact. Epiphyseal separation may also occur spontaneously in live birds during rough handling and has been referred to as traumatic epiphyseolysis.

VII. OSTEOPOROSIS

With this condition, long bones are fragile and the growth plate is irregular; thus the femur is more susceptible to breakage during necropsy. No necrosis in the femoral head is present. Otherwise, any condition causing increased bone fragility, such as rickets, or the so-called malabsorption maldigestion syndrome, might result in shattering of fragile femoral necks during necropsy.

VIII. OSTEOMYELITIS

The physis of the proximal femur affected by bacterial osteomyelitis is fragile and disarticulation during routine necropsy may break the femoral neck. Small foci of osteomyelitis (Fig. 1) will be observed in the physes and metaphyses.

IX. OSTEOMYELITIS / SYNOVITIS

Osteomyelitis is usually one manifestation of a systemic disease. It occurs when, following a bacteremic episode, there is formation of an infective focus in the bone (see previous figure). *Escherichia coli* and *Staphylococcus aureus* are most commonly isolated; less commonly, *Salmonella*, *Yersinia*, *Streptococcus*, *Pasteurella*, and *Arizona* are cultured from these lesions. More recently, an increased incidence of osteomyelitis of the thoracic vertebrae associated with *Enterococcus cecorum* has been observed in broiler chickens and broiler breeders. The epiphyses of long bones, vertebral bodies, and associated joints are usually affected. Lesions in long bones consist of focal yellow areas of caseous exudate or lytic areas. Spondylitis can result in pressure on the spinal cord and paresis. Affected joints are swollen and filled with purulent exudates (Fig. 1). Treatment is rarely effective. Prevention is based on adequate treatment of septicemic diseases.

X. OSTEOPOROSIS
(Cage Layer Fatigue)

Osteoporosis refers to a decrease in bone volume but no loss in density and affects laying hens reared in cages and is most common at the end of a laying cycle. Clinical signs are variable and include posterior paralysis (Fig. 1) or acute death with or without changes in egg production. Paralyzed hens are initially alert and may be laying on their sides (Fig. 2). On postmortem examination birds have brittle, fragile bones, thin cortices (Fig. 3), and sometimes fractures. Sternae are often deformed and there is a characteristic infolding of the ribs at the costochondral junctions (Fig. 4). Parathyroid glands can be prominent or severely enlarged (Fig. 5). In the acute death form, an egg is present in the shell gland, with the shell partially or totally calcified and no macroscopic lesion. It is hypothesized that this acute form is due to acute hypocalcemia. Osteoporosis can be caused by a vitamin D$_3$, calcium, or phosphorus deficiency, or an imbalance in the calcium and phosphorus ratio. Adding extra vitamin D and calcium to the diet may be of some benefit. Lack of activity, strain of birds, and type of housing are considered to be important risk factors. Cage layer fatigue can be a problem in a flock, but bone breakage at processing due to osteoporosis may be a more significant problem in terms of economic losses and in respect to animal welfare.

XI. PEROSIS
(Slipped Tendon)

The term perosis describes the subluxated or slipped gastrocnemius tendon, which is secondary to long bone shortening or enlarged condyles of the hock joint caused by growth plate damage (chondrodystrophy). Nutritional deficiencies affect the development of the growth plate, the classical example being manganese deficiency, but, less frequently, choline, biotin, folic acid, niacin, or pyridoxine may be involved. The deformity involves the hocks of young birds and occurs more frequently in heavier birds. Marked malposition of one or both legs is observed from the hock(s) distally. In early cases, the hock is flattened, widened, and slightly enlarged. In advanced cases, the leg from the hock distally deviates sharply from its normal position, usually laterally. Dissection usually reveals that the gastrocnemius tendon at the hock has slipped from its trochlea.

XII. RICKETS
See section on Nutritional Diseases

XIII. SHAKY LEG

This is a severe lameness of 8-18-week-old turkeys. The etiology is unknown and specific lesions are absent. Affected birds spend most time sitting and if stimulated will walk with great difficulty, on "shaky legs". It is believed that lameness is triggered by soft tissue (muscle or tendon) pain. Most turkeys recover as growth slows, but other lesions, such as breast blister, might have developed secondary to the prolonged sitting. Shaky-leg as a flock problem secondary to inactivity caused by pododermatitis from wet litter has been reported.

XIV. SPLAY LEG

This condition occurs in young birds from hatching to 2 weeks of age. There is lateral deviation of the leg (Fig. 1), usually at the knee but occasionally at the hip. Splay leg may be unilateral or bilateral. The condition results from birds being on slippery surfaces. A high incidence has been seen in poults brooded on brown paper.

XV. SPONDYLOLISTHESIS

Spondylolisthesis affects broiler chickens and is characterized by posterior paresis and paralysis due to deformation and displacement of the fourth thoracic vertebra resulting in a pinched spinal cord (Fig. 1). It is considered to be a developmental problem influenced by conformation and growth rate. Affected birds are ataxic or may assume a hock-sitting posture with their feet slightly raised off the ground, and use their wings to move (Fig. 2). Severely affected birds often become laterally recumbent and die from dehydration if not culled. This condition has been referred to in the past as "kinky back". Osteomyelitis of the thoracic vertebrae will cause similar clinical signs.

XVI. TENDON AVULSION AND RUPTURE

Normal or excessive physical stress on the intertarsal joint of growing heavy meat-type birds frequently can causes rupture of the gastrocnemius or peroneus tendons. Swelling and red to green discoloration can be noted in the affected region (Fig. 1). It has been suggested that viral-induced tenosynovitis may predispose birds to a ruptured gastrocnemius but this condition has been observed without any evidence of prior tenosynovitis.

XVII. TIBIAL DYSCHONDROPLASIA

Dyschondroplasia is a very common growth-plate cartilage abnormality of fast-growing meat-type birds (broilers and turkeys) characterized by persistence of the cartilage with failure of removal of avascular prehypertrophying chondrocytes. The cause is multifactorial, but rapid growth and electrolyte imbalance leading to metabolic acidosis are considered primary risk factors. Common dietary causes of dyschondroplasia include excessive chloride and excessive phosphorus with respect to calcium levels. Certain *Fusarium* spp. mycotoxins also produce this lesion. The pathogenesis is poorly understood, but three mechanisms have been suggested: rapidly produced prehypertrophying chondrocytes do not hypertrophy as quickly as they are produced to allow penetration by metaphyseal vessels, an inadequate vascular invasion of the cartilage cannot initiate hypertrophy, or chondrolysis is defective. Without hypertrophy and vascular penetration, degeneration and calcification do not occur and a mass of prehypertrophying cartilage remains in the metaphysis. The lesion is characterized by an abnormal mass of cartilage below the growth plate (Fig. 1). If the mass is small, the condition will be subclinical. However, lameness will be observed if the mass of remaining cartilage is very large, the weakened bone may eventually bow or fracture (Fig. 2). The lesion is most commonly observed in the proximal tibiotarsus, probably because this bone is routinely sectioned during necropsy.

XVIII. TIBIAL ROTATION
(Twisted Leg)

This condition affects young broiler chickens and turkeys. There is lateral rotation of the distal tibia on its long axis, which results in lateral deviation of the lower leg (Fig. 1 and 2). Rotation is usually unilateral and can approach 90 degrees (Fig. 3). Morbidity is low (less than 1% in broiler chickens, but occasionally up to 5% in turkeys), and the cause is unknown. Tibial rotation must be differentiated from slipped tendon, because the tendon remains in place in tibial rotation.

XIX. TWISTED TOES

Twisted toes are common in heavier breeds. Most digits or a single digit may be bent laterally or medially (Fig. 1). There is no known economic significance, but abnormal weight bearing on the toes may cause ulceration followed by pododermatitis. This is different from the ventral curling of toe due to paralysis and secondary to riboflavin deficiency.

ANGULAR BONE DEFORMITY

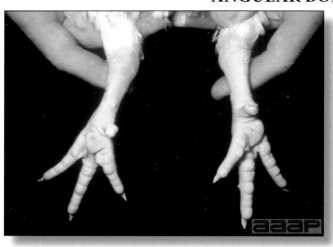

Fig. 1
Valgus deformity in a young broiler.

Fig. 2
Varus deformity in a turkey.

CHONDRODYSTROPHY

Fig. 1
Young turkeys with chondrodystrophy resulting in shortened long bones, enlargement of hock joints, and secondary valgus or varus deformity.

CONTACT DERMATITIS OF FOOT PADS

Fig. 1
Pododermatitis in a broiler breeder hen.

DEEP PECTORAL MYOPATHY

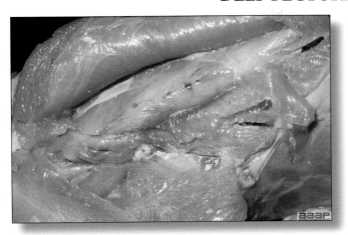

Fig. 1
Deep pectoral myopathy.

Fig. 2
Deep pectoral myopathy.

OSTEOMYELITIS

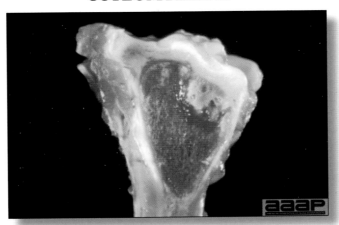

Fig. 1
Foci of osteomyelitis in the femoral head.

OSTEOMYELITIS / SYNOVITIS

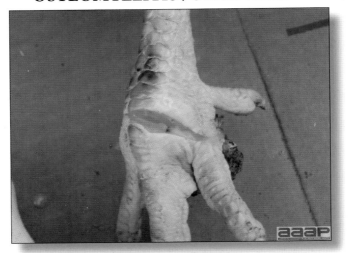

Fig. 1
Affected joint is swollen and filled with purulent exudate.

OSTEOPOROSIS

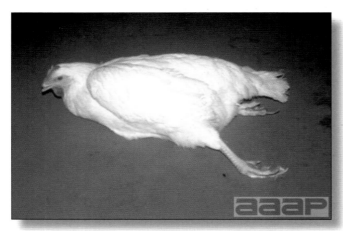

Fig. 1
Leghorn hen affected with cage layer fatigue demonstrating posterior paralysis.

Fig. 2
Acute paralysis; paralyzed hens are initially alert and may be laying on their sides.

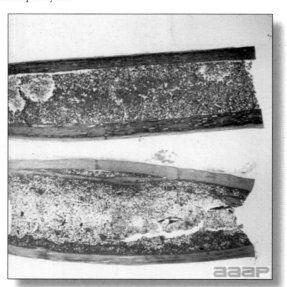

Fig. 3
Histological longitudinal section of the femur of a normal (top) and a laying hen with osteoporosis (bottom).

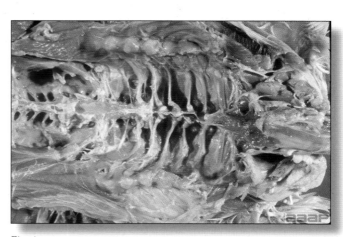

Fig. 4
Characteristic infolding of the ribs at the costochondral junctions.

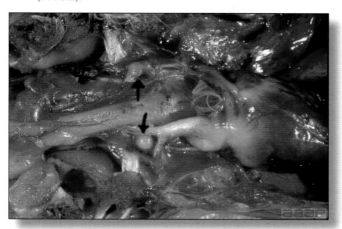

Fig. 5
Parathyroid glands can be prominent or severely enlarged (arrows).

SPLAY LEG

Fig. 1
Splay leg in a broiler chicken.

SPONDYLOLISTHESIS

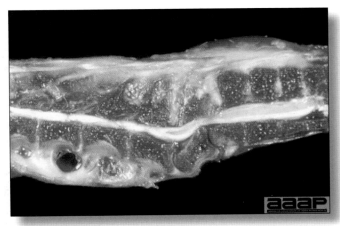

Fig. 1
Deformation and displacement of the fourth thoracic vertebra resulting in a pinched spinal cord.

Fig. 2
Affected birds are ataxic and assume a hock-sitting posture with their feet slightly raised off the ground, using their wings to move.

TENDON AVULSION AND RUPTURE

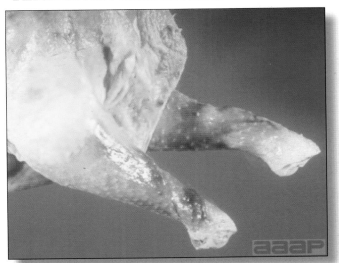

Fig. 1
Ruptured gastrocnemius tendon in a broiler chicken carcass at slaughter.

TIBIAL DYSCHONDROPLASIA

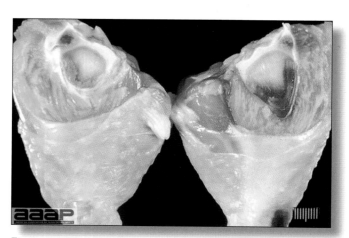

Fig. 1
Tibial dyschondroplasia is characterized by an abnormal mass of cartilage below the growth plate observed at sagittal section of the proximal tibia.

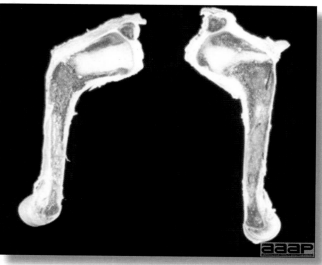

Fig. 2
If the mass of remaining cartilage is very large, the weakened bone may eventually bow or fracture.

TIBIAL ROTATION

Fig. 1
Broiler chicken with rotated tibia.

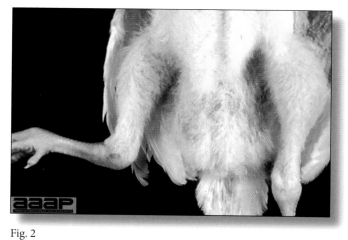

Fig. 2
Lateral rotation of the distal tibia on its long axis, which results in lateral deviation of the lower leg.

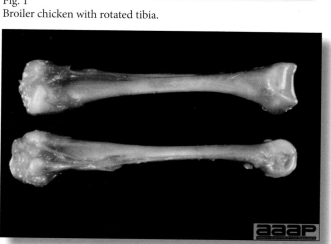

Fig. 3
Rotation is usually unilateral and can approach 90 degrees.

TWISTED TOES

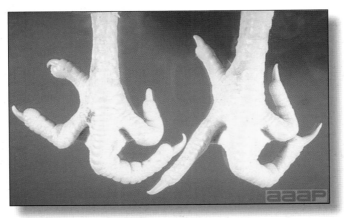

Fig. 1
Twisted toes in a 8 week-old broiler chicken.

REPRODUCTIVE DISORDERS

I. OVARIAN LESIONS

1. Atrophy and inactivity of the ovary is perhaps the most disturbing lesion in a hen of productive age. In the absence of other diseases, this finding is indicative of severe stress such as lack of feed and water and is often accompanied by neck or body molt and other evidence of difficulties including emaciation, dehydration, and so on.

2. Neoplasms affecting the ovary are fairly frequent and include involvement by Marek's disease, lymphoid leukosis, and myelocytomatosis. Adenocarcinomas, granulosa cell tumors, and arrhenoblastomas are also seen in hens. Adenocarcinomas are particularly common and are striking in their presentation because of numerous trancoelomic implants on the surfaces of abdominal organs. Ovarian neoplasms are generally readily identified by histopathological evaluation.

3. Oophoritis or follicular regression is associated with a variety of infectious diseases. The normally turgid yellow ovules become wrinkled, hemorrhagic, or discolored (green, gray-yellow, beige, etc.) (Fig. 1 and 2) and many times there is evidence of premature rupture with spillage of yolk material into the abdominal cavity. Follicular regression and rupture with abdominal yolk material is also a frequent finding in extreme dehydration. Among the infectious diseases notorious for this ovarian effect are velogenic Newcastle disease, avian influenza, fowl cholera, pullorum disease, fowl typhoid, and some strains of *Salmonella Enteritidis* and *Escherichia coli.*

4. Ovarian cysts are occasionally encountered in laying hens, sometimes in active functional ovaries. Usually these cysts are very thin walled and contain clear fluids. It is frequently difficult histologically to identify the origin of ovarian cysts but it is presumed that most are of follicular origin.

II. OVIDUCT LESIONS

1. Atresia, hypoplasia, and atrophy of the oviduct have been documented in hens of laying age. Simple atrophy associated with severe stress, chronic infection, certain intoxications, etc., is most common. Hypoplasia as a result of early infectious bronchitis in sexually immature pullets may lead to "false layer" status in hens with fully developed ovaries and partially developed oviducts. Many of these have the external appearance of active layers but yolk material is deposited in the abdominal cavity leading to the classical "yolk peritonitis". Hereditary atresia of the oviduct has been reported and the condition may affect segments or the entire oviduct. Again yolks and any surrounding albumin deposits cannot pass and are refluxed back into the abdominal cavity.

2. Neoplasms of the oviduct *per se* are rare or are rarely recognized. Adenocarcinomas of the oviduct arise predominantly in the upper magnum and tend to be highly invasive, frequently resulting in widespread peritoneal implantation. The much more common tumor occurs in the mesosalpinx arising from smooth muscle in the center of the ligament. These leiomyomas are always firm and encapsulated and vary in size from barely detectable to several centimeters in diameter. Large leiomyomas may occasionally interfere with oviduct function.

3. Salpingitis is usually seen as a sporadic individual bird problem, although, in flocks with *Mycoplasma* infection, *Salmonella* infections, and some outbreaks of pasteurellosis and colibacillosis, this lesion may occur with substantial incidence. Infections affecting the left greater abdominal air sac or the peritoneal cavity have the potential for extension into the oviduct as a descending inflammatory process in this tubular organ. In many cases, however, oviductal infections appear to originate in the distal extremity, suggesting that infection ascends from the cloacal orifice. In the early stages of salpingitis the only apparent changes may be irregularities in the mucosal surfaces including erosions or small ulcerations, edema of the mucosal folds, and accumulation of adherent fibrinopurulent exudate. As the lesion progresses the amount of exudate in the lumen increases rapidly and, ultimately, the oviduct becomes an irregular thin-walled sac filled with laminated masses of yellow cheesy exudate (Fig. 1). The oviduct becomes nonfunctional early in the infection and the ovaries of affected hens are usually atrophied. It is however possible, that some of the oviductal content in salpingitis represents impacted egg components (Fig. 2). Apart from specific bacterial infections noted above, which might be present in the early stages of salpingitis, the luminal exudate in terminal stages may yield a wide array of bacteria and even fungi.

4. Impacted or egg-bound oviducts are seen sporadically in cull hens. This condition may be more prevalent in pullets that were brought into production too early (prior to full body development) or hens that are extremely obese. Whether oviduct obstruction results from a lodged egg or an intermingled mass of broken shells, shell membranes, or aggregates of coagulated albumin and yolk, the result is the same. When impaction occurs in the uterus or vagina (which is usually the case) eggs enclosed by shell membranes may be found in the abdominal cavity. This indicates that eggs continued to form but were refluxed back into the peritoneal cavity. Hens with numerous abdominal eggs may assume a penguinlike posture.

5. Cystic oviducts are seldom clinically significant but cystic remnants of the right oviduct are very prevalent. In the chicken only the left components of a paired embryonal reproductive system develop after

hatching. Segments of the right oviduct may develop to varying degrees. These segments are closed (no anterior or posterior drainage) and, if the wall contains significant glandular tissue, a fluid secretion will accumulate resulting in production of cysts. These cysts usually occur on the right side adjacent to the cloaca and range from barely perceptible in size to massive cysts occupying most of the abdominal cavity. Affected hens appear to have ascites (water belly) but if the abdomen is opened carefully the fluid is found to be contained within a cyst. Cysts have also been described in the left oviduct but they are much less common. Left oviductal cysts may form in segments of the oviduct proper, in the wall of the oviduct or in the mesosalpinx adjacent to the oviduct. Oviductal cysts may be intriguing postmortem findings but they rarely, if ever, have any significant impact on flock performance.

6. Prolapse or eversion of the terminal oviduct can occur with alarming prevalence in some layer flocks and ranges from barely perceptible everted vaginal mucosa protruding from the vent to elongate exteriorized segments of prolapsed oviduct (Fig. 3). Circumstances in which this problem is most severe include young poorly developed pullets just coming into egg production, flocks with higher than recommended cage densities, increased levels of cannibalism associated with hyperactive flocks (hysteria) or flocks exposed to a sudden surge of increased light intensity (as in open houses in the first bright days of spring), increasing levels of obesity, and flocks that were poorly beak trimmed with substantial regrowth of beak tips. When oviposition occurs there is a normal eversion of the uterus or vaginal mucosa surrounding the egg as it is delivered. In poorly developed or obese birds this everted mucosa may be slow to retract afterwards. If there is any increased tendency toward cannibalism in the flock, cage- or pen-mates will peck at the everted mucosa causing trauma and edema, which will further slow or prevent retraction. Continuing irritation of the exposed mucosa may promote straining and overt prolapse of the oviduct. An association has also been observed between a tendency for eversion and prolapse and an increased incidence of salpingitis. Also losses from cannibalism and culling are increased in flocks with a high rate of uterine/vaginal prolapse. Control of this condition can be achieved to some degree by allowing full development of pullets before bringing them into lay, maintaining proper stocking density of cage or floor houses, careful control of lighting intensity, proper beak trimming, and maintaining feed formulations and consumption levels to avoid obesity, especially in older hens.

OVARIAN LESIONS

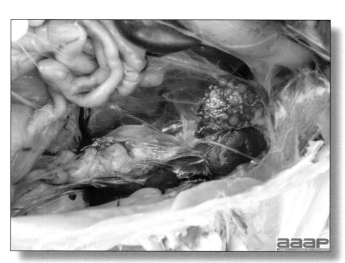

Fig. 1
Atretic ovary and oviduct.

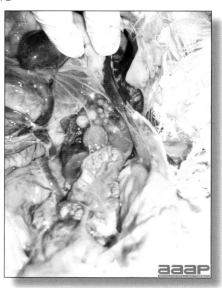

Fig. 2
Atretic ovary and oviduct.

OVIDUCT LESIONS

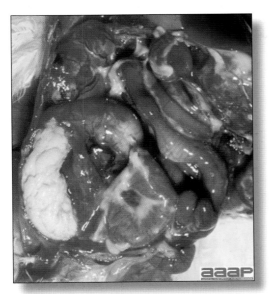

Fig. 1
Salpingitis: fibrinopurulent exudate present in the lumen of the oviduct with peritonitis.

Fig. 2
Salpingitis: compacted egg components and caseous exudate in oviduct.

Fig. 3
Laying hen dead from prolapsed oviduct.

URINARY DISORDERS

UROLITHIASIS
(Nephrosis, Renal Gout, Caged Layer Nephritis)

DEFINITION
Urolithiasis is an etiologically undefined condition seen particularly in caged laying hens and characterized by blockage of one or both ureters by urate concretions with attendant atrophy of one or more lobes of the kidney drained by the obstructed ureter.

OCCURRENCE
This condition has been recognized for years as a sporadic individual bird problem in laying flocks. More recently urolithiasis has been described as a flock problem accounting for substantial mortality in caged layers in England, the United States, and other countries throughout the world.

ETIOLOGY
A number of causative factors have been implicated in precipitating urate deposits in kidneys, joints, or on serosal surfaces throughout the body. These include dehydration, excessive dietary protein (30-40%), dietary calcium excess (3% or greater), sodium bicarbonate toxicity, mycotoxins (oosporin, ochratoxin), vitamin A deficiency, and nephrotropic strains of infectious bronchitis virus. However, the recently described urolithiasis in caged layers appears to be associated with feeding relatively high calcium levels (3% or greater) during the pullet grow-out period. Available phosphorus in the grower ration appears to be contributory in that urolithiasis is enhanced when levels are below 0.6%. Many clinicians feel that infectious bronchitis viruses, or even 'hot' vaccine strains might be involved in the process and there also is evidence that dietary electrolyte imbalances (low sodium and potassium, high chlorides) may play a role. Finally, many diagnosticians consider all current etiologic explanations of this condition to be unsubstantiated, or at best, poorly supported hypotheses.

CLINICAL SIGNS
In many cases of urolithiasis there is no consistent clinical sign other than increased mortality. Among signs associated with the condition are depression, weight loss, and an inclination of affected birds to hide. Roughened or thin eggshells may increase slightly in affected flocks and total egg production will decrease in parallel to increasing mortality. Mortality may be gradual and persistent (2-4% per month) throughout the productive lifetime of the hens or it may be more precipitous. Total mortality has approached 50% in severely affected flocks.

LESIONS
The affected ureter is usually markedly distended by cylindrical concretions surrounded by thick mucus.

Although usually unilateral, both ureters may be involved. One or more lobes of the kidney drained by the obstructed ureter often are severely atrophied (Fig. 1). The opposite functional kidney may be hypertrophied. Many affected hens will have white chalky material (urate deposits) on serosal membranes of various visceral organs.

DIAGNOSIS
Diagnosis is based on classical ureteral and renal lesions in most of dead birds necropsied. Observation of urolithiasis in an occasional dead bird is indicative of a sporadic individual bird problem and is of little consequence. Confirmation of etiologic factors noted above is usually difficult unless feed samples have been retained for analysis.

CONTROL
Until etiologic factors are better defined, it is difficult to make specific recommendations. It is advisable to respect limits of calcium and available phosphorus in rations during grow-out and to avoid electrolyte imbalance, mycotoxins and water deprivation.

Fig. 1
Nephrotropic Infectious bronchitis strain; Renal hypertrophy with atrophy of some lobes.

INTEGUMENT DISORDERS

I. KERATOCONJUNCTIVITIS
(Ammonia Burn)

Keratoconjunctivitis is an inflammation caused by excessive levels of ammonia in poorly ventilated poultry houses. Chickens appears to be more susceptible to ammonia toxicity than turkeys. Lesions include keratitis, conjunctivitis, and a corneal opacity with a possible ulceration (Fig. 1). Birds may become blind, but recovery is possible depending on the severity of damage to the cornea. Because ammonia is produced by the degradation of uric acid by bacteria in the litter, control of litter moisture and proper ventilation will prevent this problem.

Fig. 1
Broiler breeder male with corneal opacity and ulceration following exposure to high environmental ammonia levels.

II. SCABBY HIP SYNDROME

Scabby hip syndrome is a lesion observed at the slaughter plant in broiler chickens and is characterized by superficial ulceration and scabbing of skin on the thighs (Fig. 1). This is a multifactorial problem; poor feathering, high stocking density, and poor litter conditions have been incriminated. Affected carcasses are downgraded. In recent years, improvements in litter management and use of nipple drinkers have contributed to the reduction in incidence of this condition.

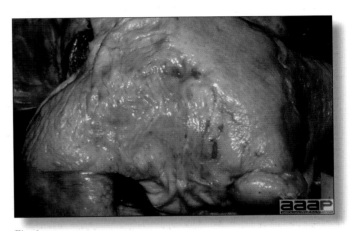

Fig. 1
Scabby hip in a broiler chicken carcass at slaughter.

III. STERNAL BURSITIS
BREAST BLISTER or BREAST
BURN or BREAST BUTTON

Sternal bursitis is a fluid-filled lesion located on the ventral aspect of the keel bone of poultry (Fig. 1). Chickens and turkeys have a synovial bursa, the sternal bursa, which under repeated trauma increases in size and may become secondarily infected. This lesion is closely associated with locomotor problems in heavy birds and increased contact time with litter. Such blisters, if not too large, are trimmed from the carcass at processing resulting in downgrading. The terms breast blister, breast button, breast burn are also used for this condition.

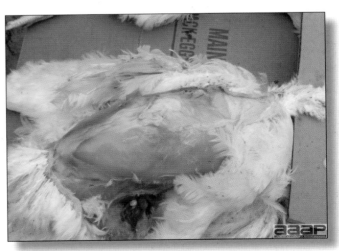

Fig.1
Sternal bursitis: fluid-filled lesion located on the ventral aspect of the keel bone of a broiler breeder following prolonged ventral decubitus.

IV. XANTHOMATOSIS

Xanthomatosis is an unusual condition characterized by the abnormal subcutaneous intracellular accumulation of cholesterol clefts mixed with multinucleated giant cells and macrophages, in chickens. Skin lesions are initially soft with fluctuating honey-colored fluid, and later become firm with marked thickening and irregularity of the surface (Fig. 1). This condition is rare and in the past was probably due to contamination of feed fat with hydrocarbons.

But xanthomatosis probably representing fat necrosis and granuloma formation can occur in various parts of the body in various species of birds including poultry.

Fig. 1
Xanthomatosis characterized by an irregular surface and thickened skin.

BEHAVIOR DISORDERS

I. CANNIBALISM

This vice can cause severe losses in some poultry flocks. Many forms are described, the most common ones being feather pulling, and vent, head (Fig. 1), and toe picking. Feather pulling occurs in any age of bird. In severe cases birds will die of hemorrhage and carcasses will be picked and eaten by pen-mates. Vent picking in cage-reared laying hens is most common in overweight birds. There is a normal eversion and prolapse of the vagina at lay. If the hen is obese, the vaginal mucosa will be exposed for a prolonged time and cage-mates will be attracted by this shiny red mucosa. The assaulted hen will bleed to death and dried blood will be present on the feathers of the pericloacal area and on the back of the legs. A pecking order has often to be established in a poultry flock and head lesions are common in turkeys, even though birds have trimmed beaks. Toe picking occurs in young chicks started on paper or in batteries, and is often initiated by hunger. Several predisposing factors such as light intensity, dense stocking, reduced animal protein feed content, lack of vitamins, amino acids, or salt in feed, sodium imbalance due to heat stress, being without feed for too long, and irritation from external parasites have all been mentioned. Recent work has shown that feather pecking may be related to low levels of insoluble fiber in the diet. Once a bird develops this habit, it will continue. Outbreaks can be sometimes controlled by using red light or reducing light intensity. In laying stock and commercial turkey flocks, trimming a third of the upper beak with laser or electrocautery is a widely used preventive measure.

II. HYSTERIA

Sporadic cases of broiler chicken and replacement pullet flocks with extremely high activity levels have been reported. The cause is unknown but tryptophan supplementation appears to alleviate the problem.

Fig. 1
Head pecking in a broiler breeder hen.

MANAGEMENT-RELATED DISORDERS

I. DEHYDRATION/STARVATION OF CHICKS/POULTS

Losses due to so-called "normal mortality" should not be more than 1% in the first 10 days of the growout period. Any mortality higher than this should be investigated

DEFINITION
Dehydration and starvation are the most common causes of mortality in chicks and poults during their first week of life. Young birds which cannot find water nor feed will eventually die of starvation and inanition, once the yolk has been absorbed i.e., before the fifth day. But dehydration and starvation can also occur at any age and in various species of birds.

ETIOLOGY
Failure to eat and/or drink can be related to farm management conditions. Since recently hatched chicks/poults are poikilothermic, optimal environmental temperatures are a must for the brooding period. A comfort zone i.e., an area where environmental temperature is ideal for the chick/poult must be established inside the barn. Check temperature charts since this temperature varies according to age and species. Feed and water must be located in the bird's comfort zone in a brightly lit area (60 to 100 lux), in order for the young bird to access them.

Severe diarrhea and high environmental temperatures can also cause dehydration.

CLINICAL SIGNS
Birds that die of dehydration or starvation do not usually show other signs of illness than weakness before death. Bear in mind that uncomfortable chicks/poults will be noisy before becoming depressed. Affected individuals are also smaller.

LESIONS
Dehydrated carcasses are light with darker feet and beak. Legs appear thinner with a prominent metatarsal vein. The skin adheres tightly to dark pectoral muscles. Upon opening the coelomic cavity, white chalky material (urates deposits) can be observed on various serosal surfaces (Fig. 1, 2 and 3) including joints (Fig. 4). Kidneys are often pale and enlarged with increased urates and ureters are often dilated with urates (Fig. 5). There is none or very little feed in the gizzard.

DIAGNOSIS
Based on history and lesions.

TREATMENT AND CONTROL
Check temperature, luminosity and bird's distribution in the brooding area as well as water and feed availability. Cold birds will huddle together while hot birds will be panting and lying on their belly, often too weak to be interested in finding the water.

II. HYPOGLYCEMIA-SPIKING MORTALITY SYNDROME IN BROILER CHICKENS

DEFINITION
Hypoglycemia-spiking mortality syndrome (HSMS) is characterized by a sudden increase in mortality (>0.5%) for at least 3 consecutive days, in a previously healthy, normal appearing broiler chicken flock aged between 7 and 21 days of age. Two clinical forms have been described; type A more severe but of short duration and type B, a milder form occurring over a longer period.

ETIOLOGY
Although the disease has been reproduced with tissue homogenates and viral particles have been identified in affected birds, the etiology is still unknown and the causal agent remains to be identified. Clinical signs and death are caused by hypoglycemia. Hypoglycemia could either be explained by a virus blocking pancreatic glucagon production or hypothetically related to melatonin deficiency and associated glycogenolysis. Melatonin deficiency could be caused by a lack of a long dark period. Stress and/or acute fasting could trigger HSMS in either situation.

CLINICAL SIGNS
Flock experiences a rapid, unexplained increase in mortality, which will decrease as quickly in a matter of a few days. Live chicks are found recumbent and uncoordinated (Fig. 1), frequently lying on their breasts with legs extended. Evidence of blindness and hyperexcitability can be seen. Death occurs rapidly, often within a few hours. Blood glucose levels are lower than <150mg/dL. Birds with very low levels from undetectable to less than 60 mg/dL frequently occur.

LESIONS
There are no specific gross or microscopic lesions. Birds appear normal and typically have food in the crop. Infrequently sinusoidal congestion or small hemorrhages are seen in the liver (Fig. 2).

DIAGNOSIS
Mortality pattern (a high spike in a mortality curve at 7-21 days of age) and low blood glucose levels in clinically affected birds are diagnostic. If a chemical analyzer is not easily accessible, glucose test strips and hand held monitor can measure fairly accurately blood glucose levels in the

field. TheraSense FreeStyle glucose meter (Abbott Labs) works with avian blood (Note: other glucometers also may work, but have not been tried or reported).

CONTROL AND TREATMENT
Although it is important to give a 24 hour period of full light to day-old chicks, a progressively decreasing day length resulting in a long daily dark period will usually prevent this problem.

III. HEAT STRESS and HYPERTHERMIA

High temperatures are stressful for poultry and frequently cause death from hyperthermia. Millions of birds die each year from hyperthermia usually because of high environmental temperature, but also because of electric power failure in closed buildings. Birds do not have sweat glands and thermoregulate via non-evaporative cooling (radiation, conduction and convection). Effect of ambient temperature on body temperature varies with body heat production which is directly related to body mass and feed intake (metabolism). If panting fails to prevent increase in body temperature birds will become depressed, then comatose, and soon die. Lethal internal high body temperature is 116°F for chicks and 117°F for adult birds. Dead birds are usually found on their breast, in good body condition. Breast muscles may have a cooked, pale appearance. Prevention of hyperthermia is based mainly on proper building insulation, optimal ventilation and evaporation techniques, feed removal early in the day to reduce metabolic heat production and adequate drinking water availability.

Panting and increased respiratory rates affect acid-base balance and cause respiratory acidosis. Higher blood pH will reduce plasma ionized calcium, which is needed for eggshell formation, hence the risk for increased thin-shelled eggs in summertime laying flocks.

IV. VACCINE REACTION (ROLLING REACTION)

DEFINITION
A normal respiratory (Newcastle disease or infectious bronchitis) vaccine reaction occurs within the week after hatchery vaccination. However, if environmental conditions are poor, or the flock is infected with vertically transmitted *Mycoplasma spp.*, this reaction might aggravate with possible secondary *Escherichia coli* or *Mycoplasma spp.* infection.

CLINICAL SIGNS AND LESIONS
Affected chicks will show head shaking, wet eyes with nasal discharge and mild coughing or sneezing (Fig. 1). Chicks will appear depressed and will huddle together or under a heat source. There will be increased mortality, growth retardation and loss of flock uniformity. At necropsy there will be serous to caseous exudates in the upper respiratory tract with airsacculitis in the case of secondary bacterial infection.

DIAGNOSIS
Diagnosis is based on poor environmental conditions, clinical signs and lesions. ELISA antibody titers to infectious bronchitis are within normal.

TREATMENT AND PREVENTION
Chicks must be provided with optimal environmental temperature and rearing conditions. Antibiotics can be administered if there is secondary bacterial infection.

DEHYDRATION/STARVATION OF CHICKS/POULTS

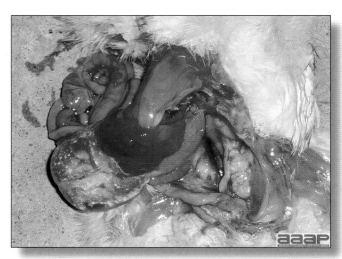

Fig. 1
Dehydrated chick: white chalky material (urates deposits) present on various serosal surfaces. Note the pale kidneys and dilated ureters.

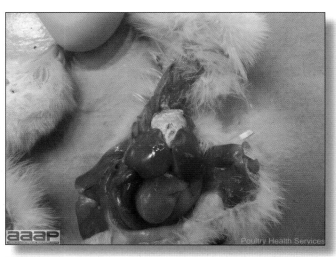

Fig. 2
Urate deposits on pericardium.

Fig. 3
Urate deposits on muscle surfaces.

Fig. 4
Presence of urate deposits in the joint.

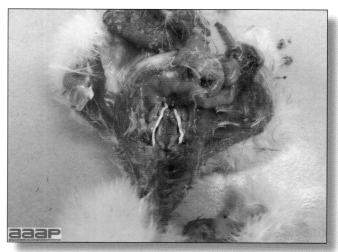

Fig. 5
Pale and enlarged kidneys with bilateral ureter dilation.

HYPOGLYCEMIA-SPIKING MORTALITY SYNDROME IN BROILER CHICKENS

Fig. 1
Hypoglycemia in live chicks.

Fig. 2
Small hemorrhages in the liver.

VACCINE REACTION

Fig. 1
Young chicken with ocular discharge.

DISEASES OF THE DUCK

Written by Peter R. Woolcock and Martine Boulianne

Name of Disease(s)	Etiology	Susceptibility	Clinical signs and lesions	Comments
Aspergillosis	Usually *Aspergillus fumigatus*.	Young birds are more susceptible than adults	Various CNS signs. Respiratory signs usually precede CNS signs and predominate. A minority develop CNS involvement.	Yellow mycotic nodules often grossly visible in brain. Lesions of aspergillosis in lungs or air sacs, perhaps in conjunctival sac or globes of eyes.
Avian botulism (limberneck)	Ingestion of toxin produced by the bacterium *Clostridium botulinum*	Most species of birds	Paralytic, often fatal disease, characterized by ascending paresis and paralysis. No gross lesions	Important environmental factors contribute to the initiation of avian botulism, outbreaks then perpetuated by the bird-maggot cycle.
Avian influenza (AI)	Orthomyxovirus (strains vary greatly in pathogenicity).	Turkeys, ducks, pheasants, quail, many wild birds, other poultry. Not a significant pathogen of ducks. Ducks are believed to be a major reservoir for avian influenza virus. H5N1 Hong Kong (2002) lethal for wild waterfowl.	Highly variable. In mild form: often swollen sinuses, ocular or nasal discharge. In severe form: hemorrhages, exudation, focal necrosis in respiratory, digestive, urogenital, cardiovascular, or multiple systems.	Enzootic forms in U.S. usually mild to moderate in severity and involve respiratory system. Egg production declines and shell abnormalities common in turkeys. Most outbreaks of AI in U.S. are in turkeys and ducks. Hong Kong 2003, first report of HPAI H5N1 causing deaths in resident and migratory water-fowl. 460 geese died from H5N1 in Xinjiang, China (2005). Now also in Europe and Africa.
Avian Paramyxovirus 1 (Newcastle Disease) and other APMV's (4, 6, 8, 9)	Paramyxovirus 1 and other APMV's (4, 6, 8, 9)	Presumably all ages but not recognized clinically other than in laying ducks. VVND seen in geese in China.	Drop in egg production, birds go into moult. Essex '70 strain will kill experimentally infected ducklings. In geese VVND pathology and high mortality (China).	Rarely diagnosed in ducks. No outbreak reported in ducks during whole of Essex '70 UK epidemic. Exotic NDV in US (Oct 2002-2003, no reports in ducks).
Avian tuberculosis	*Mycobacterium avium*	All avian species	Round nodules (granulomas) attached to serosa of gut. Focal granulomas in many other organs. Extreme emaciation in advanced cases.	Usually seen in older birds. Can be a problem is established in captive flocks. Causative bacilli readily demonstrated in acid-fast-stained smears of lesions.

Name of Disease(s)	Etiology	Susceptibility	Clinical signs and lesions	Comments
Chlamydophilosis/ Chlamydiosis (psittacosis, ornithosis)	*Chlamydophila psittaci*	Turkeys, pigeons, ducks, cage and wild birds.	Can be acute, subacute or inapparent. Pericarditis, often adhesive. Also with airsacculitis, fibrinous perihepatitis, splenomegaly, and hepatomegaly.	This is a zoonosis and a public health concern.
Colibacillosis	*Escherichia coli* septicemia.	Turkeys, chickens, commercial ducks.	Pericarditis, often adhesive.	Often with airsacculitis, fibrinous perihepatitis.
Duck viral hepatitis type I (DHV-1) Now renamed DHAV-1, DHAV-2 and DHAV-3 identified in China, South Korea & Taiwan	Picornavirus Genus: Avihepatovirus	Ducklings, typically less than 4 weeks old.	Signs: acute onset, short course, high morbidity and mortality. Liver swollen and with many hemorrhages.	Typical lesions in young ducklings almost pathognomonic. Worldwide incidence. Economically important disease for commercial duck industry.
Duck viral hepatitis type II	Astrovirus	Ducklings up to 6 weeks. Complicated by DHV type I and septicemia. Adults not susceptible.	Liver hemorrhages, swollen kidneys.	As DHV type I, deaths may be slightly slower. Only reported in UK. May see extensive biliary hyperplasia in liver sections, cf. aflatoxicosis. Also reported in China.
Duck viral hepatitis type III	Astrovirus	Ducklings up to 3-5 wks. Complicated by DHV type I and septicemia. Adults not susceptible.	Liver hemorrhages	As DHV type I, lower mortality, higher morbidity. Only reported in USA.
Duck virus enteritis (duck plague)	Herpesvirus	Mostly adult ducks but also from 2 wks of age. (Also in geese and swans).	Lesions of vascular damage on mucosa of GIT, lymphoid organs & parenchymatous organs: Widespread hemorrhages, severe enteritis. Perhaps elevated plaques in esophagus, ceca, cloaca or bursa. Hemorrhage and/or necrosis in lymphoid rings or discs of gut.	Epizootic losses in waterfowl are suggestive. Typical lesions and inclusion bodies helpful in diagnosis. An exotic, reportable disease.

Name of Disease(s)	Etiology	Susceptibility	Clinical signs and lesions	Comments
Eastern equine encephalomyelitis	Alphaviruses	Pheasants, partridges, turkeys, Pekin ducks, quail.	Circling or staggering followed by paralysis. Perhaps blindness in recovered birds. Often high morbidity and mortality.	Microscopic lesions only. Transmitted by mosquitoes.
Fowl cholera (Avian cholera or avian pasteurellosis)	*Pasteurella multocida*	Water birds, domestic poultry, game birds.	Sick birds appear lethargic or drowsy. Abnormal positions of head and neck. Ataxia, loss of equilibrium. Lesions of septicemia (hemorrhages and petechiation on serosal surfaces) and multiple foci of necrosis in liver. Birds are often in good body condition.	Acute infections are common and may result in rapid death and 'explosive' die-offs. Chronic infections with lower mortality can also occur.
Goose Parvovirus (GPV) Other names:Derzsy's Disease, Goose influenza, Goose plague,	Parvovirus	In geese & Muscovy ducks only. 0-4 wk. Younger birds more susceptible. High mortality in young goslings (1-10 days). Adults not susceptible, but do respond immunologically.	< 1 wk - anorexia, prostration & rapid death. Older - anorexia, polydipsia and weakness. Nasal and ocular discharge, head shaking. Eyelids red and swollen, profuse white diarrhea. Survivors may be stunted. Pale myocardium, congested, swollen liver, spleen & pancreas.	In China since 1956. E Europe & France in 1960's. Not thought to be present in USA - considered a foreign animal disease.
Leukocytozoonosis	*Leukocytozoon* sp.	Turkeys, ducks, geese, guinea fowl, chickens.	Pallor, splenomegaly, liver degeneration and hypertrophy in some birds. Leukocytozoons visible in blood smears. Schizonts often in liver, spleen, brain.	Outbreaks correspond with hot months when simulid flies and culicoid midges are numerous; these flies breed in and along water courses. Surviving birds (wild or domesticated) often act as carriers. Signs in birds related to anemia.
Muscovy duck parvovirus	Parvovirus	Young Muscovy ducklings only	Similar to Goose parvovirus, but skeletal muscle maybe pale and more affected.	First in W France 1989. Also in Japan and Europe. USA 1998. As yet no vaccines available in USA

Name of Disease(s)	Etiology	Susceptibility	Clinical signs and lesions	Comments
Renal coccidiosis	*Eimeria* are species-specific and many cause renal coccidiosis. Duck: *E. boschadi, E. somatarie* Geese: *E. truncata* Swan: *E. christianseni*	Chickens, turkeys, ducks, geese, perhaps in most birds.	Infected birds may be emaciated and have a prominent keel. In severe infections, kidneys may become enlarged and pale, containing multiple spots or foci of infection that coalesce into a mottled pattern. Most reports of renal coccidiosis are of asymptomatic birds or birds that show minor physiological or pathological changes due to the parasite.	Part of the parasite life cycle occurs in the kidney. Young birds and those that have been stressed by various conditions are most likely to have clinical cases of renal coccidiosis. Mortality has occurred in free-ranging wild geese, eider ducklings, and double-crested cormorants. Disease in domestic geese is usually acute, lasts only 2–3 days, and can kill large segments of the flock.
Riemerella anatipestifer New duck disease, duck septicemia	*Riemerella anatipestifer*	Ducks of any age and turkeys. Other waterfowl, chickens and pheasants may also be affected. 2-7 week-old ducks are most susceptible	Ocular and nasal discharges, mild sneezing, tremors of the head and neck, ataxia and coma. Lesions of septicemia, mostly fibrinous exudate on serosal surfaces. Mortality may vary from 2 to 75% in young ducks.	Worldwide, probably the most economically important infectious disease of farm ducks. Prevention with good biosecurity, husbandry and hygiene, control with depopulation and disinfection of the infected premises. Various antibiotics have been shown to be effective. Vaccines available in some countries.
Sarcocystis	*Sarcocystis rileyi*	Macroscopic form observed in dabbling ducks, less frequent in other species of ducks, geese and swans. Most often adult birds.	No a cause of mortality but heavy infection may cause muscle loss and be quite debilitating. Multiple cream-colored, cylindrical cysts resembling rice grains run parallel to the muscle fibers of various muscles.	Common parasitic infection of some species of waterfowl in North America. Sarcocystis is destroyed by cooking. Affected carcasses should therefore be discarded not only because of the unaesthetic appearance of the muscles but because very little is known about the health hazard to humans.

Name of Disease(s)	Etiology	Susceptibility	Clinical signs and lesions	Comments
West Nile virus	Flavivirus	Ducks, geese	Death in wild birds. Brain hemorrhages, splenitis, splenomegaly, nephritis.	Reported in New York 1999, Israel and Romania. Commercial geese in Canada 2002, ~25% mortality in goslings. Bird to bird transmission reported but mostly transmitted by mosquitoes.
Wet pox	Poxvirus	Most birds, including poultry.	1-5 mm yellow-gray plaques in mucosa of mouth, pharynx, or esophagus. Less often in sinuses or conjunctiva.	Skin lesions often on face, wattles, eyelids, comb, feet, legs, ear lobes, caruncle, snood.

DISEASES OF THE GAMEBIRDS
Written by Eva Wallner-Pendleton

Name of Disease(s)	Etiology	Susceptibility	Clinical signs and lesions	Comments
Aspergillosis (Brooder Pneumonia)	Usually *Aspergillus fumigatus*	Young birds are more susceptible than adults	Respiratory signs usually precede CNS signs and predominate. A minority develop CNS involvement.	Yellow mycotic nodules may be observed in the lungs, airsacs, tracheal bifurcation, sinuses, and occasionally in the brain or eyes.
Avian botulism (Limberneck)	Ingestion of toxin produced by the bacterium *Clostridium botulinum*	Most species of birds	Paralytic, often fatal disease, characterized by ascending paresis and paralysis. No gross lesions	Important environmental factors (failure to pick up dead birds in the flight pens) contribute to the initiation of avian botulism outbreaks then perpetuated by the bird-maggot cycle.
Avian tuberculosis	*Mycobacterium avium*	All avian species	Round nodules (granulomas) attached to serosa of gut. Focal granulomas in many other organs. Extreme emaciation in advanced cases.	Usually seen in older birds. Can be a problem is established in captive flocks. Causative bacilli readily demonstrated in acid-fast-stained smears of lesions.
Capillaria annulata (Crop threadworm)	*Capillaria annulata*	Pheasants, partridges, quail, turkeys, grouse, guinea fowl	Severe ingluvitis seen in oesophagus and crop. Drooling and weight loss. Can also cause mortality.	Earthworms also serve as intermediate hosts. Scrapings from crop mucosa examined microscopically show characteristic ova.
Colibacillosis	*Escherichia coli*	All avian species are susceptible.	Highly variable mortality and morbidity may be present. Septicemia and polyserositis are common.	A common disease secondary to stress, poor environmental conditions and previous exposure to other infectious diseases of the respiratory and GI tract.
Blackhead	*Histomonas meleagridis*	Turkeys, chickens, pheasants, chukar partridges, quail and peafowl	Yellow diarrhea and biliverdiurea, weight loss. Severe necrotic typhilitis and hepatitis seen.	Frequent worming to prevent heavy cecal worm infections. Histostat can also be used preventatively in the feed.
Eastern equine encephalomyelitis	Alphaviruses	Pheasants, partridges, turkeys, Pekin ducks, quail.	Circling or staggering followed by paralysis. Perhaps blindness in recovered birds. Often high morbidity and mortality.	Microscopic lesions only. Transmitted by mosquitoes.

Name of Disease(s)	Etiology	Susceptibility	Clinical signs and lesions	Comments
Fowl cholera (Avian cholera or avian pasteurellosis)	*Pasteurella multocida*	Water birds, domestic poultry, game birds.	Sick birds appear lethargic or drowsy. Abnormal positions of head and neck. Ataxia, loss of equilibrium. Lesions of septicemia (hemorrhages and petechiation on serosal surfaces) and multiple foci of necrosis in liver. Birds are often in good body condition at time of death.	Acute infections are common and may result in rapid death and high mortality. Chronic infections with lower mortality can also occur.
Leukocytozoonosis	*Leukocytozoon* sp.	Turkeys, ducks, geese, guinea fowl, chickens.	Pallor, splenomegaly, liver degeneration and hypertrophy in some birds. Leukocytozoons visible in blood smears. Schizonts often in liver, spleen, brain.	Outbreaks correspond with hot months when simulid flies and culicoid midges are numerous; these flies breed in and along water courses. Surviving birds (wild or domesticated) often act as carriers. Signs in birds related to anemia.
Marble Spleen Disease	Siadenovirus (Type II adenovirus)	Pheasants, can be propagated in turkey poults, guinea fowl	Sudden death in older birds, usually 12 weeks of age to adult. Depression, acute pulmonary congestion and edema, enlarged, mottled spleens.	No commercial vaccines for pheasants are available.
Mycoplasma gallisepticum (MG) Infectious sinusitis	*Mycoplasma gallisepticum*	Chickens, turkeys, gamebirds	Sinusitis, upper and lower respiratory disease observed.	Common in situations where gamebirds are raised in close proximity to infected poultry. However, the significance of other mycoplasmas found in gamebirds is poorly understood.
Pox Infections	Avipox virus	Most birds, including game birds	Skin lesions observed on the featherless regions of the head, neck, and feet. 1-5 mm yellow-gray plaques in mucosa of mouth, pharynx, trachea or esophagus.	Quailpox virus vaccine available for quail and sometimes used in other poultry.

Name of Disease(s)	Etiology	Susceptibility	Clinical signs and lesions	Comments
Quail Bronchitis	Fowl adenovirus-1 (Serotype I adenovirus)	Bobwhite quail, Japanese quail	Acute respiratory illness ("snicking", conjunctivitis, coughing, sinusitis). High mortality seen in 1-3 week old quail. Less severe illness in older birds. Increased mucus and thickened tracheal mucosa. Pneumonia, airsacculitis, hepatomegaly, splenomegaly.	Secondary bacterial infections with *E. coli* common. Older birds may be carriers. No treatment available, but maternal antibodies from recovered breeders may reduce severity of infection in progeny.
Syngamus trachea infections (gapeworms)	*Syngamus trachea*	Pheasants, turkey poults and young chickens are the most susceptible	Coughing, gasping, weight loss are seen. Characteristic red worms are seen in the tracheal lumen.	Found in birds raised outdoors when access to infected intermediate hosts (such as earthworms) is available, especially after heavy rains bring them to the surface.
West Nile virus	Flavivirus	Observed in grouse and ornamental pheasants, Corvidae and birds of prey.	Morbidity and mortality variable. Brain hemorrhages, splenitis, splenomegaly, enteritis, nephritis.	Reported in New York in 1999, Israel and Romania. Bird to bird transmission reported but mostly transmitted by mosquitoes.

APPENDIX

Revised by Drs Linnea J. Newman and Jean E. Sander

DISEASES OF CHICKENS AND TURKEYS CORRELATED WITH AGE

By knowing the species affected, salient clinical feature, and age of the flock, it is often possible to make a list of potential differential diagnoses. In the following table, some of the more common diseases are presented by age and clinical problem. Of course, this will not be absolute but can be used as a guide.

BROILERS, PULLETS, LAYERS

Typical losses to 7 weeks of age are 4-5%. Losses in the first 2 weeks account for 30-50% of total mortality.

A. BROODING PERIOD (0-2 weeks)

2. Mortality/ Poor Growth

A. Improper incubation conditions - small, weak hatchlings or increased susceptibility to infection
B. Navel and yolk sac infection (*Salmonella, Escherichia coli, Staphylococcus, Proteus*, etc.)
C. Polycystic Enteritis or Runting and Stunting Syndrome (Malabsorption Syndrome)
D. Starveout/dehydration — floor temperature, water management
E. Vaccine contamination
F. Baby chick nephropathy

2. Respiratory Disease

A. Aspergillosis (Brooder Pneumonia)
B. Vaccine Problems —Respiratory reaction

3. Musculoskeletal Disease

A. Rickets
B. Splay Leg

4. CNS Disease

A. Avian Encephalomyelitis
B. Encephalomalacia (Vitamin E Deficiency)
C. Hypoglycemia (Spiking Mortality)
D. Poor vaccine placement (*in ovo* Pox, MDV)

5. Eye Disease

A. Keratoconjunctivitis (Ammonia Burn)
B. Mycotic Keratoconjunctivitis

B. GROWING PERIOD (2-8 weeks)

1. Mortality

A. Ascites
B. Aspergillosis
C. Chicken Infectious Anemia
D. Classic Infectious Bursal Disease
E. Coccidiosis
F. Gangrenous Dermatitis
G. Histomoniasis (Blackhead)
H. Inclusion Body Hepatitis
I. Marek's Disease
J. Necrotic Enteritis
K. Ulcerative Enteritis

DISEASES OF CHICKENS AND TURKEYS CORRELATED WITH AGE

2.   Respiratory Disease

    A.   Avian Influenza
    B.   Avian Metapneumovirus (Swollen Head Syndrome)
    C.   Colibacillosis
    D.   Infectious Bronchitis
    E.   Infectious Laryngotracheitis
    F.   Mycoplasmosis
    G.   Newcastle Disease

3.   Muskoloskeletal Disease

    A.   Angular Bone Deformity (Valgus-Varus Deformities)
    B.   Infectious Synovitis
    C.   Marek's Disease
    D.   Osteomyelitis (e.g. Staphylococcosis, *Enterococcus cecorum*)
    E.   Pododermatitis
    F.   Rickets
    G.   Septic arthritides (e.g. Staphylococcosis, *Escherichia coli, Enterococcus cecorum*)
    H.   Spondylitis (*Enterococcus cecorum*)
    I.   Spondylolisthesis (Kinky Back)
    J.   Tibial Dyschondroplasia
    K.   Toxicity (Ionophore/3-Nitro, Botulism)
    L.   Viral Arthritis

4.   CNS Disease

    A.   Arbovirus infections (Eastern equine encephalitis virus)
    B.   Avian Encephalomyelitis
    C.   Encephalomalacia (Vitamin E deficiency)
    D.   Marek's Disease
    E.   Newcastle Disease

5.   Skin Disease

    A.   Exudative Diathesis (Vitamin E deficiency)
    B.   Fowl Pox
    C.   Gangrenous Dermatitis
    D.   Marek's Disease (Skin Leukosis)

6.   Other

    A.   Candidiasis (Crop Mycosis)
    B.   Cellulitis
    C.   Immunosuppression - IBD, CIA
    D.   Intestinal Parasites (Capillaria, Roundworms, Tapeworms, etc.)
    E.   Mycotoxin

C.   PULLET PERIOD (8-20 weeks)

  1.   Respiratory Disease

    A.   Avian Influenza
    B.   Infectious Bronchitis
    C.   Infectious Coryza
    D.   Infectious Laryngotracheitis
    E.   Mycoplasmosis
    F.   Newcastle Disease

   2.   Neoplastic Disease

       A. Lymphoid Leukosis
       B. Marek's Disease

   3.   Systemic Diseases

       A. Fowl Cholera (Pasteurellosis)
       B. Colibacillosis

   4.   Intestinal Disease

       A. Coccidiosis
       B. Necrotic Enteritis

D. LAYERS (>20 weeks)

   1.   Respiratory Disease

       A. Avian Influenza
       B. Infectious Bronchitis
       C. Infectious Coryza
       D. Infectious Laryngotracheitis
       E. Mycoplasmosis
       F. Newcastle Disease

   2.   Egg Production Drop

       A. Avian Encephalomyelitis
       B. Avian Influenza
       C. Hepatitis E Virus
       D. Infectious Bronchitis
       E. Infectious Coryza
       F. Mismanagement (lights, water, nutrition, etc.)
       G. Mycoplasmosis (*M. gallisepticum*)
       H. Newcastle Disease

   3.   Neoplastic Disease

       A. Lymphoid Leukosis
       B. Marek's Disease
       C. Various other tumors (Carcinoma, Sarcoma)

   4.   CNS Disease

       A. Fowl Cholera (Pasteurellosis)
       B. Newcastle Disease

   5.   Cage Layer Fatigue

   6.   Cannibalism

   7.   Fatty Liver Hemorrhagic Syndrome

   8.   Fowl Mites

   9.   Hysteria

   10. Parasitism (Capillariasis, Heterakis, Roundworms, etc.)

   11. Salpingitis/Peritonitis

DISEASES OF CHICKENS AND TURKEYS CORRELATED WITH AGE

12. Uterovaginal Prolapse

E. SPORADIC DISEASES

    1. Arbovirus Infections

    2. Avian Tuberculosis

    3. Botulism

    4. Other Parasitic Diseases

    5. Salmonellosis (Pullorum,Typhoid)

    6. Streptococcosis

TURKEYS

A. EARLY BROODING PERIOD (0-3 weeks)

    1. Mortality/Poor Growth

        A. Candidiasis
        B. Cannibalism
        C. Coccidiosis
        D. Cryptosporidiosis
        E. Omphalitis (*Salmonella, S. arizonae, E. coli, Proteus*, etc.)
        F. Mismanagement (Starveout, dehydration, poor beak trimming)
        G. Poult Enteritis andMortality Syndrome
        H. Poult malabsorption / runting-stunting syndrome
        I. Turkey Viral Hepatitis

    2. Respiratory Disease

        A. Aspergillosis (Brooder Pneumonia)
        B. Turkey Coryza (Bordetellosis)

    3. Musculoskeletal Disease

        A. Rickets
        B. Splay Leg
        C. Staphylococcosis
        D. Tibial Rotation

    4. CNS Disease

        A. Arizonosis
        B. Avian Encephalomyelitis
        C. Encephalomalacia (Vitamin E Deficiency)
        D. Mycotic Encephalitis (*Aspergillus, Ochroconis*)

    5. Eye Disease

        A. Ammonia Burns
        B. Arizonosis
        C. Injuries
        D. Mycotic Keratoconjunctivitis (*Aspergillus*)

B. LATE BROODING/EARLY GROWING PERIOD (3-12 weeks)

1. Mortality

       A. Aortic Rupture (Dissecting Aneurysm)
       B. Hemorrhagic Enteritis
       C. Histomoniasis (Blackhead)
       D. Sudden Death Syndrome of turkeys / Perirenal Hemorrhage Syndrome
       E. Leucocytozoonosis
       F. Necrotic Enteritis
       G. Round Heart Disease (Dilated Cardiomyopathy)
       H. Ulcerative Enteritis

2. Respiratory Disease

       A. Avian Influenza
       B. Bordetellosis (Turkey Coryza)
       C. Colibacillosis
       D. Fowl Cholera (Pasteurellosis)
       E. Mycoplasmosis (MM, MS, MG)
       F. Newcastle Disease
       G. Avian Metapneumovirus infection (Turkey rhinotracheitis)

3. Musculoskeletal Disease

       A. Bacterial Arthritides (*Staphylococcus, Escherichia coli*)
       B. Contact dermatitis of foot pads
       C. Spondylolisthesis ("Kinky Back")

4. Other
       A. Mycotoxins
       B. Roundworms

C. FINISHING PERIOD (>12 weeks-market)

1. Mortality

       A. Aortic Rupture (Dissecting Aneurysm)
       B. Cannibalism
       C. Erysipelas

2. Respiratory Diseases

       A. Aspergillosis
       B. Avian Influenza
       C. Chlamydiosis
       D. Fowl Cholera (Pasteurellosis)
       E. Newcastle Disease
       F. Ornithobacterium infection (ORT)

3. Musculoskeletal Disease

       A. Angular Bone Deformity (Valgus-Varus Deformities)
       B. Bacterial Arthritides (*Staphylococcus, E. coli*, Erysipelas, *Pasteurella multocida*)
       C. Osteomyelitis
       D. Scoliosis
       E. Tibial Dyschondroplasia

<u>DISEASES OF CHICKENS AND TURKEYS CORRELATED WITH AGE</u>

4. Other

    A.   Breast Buttons/Blisters
    B.   External Parasites (Mites, Lice)
    C.   Internal Parasites (Round Worms, Cecal Worms)
    D.   Pendulous Crop
    E.   Turkey Pox

D. BREEDERS (>30 weeks). Diseases of the finishing period can also occur during the laying period.

1. Mortality

    A.   Aspergillosis
    B.   Fowl Cholera (Pasteurellosis)
    C.   Salpingitis/Peritonitis

2. Neoplasia

    A.   Lymphoproliferative Disease
    B.   Reticuloendotheliosis

3. Egg Production Drop

    A.   Arbovirus infections (EEEV, WEEV, HJV)
    B.   Avian Influenza
    C.   Avian Metapneumovirus infection
    D.   Mismanagement (lights, water, nutrition, etc.)
    E.   Mycoplasmosis
    F.   Newcastle Disease

DISEASES WITH LESIONS IN THE CARDIOVASCULAR SYSTEMS

Name of Disease(s)	Etiology	Species Affected	Lesions	Comments
Ascites	Metabolic; related to rapid growth rate and high yield in broiler chickens.	Young, fast-growing chickens (males > females).	Enlarged heart. Ascites. Enlarged or cirrhotic liver. Fibrin exudation in severe cases.	Exacerbated by low oxygen conditions at hatch or brooding, high altitude, heavy dust, lung pathology. Controlled via lighting programs to slow growth.
Chlamydiosis	*Chlamydophila psittaci*	Turkeys, pigeons, ducks, cage and wild birds.	Pericarditis, often adhesive.	Often with airsacculitis, fibrinous perihepatitis, splenomegaly, and hepatomegaly.
Colibacillosis	*Escherichia coli* septicemia.	Turkeys, chickens, commercial ducks.	Pericarditis, often adhesive.	Often with airsacculitis, fibrinous perihepatitis.
Dissecting aneurysm (aortic rupture)	Unknown. Non-infectious. A strong nutritional influence, especially copper metabolism.	Turkeys. Occasionally, chickens.	Ruptured artery, usually abdominal aorta. Rarely, aortic arch. Extensive internal hemorrhage.	Sudden deaths in rapidly growing, highly conditioned birds. Losses can be extensive. Usually in males.
Endocarditis	Various bacteria: Erysipelas, *Pasteurella, Staphylococcus, Streptococcus.*	Chickens and turkeys.	Yellow, irregular masses on the heart valves.	Low incidence.
Marek's disease	Herpesvirus	Chickens	Focal or multifocal tumors in the myocardium.	May be associated with tumors in other organ systems.
Mycoplasma gallisepticum infection	*M. gallisepticum*	Turkeys, chickens, other poultry, birds.	Pericarditis, often adhesive.	Often with airsacculitis, fibrinous perihepatitis.
Pullorum disease, fowl typhoid, possibly paratyphoid	*Salmonella* Pullorum, S. Gallinarum, other salmonellae.	Chickens, turkeys, and geese.	Nodules in myocardium. Adhesive pericarditis.	Oophoritis or orchitis may occur in some adult birds. Enteric or septicemic diseases with diarrhea in young birds.
Round heart disease	Possibly toxic agents in turkeys (antitrypsin and furazolidone implicated).	Chickens and turkeys.	A greatly enlarged, round heart. Ascites. Fibrin exudation in severe cases.	Uncommon in chickens. Most outbreaks associated with built-up litter. Variable mortality. Potentiated by furazolidone treatment in turkeys.

DISEASES WITH SIGNS SUGGESTIVE OF CNS DISEASE[A]

Name of Disease(s)	Etiology	Species Affected	Lesions	Comments
Aspergillosis	Usually *Aspergillus fumigatus*.	Usually young chicks and poults or captive game birds.	Various CNS signs. Respiratory signs usually precede CNS signs and predominate. A minority develop CNS involvement.	Yellow mycotic nodules often grossly visible in brain. Lesions of aspergillosis in lungs or air sacs, perhaps in conjunctival sac or globes of eyes.
Avian encephalomyelitis (epidemic tremor)	Hepatovirus (Picornaviridae)	Chicks, young poults, and pheasants.	Tremors of head and neck and legs. Paresis progressing to paralysis and prostration.	Microscopic lesions in the CNS. Survivors often develop cataracts. Invert the birds to accent the tremors for diagnosis.
Bacterial encephalitis	*Salmonella, S. arizonae,* paratyphoid species, *Escherichia coli.*	Turkey poults, chicks.	Various CNS signs. Ophthalmitis and omphalitis often present in some of the poults.	Exudate grossly visible in meninges and ventricles. Confirm by culture.
Botulism	Preformed toxin of *Clostridium botulinum.*	Usually chickens	Paresis progressing to paralysis of legs, neck, wings, nictitating membrane. Loose feathers.	No gross or microscopic lesions of value. Perhaps putrid feed or maggots in crop.
Encephalomalacia, (Crazy chick disease)	Vitamin E deficiency	Chicks (usually less than 8 weeks old), turkeys (usually 2-4 weeks old).	Ataxia, falling and flying over backwards, loss of balance; prostration with legs outstretched, toes flexed, and head and neck back.	Hemorrhage and malacia of cerebellum often grossly visible. Confirm by microscopy. Perhaps exudative diathesis along ventrum or muscle necrosis.
Enterococcus cecorum	*Enterococcus cecorum*	Chickens	Incoordination, abnormal gait	Associated with spondylitis, femoral head necrosis, and osteomyelitis in broiler chicken and broiler breeder flocks.
Equine Encephalomyelitis Virus	Alphaviruses	Pheasants, partridges, turkeys, Pekin ducks, quails, chickens.	Circling or staggering followed by paralysis. Perhaps blindness in recovered birds. Often high morbidity and mortality.	Microscopic lesions only. Transmitted by mosquitoes.
Fowl cholera	*Pasteurella multocida*	Turkeys, chickens, perhaps other species.	Abnormal positions of head and neck. Ataxia, loss of equilibrium.	A localized form of chronic fowl cholera. May/may not accompany acute outbreak. Lesions may be present in cranial bones or inner ear.

DISEASES WITH SIGNS SUGGESTIVE OF CNS DISEASE[A]

Name of Disease(s)	Etiology	Species Affected	Lesions	Comments
Hypoglycemia-spiking mortality syndrome (HSMS)	Unknown. Arenavirus, rotavirus, thiamine deficiency, mycotoxins have been implicated.	Young chicks and poults (< 3 weeks).	Ataxia and death. Dead often found ventrally recumbent with head and neck outstretched.	Sudden, high mortality that may last 1- 2 days. Often associated with a stress event such as movement from brood chamber to whole house. Capsular or parenchymal liver hemorrhage may be present. Blood sugar levels are very low in affected compared to normal birds.
Marek's disease	Alpha herpesvirus	Chickens usually 6•20 weeks old.	Paresis progressing to paralysis of a leg or wing. Often one leg is held forward and one leg is held backward in the recumbent bird.	Microscopically there are infiltrating neoplastic cells in affected nerve trunks and the CNS. Grossly, lesions may be visible in affected nerve trunks.
Newcastle disease	Paramyxovirus	Usually chickens but most birds susceptible.	In chicks CNS signs, usually proceeded by respiratory signs. Progressive paresis followed by paralysis and death.	Microscopic lesions are present and helpful in diagnosis. Only a small part of birds with respiratory signs have CNS signs.
Ochroconosis (Previously known as Dactylariosis)	Ochroconis gallopava (fungus) (previously known as Dactylaria gallopava)	Turkey poults, chicks.	Incoordination, tremors, torticollis, paralysis, perhaps ocular opacities.	Focal gross brain lesions, often pulmonary nodules or airsacculitis. Fungus often in sawdust litter.
Pox vaccination reaction	Fowl pox vaccine	Chickens < 2 weeks. Low incidence for 5•7 days.	Ataxia, mild extensor rigidity. Head drawn back, wings drawn up, tiptoe gait.	In ovo administration of pox vaccine; errors in subcutaneous injection. May relate to vaccine titer.

[A]Signs suggestive of CNS disease in birds include ataxia, paralysis, circling, trembling, twisting of the head and neck, falling over backward, and loss of balance. Any combination of CNS signs may occur.

DISEASE WITH LESIONS IN THE MOUTH, PHARYNX, ESOPHAGUS, CROP, PROVENTRICULUS, GIZZARD

Name of Disease(s)	Etiology	Species Affected	Lesions	Comments
Candidiasis (crop mycosis)	*Candida albicans*	Poultry, game birds, perhaps other birds.	Gray, thin, pseudomembranous patches on the mucosa. Little inflammation.	Often secondary to parasitism, malnutrition, poor sanitation, impaction, antibiotic usage, other disease. Affects any or all organs listed in title.
Capillariasis	*Capillaria contorta, C. annulata*	Chickens, turkeys, game birds.	Worms sewn into inflamed, thickened mucosa.	In the esophagus and crop. Common in game birds. Scrapings usually necessary for identification.
Duck plague (Duck virus enteritis)	Herpesvirus	Ducks, geese, swans.	Hemorrhage and necrosis of the esophageal and cloacal tissue. Liver has petechial hemorrhages.	Intranuclear inclusions produced in infected tissue.
Mycotoxicosis	Trichothecenes	All poultry	Oral ulcerations	Produced by *Fusarium* species of mold.
Pendulous crop	If epizootic, influenced by coarse roughage; or by genetics in turkeys.	Turkeys, chickens, perhaps others.	Crop and esophagus enlarged, perhaps impacted.	Secondary mycosis often present in atonic crop or esophagus. Sporadic cases sometimes from vagal paralysis.
Trichomoniasis (canker in pigeons; frounce in falcons)	*Trichomonas gallinae*	Raptors, doves, pigeons, turkeys, chickens.	Raised conical masses in mucosa of mouth, pharynx, esophagus, crop.	Many trichomonads in oral fluids. Lesions sometimes in proventriculus. Also in the liver of pigeons and some raptors. Lesions often invasive.
Vitamin A deficiency	Inadequate vitamin A	Chickens, turkeys.	Pustule-like lesions in esophagus, perhaps mouth and pharynx. Variable rhinitis, sinusitis, conjunctivitis. Perhaps excessive urates in urinary tract or cloaca.	Sticky eyelids and ataxia often the only gross lesions and signs in young birds. Squamous metaplasia of columnar epithelium in esophageal mucous glands and nasal epithelium.
Wet pox	Avipoxvirus	Most birds, including poultry.	1-5 mm yellow-gray plaques in mucosa of mouth, pharynx, or esophagus. Less often in sinuses or conjunctiva.	Skin lesions often on face, wattles, eyelids, comb, feet, legs, caruncle, snood.

DISEASES WITH LESIONS IN THE INTESTINE

Name of Disease(s)	Etiology	Species Affected	Lesions	Comments
Arizonosis	*Salmonella arizona*	Poults, chicks.	Enteritis, unabsorbed yolk. Large mottled liver, perhaps peritonitis or intraocular turbidity.	Poults often unthrifty and with excessive mortality. Perhaps diarrhea.
Avian Tuberculosis	*Mycobacterium avium*	Chickens, most other poultry and birds.	Round nodules (granulomas) attached to serosa of gut. Focal granulomas in many other organs. Extreme emaciation in advanced cases.	Usually seen in old chickens kept beyond one laying season. Causative bacilli readily demonstrated in acid-fast-stained smears of lesions.
Coccidiosis	Many species of *Eimeria*	Chickens, turkeys, ducks, geese, perhaps in most birds.	Enteritis of variable severity and location. *E. acervulina* – upper small intestine. *E. necatrix* and *E. maxima* ••mid-small intestine. *E. tenella* ••ceca. *E. brunetti* – posterior gut.	Five major pathogens for chickens are listed. For details see text. Coccidiosis often occurs concurrently with other diseases. Uncommon in ducks, geese.
Colibacillosis	*Escherichia coli*	Chickens, turkeys.	Enteritis, omphalitis, salpingitis, peritonitis, arthritis, and panophthalmitis are frequent lesions. In respiratory disease often associated with pericarditis, perihepatitis, and airsacculitis. See notes.	Many syndromes identified. Seen in very young and in adults. Three serotypes of the agent account for most outbreaks. Often a secondary infection.
Coligranuloma	A mucoid coliform	Chickens, turkeys.	Granulomas along cecum, duodenum, in mesentery and liver.	Resembles avian tuberculosis but no acid-fast bacteria in lesions. Must be differentiated from avian tuberculosis. An uncommon disease.
Duck virus enteritis (Duck plague)	Herpesvirus	Wild or domestic ducks, geese, swans.	Widespread hemorrhages, severe enteritis. Perhaps elevated plaques in esophagus, ceca, cloaca or bursa. Hemorrhage and/or necrosis in lymphoid rings or discs of gut.	Epizootic losses in waterfowl are suggestive. Typical lesions and inclusion bodies helpful in diagnosis. An exotic, reportable disease.
Fowl typhoid	*Salmonella gallinarum*	Chickens, turkeys, occasionally other poultry.	In recently hatched, same as pullorum (below). In older birds, pale cadaver, marked enteritis, splenomegaly, gray foci in liver, bile-stained (bronze) liver.	Closely resembles pullorum in recently hatched chicks and poults but mortality persists into adulthood. Diarrhea and anemia in older birds.

DISEASES WITH LESIONS IN THE INTESTINE

Name of Disease(s)	Etiology	Species Affected	Lesions	Comments
Hemorrhagic enteritis of turkeys (HE)	Adenovirus (Siadenovirus)	Turkeys often 6-12 weeks old.	Severe enteritis in small intestine with much blood in the gut. Spleen enlarged and mottled early in the course.	Bloody feces often noted. Mortality may be high. A similar virus causes marble spleen disease in pheasants. Subclinical HE may be associated with immunosuppression and secondary *Escherichia coli* infection.
Hexamitiasis (new name Spironucleosis)	*Spironucleus meleagridis.* In pigeons *S. columbae.*	Turkey, poults, game birds, pigeons.	Catarrhal enteritis in upper half of small intestine. Local bulbous dilations in affected gut. *Spironucleus* in crypts of Lieberkuhn.	*Escherichia coli* infection birds have watery diarrhea. They often die in convulsions. Closely resembles paratyphoid or transmissible enteritis.
Histomoniasis (blackhead)	*Histomonas meleagridis*	Turkey poults, game birds, chickens, replacement pullets.	Ceca swollen and usually with cecal cores. Circular or oval recessed lesions in liver.	Typical lesions in ceca and liver are pathognomonic.
Infectious bursal disease	Avibirnavirus	Chickens typically 3•6 weeks old.	Marked inflammation of the bursa of Fabricius, which is swollen early but atrophied later. Enlarged spleen. Hemorrhages common in heavy muscles.	Diarrhea, incoordination, dehydration, vent picking are usual signs. Course is about 1 week. Immune system damaged. May be followed by secondary infection such as inclusion body hepatitis, gangrenous dermatitis, ulcerative or necrotic enteritis, etc...
Internal parasitism	Ascarid tapeworms, cecal worms, *Capillaria* spp.	Many birds, including poultry.	Parasites usually visible in appropriate location. Variable degree of enteritis. Emaciation may be marked.	Microscopic exam of scrapings necessary to identify *Capillaria.*
Marek's disease	Alpha herpesvirus	Chickens.	Diffuse neoplasia involving an area of gut.	Focal of diffuse neoplasia usually apparent in other visceral organs, e.g., liver, spleen, gonads, kidneys, lungs.

DISEASES WITH LESIONS IN THE INTESTINE

Name of Disease(s)	Etiology	Species Affected	Lesions	Comments
Necrotic Enteritis	*Clostridium perfringens*	Chickens.	Focal or diffuse necrosis of intestinal mucosa, particularly the ileum.	Most common in flocks raised without antibiotics, in the presence of coccidiosis, intestinal irritants such as non-starch polysaccharides (wheat, barley, rye diet), biogenic amines.
Nonspecific enteritis	Many infectious agents.	Poultry and other birds.	Enteritis accompanies many infectious diseases that have lesions of greater diagnostic value in other systems.	Other diseases that may have enteritis include: cholera, erysipelas, salmonellosis, vibrionic hepatitis, spirochetosis, botulism, aflatoxicosis, influenza, candidiasis, and others.
Paratyphoid	*Salmonella sp.* (about 20 major species).	Poults, chicks, other young birds. Sometimes in adult poultry.	Severe enteritis. Often mucosal plaques and/or cheesy cecal cores. Occasionally the other lesions described for pullorum disease may occur.	Usually in birds less than 8 weeks old but occasionally in older birds. Occurs frequently in young turkey poults. Interspecies transmission possible.
Pullorum disease	*Salmonella* Pullorum	Chickens, turkeys, occasionally other poultry.	Enteritis, dehydration, unabsorbed yolk. Nodules in lungs, heart, or gizzard. Perhaps mucosal plaques, cecal cores, and focal necrotic hepatitis. May present as synovitis in broilers.	Typically epizootic in birds up to 4 weeks old. Begins shortly after hatching. White adherent diarrhea common. Persists in some adults as oophoritis, orchitis, or myocarditis.
Turkey coronavirus enteritis	Coronavirus	Turkeys, especially poults.	Marked mucoid enteritis, dehydration.	Diarrhea present. Mortality may be very high with young poults. Less severe in older turkeys.
Ulcerative enteritis (quail disease)	*Clostridium colinum*	Captive game birds, turkeys, chickens.	Acute enteritis early. Usually many deep ulcers along the intestine. Enlarged spleen. Focal and/or diffuse yellow areas in liver.	A common disease of game birds. Resembles coccidiosis in chickens. Often secondary in chickens.

DISEASES WITH LESIONS IN LIVER[A]

Name of Disease(s)	Etiology	Species Affected	Lesions	Comments
Aflatoxicosis (mycotoxicosis)	Toxin usually from *Aspergillus flavus*.	Poults, pheasants, chicks, ducklings.	Liver pale and mottled and with bile duct hyperplasia. Catarrhal enteritis.	Toxin in feeds, especially peanut meal; in spilled feed in litter; in litter alone. Signs often include ataxia, convulsions, opisthotonos. The only signs may be unthriftiness, poor weight gain, low production.
Avian Tuberculosis	*Mycobacterium avium*	Usually in chickens. Also, other poultry and wild birds.	Focal granulomas in the liver. Lesions often are numerous. Nodules (granulomas) along the periphery of the gut. Lesions in most organs and marrow in advanced cases. Emaciation.	Usually encountered in older chickens – those over 1 year. Extreme emaciation is the hallmark of tuberculosis.
Duck Hepatitis Virus type I (DHV-1)	Picornavirus	Ducklings, typically less than 4 weeks old.	Liver swollen and with many hemorrhages	Signs: acute onset, short course, high morbidity and mortality. Typical lesions in young ducklings almost pathognomonic.
Duck Hepatitis Virus type 3 (DHV-3)	Astrovirus	Ducklings up to 5 weeks of age	Liver swollen and with many hemorrhages	Death within 1 to 2 hours after onset of clinical signs. Convulsions, opisthotonos. Mortality rarely > 30% but high morbidity.
Fowl cholera	*Pasteurella multocida*	Poultry, wild birds, especially waterfowl.	Diffuse streaking of the liver in acute cases. Later there may be 1-3-mm focal areas of hepatic necrosis.	Focal hepatic lesions closely resemble those of salmonellosis, tuberculosis, and listeriosis. Fowl cholera often septicemic.
Histomoniasis (Blackhead)	*Histomonas meleagridis*	Turkeys, game birds, chickens.	Recessed round to oval focal hepatic lesions up to 2.0 cm. Typhlitis with/without cecal cores.	Classical hepatic and cecal lesions together are pathognomonic. Frequently occurs in turkeys raised with or after chickens. Agent transmitted in ova of cecal worms and in earthworms.
Inclusion body hepatitis	Adenovirus	Chickens typically 5-8 weeks old.	Yellow-tan hepatic areas with hemorrhages. Intranuclear inclusions. Perhaps icterus. Hemorrhages at many sites (skin, muscles, subserosa).	Often follows infectious bursal disease, which damages immune system.

[A]This table contains only diseases that are more frequently encountered, are more significant economically, or have hepatic lesions clearly of diagnostic value. Many other avian diseases may have hepatic lesions, e.g., psittacosis, Arizona (paracolon) infection, turkey viral hepatitis, synovitis, *Riemerella anatipestifer* infection, staphylococcosis, and listeriosis.

DISEASES WITH LESIONS IN LIVER[A]

Name of Disease(s)	Etiology	Species Affected	Lesions	Comments
Leukosis complex	Retroviruses	Chickens, perhaps other species.	Focal or diffuse neoplastic lesions in the liver. Other organs frequently affected with focal or diffuse neoplastic lesions.	If epizootic in chickens less than 5 months old the disease may be Marek's disease. In older birds, probably lymphoid leukosis. See section on viral tumors.
Salmonellosis, Pullorum disease, fowl typhoid, paratyphoid	*Salmonella* Pullorum, S. Gallinarum, other *Salmonella*.	Chickens and turkeys. Many kinds of poultry, birds, and mammals have paratyphoid.	Sometimes 1-3 mm focal areas of hepatic necrosis. Often splenomegaly. Enteritis, sometimes with raised plaques in the mucosa and with cheesy plugs in the gut or cecum. May be septicemic with few lesions, especially in the very young.	Salmonella infections predominate in the young. Many are transmitted through the egg. Interspecies transmission possible with paratyphoid. Salmonella infections often are septicemic.
Ulcerative enteritis	*Clostridium colinum*	Captive game birds, turkeys, chickens.	Focal and/or diffuse yellow areas in the liver. Deep ulcers scattered throughout the intestine.	A common disease of captive game birds. Increasing in poults and chickens.
Vibrionic hepatitis	*Campylobacter fetus* spp. *jejuni*	Chickens, usually well started or adults.	Asterisk or cauliflowerlike foci of hepatic necrosis. Sometimes hemorrhages, subcapsular hematocyst, ascites, or hydropericardium.	Only about 10% of affected birds have hepatic lesions. *Campylobacter* often cultured from bile.

DISEASES WITH LESIONS IN HEMOPOIETIC SYSTEM

Name of Disease(s)	Etiology	Species Affected	Lesions	Comments
Chicken infectious anemia (CIA)	Gyrovirus (Circoviridae)	Chickens (2-4 weeks old).	Anemia, thymic atrophy. Pale pink to yellow bone marrow.	Often associated with gangrenous dermatitis. Particularly on wings (blue wing disease). Course is usually about 1 week. Affected broilers can be traced to a CIA-shedding breeder flock.
Duck virus enteritis (Duck plague)	Herpesvirus	Wild or domestic ducks, geese, swans.	Widespread hemorrhages. Severe enteritis. Perhaps elevated plaques in esophagus, ceca, rectum, cloaca, or bursa. Hemorrhage and/or necrosis in lymphoid rings (discs) of gut.	Epizootic losses in waterfowl are suggestive. Typical lesions and inclusion bodies helpful in diagnosis. An important reportable disease.
Infectious bursal disease	Avibirnavirus	Chickens typically 3-6 weeks old.	Marked inflammation of the bursa of Fabricius, which is swollen early but atrophied later. Enlarged spleen. Hemorrhages common in heavy muscles.	Diarrhea, incoordination, dehydration, vent picking are usual signs. Course is about 1 week. Immune system damaged. May be followed by inclusion body hepatitis or gangrenous dermatitis.
Leukocytozoonosis	*Leukocytozoon* sp.	Turkeys, ducks, geese, guinea fowl, chickens.	Pallor, splenomegaly, liver degeneration and hypertrophy in some birds. Leukocytozoons visible in blood smears. Schizonts often in liver, spleen, brain.	Outbreaks correspond with hot months when simulid flies and culicoid midges are numerous; these flies breed in and along water courses. Surviving birds (wild or domesticated) often act as carriers. Signs in birds related to anemia.
Lymphoid leukosis	Retroviruses	Chickens, perhaps other species.	Internal neoplasms. Neoplastic lymphoblastic cells in bursa of Fabricius and many other organs.	May resemble Marek's disease. Usually observed in chickens over 4 months old. Neoplasia frequently in liver, spleen, kidneys; nodular tumors in bursa of Fabricius.
Marek's disease	Alpha herpesvirus	Chickens, perhaps other species.	Internal neoplasms. Neoplastic pleomorphic lymphoid cells infiltrate, CNS, spleen, liver, kidney, nerve trunks, iris, ovary, and many other organs.	Often epizootic in chickens 6-20 weeks old. Persists in older birds. May closely resemble lymphoid leukosis. Bursa of Fabricius seldom neoplastic.

DISEASES WITH LESIONS IN THE MUSCULOSKELETAL SYSTEM

Name of Disease(s)	Etiology	Species Affected	Lesions	Comments
Biotin deficiency	Biotin deficiency	Turkey poults.	Large twisted hocks and bowed shanks. Hyperkeratosis of skin on soles of feet and toes.	Lesions include hyperkeratotic skin at corners of mouth and on eyelids.
Cage layer fatigue (osteoporosis, adult rickets)	Controversial etiology. Possibly low phosphorus, low calcium, or imbalance. Caging a factor.	Chickens (caged layers).	Bones that easily break and splinter. Fractures, sometimes vertebral. Depletion of bone mineral.	Leg weakness or inability to stand. May recover if promptly removed and kept on floor.
Contact dermatitis of foot pads (Pododermatitis or Bumblefoot)	Trauma with secondary bacterial infection of feet.	Chickens, turkeys, falcons, other birds.	Swollen foot or toepads often with open lesions. May involve joints or toes or feet.	Usually sporadic. Concrete floors without litter may result in many cases. In falcons related to small, hard perches.
Crooked (twisted) toes	Unknown	Chickens, turkeys.	Affected toes deviated laterally or medially.	Incidental finding. Don't identify as cause of lameness or confuse with riboflavin deficiency.
Deep pectoral myopathy (green muscle disease)	Ischemia following exertion.	Chickens, turkeys.	Unilateral or bilateral green areas of necrosis in deep pectoral muscles.	An aseptic necrosis of muscle that follows stress leading to muscle swelling.
Enterococcus Cecorum	*Enterococcus Cecorum*	Chickens	Incoordination, abnormal gait	Associated with spondylitis, femoral head necrosis, and osteomyelitis in broiler chicken and broiler breeder flocks.
Gout (articular gout)	A metabolic disease with deposition of uric acid crystals.	Chickens, turkeys, possibly other birds.	White to gray semisolid tophi deposited in and around joints. Affected joints enlarged and distorted. Most obvious on feet and legs. Emaciation.	Similar uric acid crystals may be deposited on viscera, especially the pericardium and capsule of liver. Perhaps in the ureters.
Infectious synovitis	*Mycoplasma synoviae*	Chickens, turkeys.	Joints and tendon sheaths swollen, most apparent on hocks, shanks, feet. Sticky synovial exudate. Sometimes the liver is green.	Many birds are lame and squat on floor. Breast blisters a common sequel. Other mycoplasmas may produce similar lesions

DISEASES WITH LESIONS IN THE MUSCULOSKELETAL SYSTEM

Name of Disease(s)	Etiology	Species Affected	Lesions	Comments
Nutritional myopathy	Deficiency of vitamin E, selenium, and sulfur-containing amino acids.	Chickens, turkeys, ducks.	In chickens and ducks necrotic muscle fibers in breast muscles or legs appear as white streaks or masses. In turkeys gray-white patches in gizzard musculature.	In chicks muscle lesions may be accompanied by encephalomalacia or exudative diathesis.
Osteomyelitis	Various bacteria metastasize to marrow.	Poultry and other birds.	Osteomyelitis at various sites. Commonly in femur, tibia, or vertebra.	Sporadic. Nonspecific.
Osteopetrosis	Associated with retrovirus infections.	Chickens.	Shanks and other long bones thickened, heavy, dense and with small marrow cavity.	Higher incidence in roosters. Bone lesions non-neoplastic.
Chondrodystrophy (perosis or slipped tendon)	Usually manganese or choline deficiency. Sometimes niacin, biotin.	Poultry and game birds in captivity.	Gastrocnemius tendon often slips off trochlea at hock and leg deviated (usually laterally) at hock. Bizarre position of leg distal to hock.	May be unilateral or bilateral. A disease of rapidly growing, young birds. Often in flocks fed mostly corn.
Rickets	Imbalance or deficiency of Ca/P/vitamin D_3.	Poultry and other birds raised in captivity.	In young birds - soft beaks and bones, beaded ribs, crooked keels, enlarged epiphyses, and enlarged paratyphoids.	Usually seen in birds a few weeks old; deficiency of vitamin D_3 frequently the cause.
Splay leg	Slippery surface under birds.	Young chickens, turkeys.	Lateral deviation of legs at hips.	High incidence in flocks brooded on slippery paper.
Staphylococcal arthritis	*Staphylococcus aureus* infection with localization in various joints.	Turkeys, chickens.	Affected joints swollen, painful, usually with exudate. Lesions may be generalized but usually involve hocks and feet. May involve thoracolumbar junction.	Affected birds often severely crippled. Arthritis often preceded by septicemia. Common disease in turkeys.
Tibial dyschondroplasia	Possibly influenced by genotype. May be related to excess phosphorus or other dietary factors.	Broiler chicks, young turkeys.	Anterolateral bowing of tibias. Abnormal mass of cartilage in proximal tibia and/or metatsus. Perhaps fracture at site. Normal parathyroids. Osteomyelitis often occurs concurrently.	Squatting, reluctance to move, abnormal posture and gait. Retarded growth. Similar syndrome in ducks.
Viral arthritis (reovirus infection)	Reovirus	Chickens (meat type), turkeys.	Swelling of tendons and tendon sheaths, mostly near hocks.	Sometimes leads to rupture of gastrocnemius tendon(s).

DISEASES WITH LESIONS IN THE REPRODUCTIVE TRACT

Name of Disease(s)	Etiology	Species Affected	Lesions	Comments
Egg yolk peritonitis	Yolk ovulated into peritoneal cavity. Exact cause unknown.	Layers.	Yolk material among abdominal organs. Peritonitis of variable severity.	Rarely associated with abnormal ovary or oviduct. Occasionally accompanies many acute diseases.
Internal laying	Eggs regurgitated into peritoneal cavity. Cause unknown.	Layers.	Hard- or soft-shelled eggs in peritoneal cavity. Peritonitis.	Perhaps some cases are a sequel to oviduct damage from infectious bronchitis.
Lymphoid leukosis	Retrovirus	Chickens, perhaps other species.	Neoplasia of ovary. Focal or nodular neoplasia of the bursa of Fabricius.	Accompanied by many other neoplastic lesions. Usually in sexually mature birds. See notes on lymphoid leukosis (Avian leukosis (ALV)/sarcoma viruses).
Marek's disease	Alpha herpesvirus	Chickens, perhaps other species.	Neoplasia of ovary. Atrophy of bursa of Fabricius.	Often accompanied by other neoplastic lesions. Usually in sexually immature birds. See notes on Marek's disease.
Prolapse of oviduct	Unknown. Often accompanies obesity.	Layers.	Oviduct prolapsed and often cannibalized.	Occurs frequently in young, fat pullets beginning to lay.
Pullorum disease fowl typhoid, paratyphoid	Salmonella Pullorum, S. Gallinarum, Salmonella sp.	Adult chickens and turkeys.	Oophoritis with bloody, cheesy or atrophic follicles. Orchitis.	Sometimes accompanied by focal myocarditis and pericarditis.
Salpingitis	May accompany mycoplasmosis or follow infectious bronchitis. Usually nonspecific. E. coli commonly cultured.	Layers.	Salpingitis of variable extent. Oviduct may be dilated with exudate.	A common cause of routine daily mortality in laying hens. May be an incidental finding in broilers with a history of airsacculitis.

DISEASES WITH LESIONS IN THE RESPIRATORY TRACT[A]

Name of Disease(s)	Etiology	Species Affected	Lesions[B]	Comments
Aspergillosis	*Aspergillus fumigatus*	Most poultry and birds susceptible. Commonly in chickens, turkeys, waterfowl, penguins, captive game birds.	Mycotic granulomas usually in lungs, perhaps along airways. Mycotic plaques or fuzzy mycelium often in air sacs. Lesions may transplant or metastasize to internal organs, brain, globes of eyes. Rarely in conjunctival sac.	Usually transmitted via moldy feed, litter, or hatchery contamination. *Ochroconis gallopava* and other fungi may produce similar signs and lesions.
Avian chlamydiosis	*Chlamydophila psittaci*	Occasionally turkeys, ducks. More often in wild exotic birds.	Variable, but often airsacculitis, pericarditis, fibrinous perihepatitis. Splenomegaly may be the only lesion in chronic cases.	Among poultry, most often seen in turkeys. May infect humans processing infected poultry, resulting in flu-like symptoms.
Avian influenza (AI)	Orthomyxovirus (strains vary greatly in pathogenicity).	Turkeys, ducks, pheasants, quail, many wild birds, other poultry.	Highly variable. In mild form: often swollen sinuses, ocular or nasal discharge. In severe form: hemorrhages, exudation, focal necrosis in respiratory, digestive, urogenital, cardiovascular, or multiple systems. See AI section.	Enzootic forms in U.S. usually mild to moderate in severity and involve respiratory system. Egg production declines and shell abnormalities common in turkeys. Most outbreaks of AI in U.S. are in turkeys and ducks.
Avian Metapneumovirus infections	Metapneumovirus	Chickens and turkeys of any age, pheasants, guineas	Upper respiratory signs with nasal and ocular exudate, swollen heads and sinuses.	Variations in clinical presentation dependent on secondary infections. Egg production drops.
Bordetellosis (Turkey coryza)	*Bordetella avium*	Turkeys, chickens.	Nasal, ocular, and sinus exudation. Tracheitis. Perhaps flattened trachea. Pneumonia uncommon in uncomplicated cases.	Causes deciliation of trachea leading to increased susceptibility to other respiratory diseases.

[A]Respiratory signs include one or more of the following: snicks, sneezes, dyspnea, gasping, rapid breathing, rales, ocular or nasal discharge, swollen sinuses. Signs in broilers may be severe, even with the milder strains. Secondary *Escherichia coli* infection is common.

[B]Unless stated otherwise lesions are those seen in turkeys or chickens.

DISEASES WITH LESIONS IN THE RESPIRATORY TRACT[A]

Name of Disease(s)	Etiology	Species Affected	Lesions[B]	Comments
Infectious bronchitis (IB)	Coronavirus	Chickens.	Mild to moderate inflammation of respiratory tract. Occasional nephropathic outbreaks. When complicated by mycoplasmosis, severe airsacculitis, pericarditis, fibrinous perihepatitis.	Signs predominantly respiratory. Little mortality unless complicated and mostly in chicks less than 10 weeks old. In layers, egg production may drop as much as 50%. May result in the production of wrinkled egg shells and/or cystic salpingitis.
Infectious laryngotracheitis	Iltovirus (Herpesviridae)	Chickens. Occasionally pheasants.	Usually severe hemorrhagic laryngotracheitis with bloody exudate and possibly fibrinous casts or cheesy plugs. Occasionally only mild laryngotracheitis with conjunctivitis.	Usually in semimature or mature chickens. Spreads more slowly than other viral diseases. May cause high mortality. Severe dyspnea with loud gasping and expectoration of bloody exudate. Mild form has few signs and lesions.
Mycoplasma gallisepticum infection (MG)	*M. gallisepticum*	Primarily chickens, turkeys, but also in many other poultry/birds.	Airsacculitis of variable severity. Usually pericarditis and, perhaps, fibrinous perihepatitis.	Often secondary to other diseases and vaccinations (especially Newcastle and infectious bronchitis) and stresses. Usually a chronic respiratory disease.
Newcastle disease (ND)	Paramyxovirus (strains vary greatly in pathogenicity).	Most poultry and birds susceptible. Usually seen in chickens.	Highly variable with various viral strains. Enzootic ND produces few/no gross lesions. Exotic ND produces many but variable lesions: swelling around eyes and on neck, hemorrhages at many sites including intestinal mucosa, severe tracheitis. See ND section.	Enzootic ND in U.S. usually manifested as a respiratory disease but a modest number of young birds may show concurrent or closely following CNS signs. Exotic ND rarely in the U.S. In layers with ND egg productions drops drastically or ceases.

[A]Respiratory signs include one or more of the following: snicks, sneezes, dyspnea, gasping, rapid breathing, rales, ocular or nasal discharge, swollen sinuses. Signs in broilers may be severe, even with the milder strains. Secondary *Escherichia coli* infection is common.

[B]Unless stated otherwise lesions are those seen in turkeys or chickens.

DISEASES WITH LESIONS IN SKIN

Name of Disease(s)	Etiology	Species Affected	Lesions	Comments
Biotin deficiency	Biotin deficiency	Chicks, turkey poults.	Dermatitis on feet and shanks. Crusty, scab like lesions at the commissures of the mouth, perhaps of the eyelids.	In turkeys enlarged hocks and bowing of the metatarsus usually are more obvious and may precede or accompany cutaneous lesions. Sometimes implicated in perosis, especially in poults.
Cellulitis	*Escherichia coli*	Young chickens.	Yellow, thickened skin in the ventral area with subcutaneous exudates that may be edematous to caseous.	Bacterial infection secondary to skin scratches.
Erysipelas	*Erysipelothrix rhusiopathiae*	Turkeys, usually older males	Acute septic disease with serosal, cutaneous, and muscular hemorrhages and splenomegaly.	The bacteria reside in the soil and enter skin wounds created when males fight. Most common in toms housed outside.
External Parasites	Mites, lice, etc.	All poultry types.	Ectoparasites move fast but may be visible around the vent. Feces and/or eggs may be seen on the skin or feather shafts.	Ectoparasites may cause discomfort adversely affecting performance.
Fowl pox, pigeon pox, canary pox, turkey pox	Poxvirus	Perhaps all birds.	Yellow pustules or dark brownish-red scabs on any unfeathered skin (face, wattles, eyelids of chickens; caruncle, snood of turkeys; feet and legs of cage and wild birds).	Yellow or gray plaques may occur in oral cavity, pharynx, conjunctiva, sinuses. Can be transmitted by mosquitoes. Intracytoplasmic inclusion bodies in epithelium of lesions.
Gangrenous dermatitis	Skin wounds with secondary *Staphylococcus*, *Clostridia*, etc.	Chickens.	Traumatized skin with an underlying cellulitis.	Severe losses have occurred in 4-16 week-old chickens and turkeys. Some outbreaks follow infectious bursal disease or adenoviral infection.

DISEASES WITH LESIONS IN THE SKIN

Name of Disease(s)	Etiology	Species Affected	Lesions	Comments
Pantothenic acid deficiency in chicks	Deficiency of pantothenic acid.	Chicks, perhaps other poultry.	Dermatitis on feet and shanks. Crusty scab like lesions at the commissures of the mouth, perhaps of the eyelids.	Ataxia followed by inability to stand. Myelin degeneration in much of cord.
Riboflavin deficiency	Riboflavin deficiency	Turkey poults and chicks.	Poults: Dermatitis on feet, shanks. Crusts at corners of mouth, on eyelids, vent. Chicks: drooping wings. Both: hypertrophy and myelin degeneration of nerve trunks.	Poults and chicks may walk on hocks or sprawl with toes curled. A common deficiency.
Vesicular dermatitis (photosensitization)	Photosensitizing seeds, plants, feeds with/without fungus; phenothiazine. Sunlight required on skin.	Chickens, perhaps other birds.	Vesicles or scabs on unfeathered skin (comb wattles, face, legs caruncle, feet).	Affected skin wrinkled, ulcerated, and shrunken after vesicles rupture.
Xanthomatosis	Uncertain etiology. Possibly toxic materials in animal fats.	Chickens.	Thickened, roughened yellow areas of skin. Swelling of the wattle(s) of intermandibular area	Skin lesions are permanent. Affected birds condemned at slaughter. Cholesterol crystals and large foamy macrophages in affected skin.

DIFFERENTIAL DIAGNOSIS OF COMMON RESPIRATORY DISEASE OF CHICKENS[A]

Diagnostic Aid	Mesogenic Newcastle Disease	Infectious Bronchitis	Infectious Laryngotracheitis	Mycoplasma gallisepticum infection	Infectious Coryza
Speed of spread in the flock	Rapid	Rapid	Moderate	Slow; persistent	Rapid
Duration of flock signs	2 weeks	2 weeks	2-4 weeks	Weeks to months	Weeks to months
Egg production drop	Nearly complete cessation of lay.	Up to 50%	1-20%	1-20%	1-20%
Mortality in chicks less than 3 weeks old	25-90%	5-60%	Rarely occurs in chicks.	5-40% (seldom occurs in chicks)	Seldom occur in chicks.
Mortality in adults	0-5% (very high with exotic Newcastle).	Usually 0	Up to 50%	Low; many culls	Low; many culls.
Egg-borne transmission	No	No	No	Yes	No
Etiology	Paramyxovirus	Coronavirus	Iltovirus (Herpesvirus)	Mycoplasma gallisepticum	Avibacterium paragallinarum
Vaccines available	Yes	Yes	Yes	Yes	Yes

[A] Aspergillosis, avian influenza, and chlamydiosis have been omitted from this table although they may be classified as respiratory diseases. Occasional respiratory noises may be heard with certain other diseases if the bird is sick enough that mucus secretion accumulates in the respiratory tract.

POULTRY DRUG USE GUIDE

Diagnostic Aid	Mesogenic Newcastle Disease	Infectious Bronchitis	Infectious Laryngotracheitis	Mycoplasma gallisepticum infection	Infectious Coryza
Natural carrier chickens	Rarely	Yes - up to 2 weeks (vaccinates also).	Yes – probably lifelong (vaccinates also).	Yes	Yes
Clinical diagnosis	Adults cease lay within 3 days. In chicks, acute respiratory disease with CNS signs in some and high mortality.	Acute epornitic respiratory disease without CNS signs but with sharp drop in egg production and shell quality.	Severe dyspnea with bloody mucus from the nares and high mortality in adult birds.	Chronic respiratory signs influenced by weather and season.	Respiratory signs with facial edema ocular and nasal discharge. Exudate has foul odor.
Unique features of agent	Produces hemorrhagic skin and death in chick embryos. Hemagglutinates.	Curling, stunting of embryos with passaging. Does not hemagglutinate.	Pocks on CAM, intranuclear inclusions in CAM and tracheal epithelium.	Growth in special broth; but not on blood agar initially. Hemagglutinates.	Tiny dewdrop colonies on blood agar in candle jar (needs Staph nurse colony). Pleomorphic, Gram negative.
Serologic tests	HI test of value after 5 days; VN test of value after 10-21 days. ELISA test available	VN test of value. ELISA test available	ELISA test available. Seroconversion is slow.	Plate and tube agglutination tests. HI test. ELISA test available. Seroconversion is slow.	No test.
PCR	Yes (RT-PCR)	Yes (RT-PCR)	Yes (PCR/RFLP)	Yes (commercial kits available)	Yes (ERIC-PCR)
Gross lesions	Often none. Possibly mild airsacculitis and airway inflammation.	Possibly none. Often airsacculitis and tracheitis.	Bloody mucus on face, beak. Severe tracheitis with cheesy plugs in dead bird.	Marked airsacculitis. Fibrinous perihepatitis, adhesive pericarditis.	Facial edema. Eyelids adhered. Mucoid ocular and nasal discharge.

POULTRY DRUG USE GUIDE

Revised by Drs Linnea J. Newman and Jean E. Sander

The following listing of approved poultry drugs for United States use is intended to provide a general guide of dose and preslaughter withdrawal time. When calculating withdrawal time, each day is a full 24 hours long, starting with the hour the bird last received the drug. The listing is neither inclusive nor exclusive and may change. Many of the listed drugs are approved for use in combination with other drugs. Drug approval does not indicate cross-clearance. For information regarding approved combinations and specific indications of use, the reader should consult one of the references listed at the end of this section. Italicized compounds are not for use in laying hens.

When the decision is reached to use antimicrobials for therapy, veterinarians should strive to optimize therapeutic efficacy and minimize resistance to antimicrobials to protect public and animal health.

CHICKEN DRUG LIST

Active Ingredients	Route	Withdrawal Time (days)	Dose
Amprolium	Water	0	0.006 - 0.024%
Amprolium[A]	Feed	0	36.3 – 227 g/ton
Bacitracin methylene disalicyclate	Water	0	27.5 – 158 mg/L
Bacitracin methylene disalicyclate[B]	Feed	0	4-200 g/ton
Bacitracin zinc	Water	0	100-400 mg/gal
Bacitracin zinc[B]	Feed	0	4 - 50 g/ton
Bambermycins	Feed	0	1-2 g/ton
Ceftiofur sodium[C]	SQ	0	0.08-0.20 mg/chick
Chlortetracycline	Water	1	100-1,000 mg/gal
Chlortetracycline	Feed	1	10¥500 g/ton
Clopidol	Feed	5	113.5 – 227 g/ton
Cyromazine[D]	Feed	3	0.01 lb/ton
Decoquinate	Feed	0	27.2 g/ton
Diclazuril	Feed	0	0.91 g/ton
Erythromycin	Water	1	0.500 g/gal
Gentamicin sulfate	Inject	35	0.2 mg

[A] 36.3 - 113.5 g/ton for laying hens.

[B] 10-25 g/ton for laying hens.

[C] For use in day-old chicks only. Extralabel use of cephalosporins in chickens is prohibited by the FDA.

[D] For US in layers or breeders only.

Active Ingredients	Route	Withdrawal Time (days)	Dose
Halofuginone Hydrobromide	Feed	4	2.72 g/ton
Lasalocid	Feed	0	68-113 g/ton
Lincomycin	Feed	0	2 – 4 g/ton
Lincomycin HCl	Water	0	16 mg/L
Lincomycin/ Spectinomycin[E]	Water	3	833 mg antibacterial action/L
Monensin	Feed	0	90-110 g/ton
Narasin	Feed	0	54-72 g/ton
Narasin/nicarbazin	Feed	5	54 - 90 g/ton of combination
Neomycin/Oxytetracycline	Feed	3	10 – 50 g/ton
Nicarbazin	Feed	4	113.5 g/ton
Nitarsone	Feed	5	170.1 g/ton
Oxytetracycline hydrochloride	Water	0	200 - 800 mg/gal
Oxytetracycline	Feed	0 - 3	10-500 g/ton
Penicillin (from procaine penicillin)	Feed	0	2.4 – 50 g/ton
Piperazine	Water	14	50 – 100 mg/bird
Robenidine hydrochloride	Feed	5	30 g/ton
Roxarsone	Water	5	5 ml / gal
Roxarsone	Feed	5	22.7-45.4 g/ton
Salinomycin	Feed	0	40-60 g/ton

[E] Use only up to 7 days of age.

Active Ingredients	Route	Withdrawal Time (days)	Dose
Semduramicin	Feed	0	22.7 g/ton
Spectinomycin dihydrochloride	Water	5	0.5 – 2 g/ gal
Spectinomycin dihydrochloride[c]	Inject	0	2.5 mg/bird
Steptomycin Sulfate	Water	4	0.5-1.5 g/ gal
Sulfadimethoxine	Water	5	1.875 g/gal
Sulfadimethoxine/ ormetoprim	Feed	5	113.5 g/ton/ 68.1 g/ton
Sulfamethazine sodium	Water	10	61-89 mg/lb BW/day
Sulfaquinoxaline	Water	10	0.025-0.04% 10 – 45 mg/lb/day
Tetracycline hydrochloride	Water	5	89 mg/L
Tylosin tartrate	Water	1	100 - 500 mg/L
Tylosin[F]	Feed	0 - 5	4 - 1000 g/ton
Virginiamycin	Feed	0	5-20 g/ton
Zoalene	Feed	0	36.3 – 113.5 g/ton

[F] For layers use 20-50 g/ton dose. Highest dose level (1,000 g/ton) requires 5-day withdrawal.

TURKEY DRUG LIST

Active Ingredients	Route	Withdrawal Time (days)	Dose
Amprolium	Water	0	4-16fl oz/50 gal
Amprolium	Feed	0	113.5 – 227 g/ton
Bacitracin methylene disalicyclate	Water	0	400 mg/gal
Bacitracin methylene disalicyclate	Feed	0	4-200 g/ton
Bacitracin zinc	Feed	0	4-50 g/ton
Bambermycins	Feed	0	1-2 g/ton
Ceftiofur sodium [A]	SQ	0	0.17 – 0.5 mg/poult
Chlortetracycline	Water	1	100-400 mg/gal
Chlortetracycline	Feed	1	10-400 g/ton
Clopidol	Feed	5	113.5 – 227 g/ton
Diclazuril[B]	Feed	0	0.91 g/ton
Erythromycin	Water	1	57.8 – 115.6 mg/L
Fenbendazole[B]	Feed	0	14.5 g/ton
Gentamicin	Inject	63	1 mg/bird
Halofuginone Hydrobromide	Feed	7	1.36 – 2.72 g/ton
Lasalocid[B]	Feed	0	68-113 g/ton
Monensin	Feed	0	54-90 g/ton
Neomycin/ Oxytetracycline	Feed	5	10 – 200 g/ton
Neomycin sulfate	Water	0	10 mg/lb BW/day
Nitrasone	Feed	5	170.1 g/ton
Oxytetracycline hydrochloride	Water	5	200 - 400 mg/gal
Oxytetracycline	Feed	0	10-200 g/ton
Penicillin (G potassium)	Water	1	1.5 Million units/gal
Penicillin (from procaine penicillin)	Feed	0	2.4 - 50 g/ton
Piperazine sulfate	Water	14	100-200 mg/bird
Ractopamine Hydrochloride	Feed	0	4.6-11.8 g/ton
Roxarsone	Feed	5	22.7-45.4 g/ton
Roxarsone	Water	5	5 ml /gal
Spectinomycin dihydrochloride	Inject	0	1.0-5 mg/bird
Sulfadimethoxine	Water	5	0.938g/gal

Active Ingredients	Route	Withdrawal Time (days)	Dose
Sulfadimethoxine/ Ormetoprim	Feed	5	56.7 g/ton/ 34.05 g/ton
Sulfamethazine sodium	Water	10	53-130 mg/lb BW/day
Sulfaquinoxaline	Water	10	0.025 – 0.04% 3.5 – 55 mg/lb/day
Triple Sulfa (sulfamethazine, sulfamerazine, sulfaquinoxaline)	Water	14	0.025% solution
Tylosin tartrate	Water	3	500 mg/L
Virginiamycin	Feed	0	10-20 g/ton
Zoalene[B]	Feed	0	113.5 – 170.3 g/ton

[A] For use in day-old poults only. Extralabel use of cephalosporins in turkeys is prohibited by the FDA.

[B] Do not use in breeding turkeys.

REFERENCES

1. Arriuja-Dechert, A. (ed) 1999. Compendium of Veterinary Products, 5th ed. Adrian J. Bayley, Pub., Port Huron, MI.

2. FDA Website: www.accessdata.fda.gov/scripts/animaldrugsatfda/index.cfm?gb=1&showtype=adv (last consulted July 20th, 2012).

3. Lundeen, T. (ed.) 1999. 2011. Feed additive Compendium, The Miller Publishing Co., Minnetonka, MN

NECROPSY OF THE FOWL

Written by Drs Richard J. Julian and Martine Boulianne

History

As with disease investigation in other species, a good history will often provide clues that will help solve the problem. Obtain information on the type of bird, age, feed and water source and consumption rate, growth, production, morbidity and mortality, the owner's description of the case, vaccination program, drugs being used etc…

In any poultry operation, many of the problems can relate to management, environmental factors, and stress rather than to infection so if you are on the premises, examine carefully the housing conditions. Is the ventilation adequate? Are ammonia fumes a problem? Is it too hot or too cold? Is the litter wet or is it too dry and dusty? Is the room too light? Are there sufficient hours of light for best production? Is the nest area darkened? Are the roosts too high? Do the birds appear comfortable? Chickens can talk and the sounds they make can indicate comfort, hunger, pain, panic, or disease.

Examination of Live Specimens

Check the general appearance of the individual or group and try to determine which organ or system is involved in the illness. Note any signs or lesions that might point to a diagnosis. If the birds show lameness or paralysis, is the lesion in the nervous system, bones, joints, muscles or skin? Some conditions, particularly those affecting locomotion, are easier to diagnose in live birds. Botulism which produces neck paralysis in chickens (leg and wing paralysis are more obvious in turkeys, ducks, and pheasants) is an example.

Examine the skin of the head, body, and legs for lice and mites, injury (particularly cannibalism), molting, swellings, cyanosis, or staphylococcal or clostridial dermatitis. Listen for unusual breathing sounds (snicking, gurgling) and look for gasping or head-shaking that might indicate respiratory distress. Mouth-breathing (panting) is normal in chickens in hot weather. Exudate from nostrils and eyes and dirty feathers also suggest respiratory infection. Examine the droppings for evidence of diarrhea and the presence of blood.

If a post-mortem examination is to be carried out, birds that are representative of the problem in the flock must be selected. If there has been mortality, both sick and dead birds should be opened. A couple of cull birds will not provide the answer. If the problem is a drop in production, try to find birds that look like they have recently stopped laying. It is important to do both an external and an internal examination and to follow a specific pattern to avoid missing important lesions.

If alive, the fowl may be killed by any of three methods.

A. Administration of CO_2 gas in an appropriate closed container.

B. Disarticulate the head at the atlantooccipital joint (Fig. 1).

C. Intravenous injection of barbiturate (Fig. 2).

Moisten the feathers with water containing detergent. If psittacosis is suspected, the bird should be soaked in 5% Lysol solution and a laminar flow hood should be used for the necropsy.

If swabs are desired for culturing, use sterile instruments to cut the infraorbital sinus, a joint, or to remove any organs. All unnecessary manipulations and delays prior to culture increase the probability of contamination. Take intestinal cultures last.

Necropsy

1. Examine the head, including the eyes, ears, nostrils, comb, wattles, mouth, and beak (Fig. 3).

   Infected eyes and conjunctivitis are often seen in case of a respiratory disease such as infectious sinusitis (*Mycoplasma gallisepticum* infection), infectious coryza, infectious laryngotracheitis (ILT), infectious bronchitis (IBV), and Newcastle Disease (ND). Keratoconjunctivitis with central corneal ulcers suggests ammonia burn.

   Swollen sinuses are seen in infectious coryza in chickens, infectious sinusitis, *Cryptosporidium* and *Bordetella* infection in turkeys and pheasants. They may also be seen as the result of other infections such as avian influenza (AI).

   Chronic fowl cholera causes swollen wattles in adult chickens. Fowl pox causes scabs on the comb, eyelid, and wattle, but must be differentiated from injury. Pox also occurs in turkeys, pigeons, doves, canaries and other avian species. A frozen comb or wattles is usually mottled red and white before becoming gangrenous.

   If the bird has had its beak trimmed, check for proper healing, overgrowth of the lower beak or over-trimming (cut too short).

   Injury on the head, neck, or breast or back may indicate predators. Dermatitis and scabby or crusty lesions around the mouth and eyes suggest vitamin B deficiency.

2. Cut across the upper beak (Fig. 4) and examine the sinuses (Fig. 5). Then insert one blade of a sterile scissors into the infraorbital sinus. Make a longitudinal lateral incision through the wall of each

sinus and examine them (Fig. 6). Cut through one lateral commissure of the mouth (Fig. 7) and down the esophagus into the crop (Fig. 8). Examine the oral cavity, note the content and odor of the esophagus and crop.

White plaques in the mouth, esophagus, or crop may be caused by capillaria worms, yeast infection (candidiasis), or possibly trichomoniasis or vitamin A deficiency, but most frequently by fowl pox (wet form). Single white plaques in the mouth are common in hens and turkey breeders and occur where salivary ducts open into the mouth. Tricothecene mycotoxicosis may produce similar lesion in young chickens and turkeys.

If the crop is enlarged and full, it may be an impacted or a sour crop (pendulous crop). The problem may be caused by excess water intake, defects in the crop itself, partial blockage of the proventriculus or gizzard or Marek's disease. Necrosis of the crop in sparrows and other songbirds feeding around bird feeders is caused by *Salmonella* infection.

3. Examine the soft palate and larynx and cut down the trachea (Fig. 9).

Wet fowl pox lesions are seen on the roof of the mouth and on the larynx.

Granulation, congestion, and mucus in the trachea are seen in IBV, *Escherichia coli* infection, infectious coryza and bordetellosis. Hemorrhage and blood clots occur in ILT and occasionally in ND or infectious coryza and may cause severe gasping.

Gapeworms in pheasants, quail, and other birds are caused by *Syngamus* and cause gasping. *Cyanthastoma* cause similar infection in waterfowl and tracheal flukes may be found in waterfowl in the tropics.

4. Check under the wings (Fig. 10) and on the abdomen for lice and mites, and the vent for injury.

Incise the loose skin between the medial surface of each thigh and the abdomen (Fig. 11), reflect the legs laterally and dislocate the hip joints (Fig. 12). Connect the lateral skin incisions with a transverse skin incision across the middle of the abdomen and reflect the skin of the breast anteriorly (Fig. 13). Tightly adhering skin and dark tissues indicate dehydration. Examine the skin, integument, and muscles.

Emaciation, along with small organs, suggests malnutrition, beak injury (poor trimming), peck order (behavioral) problems, chronic disease (coccidiosis), bumblefoot or other lameness, or chronic poisoning (lead, insecticide, etc.).

Muscular degeneration due to vitamin E-selenium deficiency can cause lameness, particularly in ducks. Sarcosporidial cysts produce small, white lesions in the muscle of waterfowl.

Gangrenous dermatitis is caused either by *Staphylococcus* or *Clostridium septicum* infection. The clostridial infection shows fairly typical lesions of emphysematous or serosanguineous cellulitis in affected areas. Scabby hip syndrome can be observed on the thighs and is usually associated with overcrowding, poor litter conditions, or poor feathering. Cellulitis is characterized by the presence of a subcutaneous caseous plaque on the side of the lower abdomen, often near the vent.

Skin leukosis is Marek's disease virus causing viral dermatitis in the feather follicles. At processing this can be confused with scabby hip or other causes of dermatitis.

5. Examine the feet, their plantar surfaces, then bones and joints of the lower limbs for abnormality and deformity. Break the metatarsal bone to verify for strength (Fig. 14). Using a sharp knife or scalpel, open each tibiotarsal joint and examine the joint fluid for signs of exudate (Fig. 15). Make a longitudinal cut through the anterio-medial aspect of the tibial head to expose the growth plate of immature birds (Fig. 16). With an osteotome split one femur longitudinally and examine the bone marrow.

Angular bone (valgus-varus) deformity of the intertarsal joint is caused by lateral or medial bending of the tibio-tarsal and metatarsal bones and is a common problem in meat-type chickens. It has a variety of possible causes (nutritional, genetic, management, etc.). Slowing growth in young broilers will help prevent leg deformity.

Other types of hock and stifle lameness are frequent in heavy roasters and turkeys and may be mainly due to injury as the result of heavy weight and fast growth.

Check for poor bone-breaking strength (osteoporosis or cage layer fatigue) or, in young birds, for rubbery bones, soft beaks, and beaded ribs (rickets) which may indicate calcium, phosphorus or vitamin D_3 imbalance. While cage layer fatigue maybe due to nutritional factors, it is also associated with continuous high production.

Curled-toes in young birds may indicate a riboflavin deficiency, but in older birds and turkeys it may be due to genetic factors or a lack of roosts. Cracked feet and foot dermatitis may be pantothenic acid or biotin deficiency, but scaly leg in cage birds is likely parasitic (mites). Toe injury in young birds may be cannibalism or mechanical injury.

Swelling of the feet, hocks or other joints suggests articular gout, infectious synovitis (*Mycoplasma synoviae*) or other localized bacterial infection. These bacterial infections often extend into osteomyelitis. Infection in the wing joints in pigeons is usually due to Salmonella. In broilers, roasters and broiler breeders viral arthritis may cause lameness or ruptured tendons. If growing birds are lame and there is no evidence of infection or rickets, split some bones and look for necrosis due to osteomyelitis or dyschondroplasis. Lameness may require splitting the spine at T4 to look for spondylolisthesis. Also, consider Marek's disease and examine the sciatic nerve (Fig. 17) by separation or section of the adductor muscles.

6. At this point, remove the breast carefully by cutting through the pectoral muscles on each side of the keel and over the costrochondral junctions (Fig. 18). Examine the organs, paying particular attention to the air sacs (Fig. 19), lungs, and liver. Carefully push the liver and gizzard to the left, incise the peritoneum (Fig. 20) and cut the esophagus cranially to the proventriculus (Fig. 21) to exteriorize the whole gastrointestinal tract and remove it after transecting the rectum.

If there is fibrin on the liver and/or in the pericardial sac, suspect secondary *E. coli* infection. These lesions in turkeys and cage birds might also be due to *Chlamydophila psittaci,* or in ducks due to *Riemerella anatipestifer* infection.

Peritonitis in layers is usually "egg peritonitis" frequently with secondary *E. coli* infection, although acute fowl cholera may also causes peritonitis.

White crystals on the heart sac, liver, and other tissues and organs are uric acid crystals (visceral gout) and are secondary to hyperuricemia from nephritis or water deprivation.

Tumors in or on the organs may be due to Marek's disease, lymphoid leukosis, or other tumors. Multiple small tumors on organs and peritoneum in adult hens are frequently metastasis from a carcinoma of the oviduct or pancreas. This may result in ascites.

Ascites may also result from heart or liver disease or from ingestion of some toxic material.

Focal white lesions on organs may be due to tuberculosis (*Mycobacterium avium*) or other bacterial septicemia. Histomoniasis causes large irregular circumscribed lesions on the liver in turkeys, pheasants and peacocks but less so in chickens.

In young birds, examine the lungs and air sacs for yellow-white foci or plaques caused Aspergillosis (brooder pneumonia). Gasping in young birds is a sign of tracheal or bronchial injury, obstruction, irritating fumes or an infectious agent.

In baby chicks, look for yolk sac infection (omphalitis), peritonitis or pericarditis in which the abdomen can be swollen, wet, and discolored. The yolk sac may be infected with *E. coli, Salmonella, Staphylococcus, Pseudomonas,* etc…

Baby birds also die because they did not learn how to eat (starve-outs) or drink (dehydration). Deaths occur mainly between 4 and 6 days of age once the yolk sac has been absorbed.

Broiler chickens that die suddenly from flip-over (sudden death syndrome, heart attack), heat stroke, or suffocation (piling-up) have congested, edematous lungs, a full digestive tract and congested mottled breast muscle.

Turkeys (6-18 wks old) die suddenly of unknown cause and have perirenal hemorrhage, swollen spleen and congested liver and lungs. Turkeys with aortic rupture will have a large blood clot in the abdomen and chickens or turkeys with hepatic lipidosis may also have hemorrhage and hematoma of the liver. .

Birds that die from anemia are pale and the blood is watery. Birds may have bled to death (pick-outs, ruptured fatty liver, acute cecal coccidiosis, ruptured aorta, hemorrhagic enteritis in turkeys, etc…) and are pale.

To identify anemia due to parasites in the red blood cells (*Leukocytozoon* for duck, *Plasmodium* for canary), a blood smear needs to be collected from a live, sick bird and examined by Wright's stain or Diff Quick.

Sulfa poisoning also produces anemia with widespread hemorrhage in the tissues. Chicken infectious anemia virus (CIA) produces similar lesion.

7. Remove and examine the heart (Fig. 22) and its pericardial sac. Open the heart to visualize the valves.

Right ventricular dilation and hypertrophy can sometimes be observed in broiler chickens whereas a bilateral dilated cardiomyopathy can be seen in turkey poults, both often accompanied by secondary ascites.

Pericarditis and endocarditis will be sometimes observed as lesions of septicemia.

8. Examine the lymphoid system; the spleen (located at the junction of the proventriculus and gizzard, (Fig. 23), bursa (dorsally to the cloaca, (Fig. 24) and lymphoid tissue of the neck (thymus). Lymphoid tissues of the intestines (peyers's patches and cecal tonsils) will be examined upon opening the gut.

Infectious bursal disease (IBDV) will affect the bursa of Fabricius and initially cause bursal edema (4-5 days post-infection) before atrophy (7 days post-infection). This disease causes immunosuppression and increases the susceptibility to secondary infections.

Swelling, congestion and hemorrhage with or without focal necrosis in the spleen, liver and lymphoid tissue suggest septicemia (fowl cholera, fowl typhoid, streptococcosis, colisepticemia or erysipelas) or viremia (ND and highly pathogenic AI).

Marek's disease and lymphoid leukosis produce tumors in lymphoid tissue except for the bursa which is only affected by lymphoid leukosis (occasional Marek's lesions may occur in the stroma of the bursa).

9. With a scissor or an enterotome, make a longitudinal incision through the proventriculus (Fig. 25). Open the proventriculus, gizzard (Fig. 26), small (Fig. 27) and large intestines to the cloaca. Check the cloaca carefully for evidence of picking injury.

If hens are not properly beak-trimmed or are too fat, mortality from "pick-out" is common. The whole intestine may be picked out through the cloaca. Prolapse of the vagina (and cloaca) may occur from straining, secondary to injury or inflammation.

Examine the digestive tract for lesions and the various kinds of enteritis (hemorrhagic, necrotic, ulcerative, etc...), parasites (roundworms, capillary worms, tapeworms, cecal worms, and coccidia), and gastrointestinal accidents. A large proventriculus in broilers may be from lack of fibre in the diet resulting in poor development of the gizzard.

Green staining of the digestive tract is just bile and indicates that the bird is not eating. The liver and spleen may be small (if the bird is thin) and the gallbladder full.

Check the ceca, intestine and liver for lesions of histomoniasis, avian tuberculosis, coccidiosis or tumor, the liver for other varieties of bacterial, viral or protozoal hepatitis, cholangiohepatitis etc., and the pancreas for tumors.

A large, yellow liver may be normal fat storage in a laying bird (estrogen stimulation) but layers can die from a ruptured fatty liver.

A yellow or hemorrhagic liver particularly with focal necrosis may be viral hepatitis which is seen in broiler chickens, pigeons, ducks, raptors, owls and psittacines.

10. Examine the testicles in males (Fig. 28) or the single left ovary in females (Fig. 29) and cut the oviduct longitudinally in adult females.

The development of the ova helps to identify length of illness. Shrinking ova indicate illness of several days' duration or one from which the bird may be recovering. Small, sac-like ova indicate that the bird has been out of lay for a week or more and may be in a molt.

Semi-solid (cooked) ova indicate bacterial infection.

An impacted oviduct may be secondary to vent-picking, egg material left in the oviduct, or the bird may be egg-bound. Infection (Mycoplasma, IBV, E. coli) can cause salpingitis as well.

A large or small fluid-filled cyst in the right abdomen beside the cloaca is the cystic remnant of the right oviduct.

A drop in production may be related to systemic disease (IBV, MG, avian encephalomyelitis, influenza-like, ND, etc...) or management failure (lack of light, temperature change, lack of water) or nutritional problems etc.

Deformed shells suggest IBV, and soft shells (higher than 1-2%) may indicate calcium or vitamin D deficiency.

A normal-appearing dead bird with an egg in the shell gland or a recently laid egg may have died from acute hypocalcemia. These birds often have fragile bones.

Hard (fibrotic) or swollen testes indicate bacterial infection (Paratyphoid).

11. Examine the kidneys and ureters (Fig. 30).

Ureters plugged with urates or hard stony material (urolithiasis) and swollen pale kidneys indicate hyperuricemia and nephrosis. This may be due to lack of water, or a Calcium/Phosphorus imbalance.

Swollen kidneys and nephritis may be due to IBV (nephrotrophic strain) or E. coli infection and usually cause death from uricemia as well.

12. Examine the lungs by reflecting them medially from their attachment to the rib cage (Fig. 31 and 32).

Pneumonia in turkeys is caused by a Pasteurella multocida infection (fowl cholera) and the lungs may be quite solid. ND, AI and *E. coli* infection can also cause pneumonia.

13. Disturbances of the nervous system may cause incoordination, staggering, paralysis, walking backwards (with wings flapping for balance), tremors, stargazing, and other odd behaviors. In case of lameness, paresis, paralysis, or incoordination, examine the brachial plexus, intrapelvic portion of the sciatic nerves that you can expose by removal of the overlying portion of the kidneys, spinal cord, and brain. To examine the brain, disarticulate the head and skin it. Remove the calvarium with strong scissors using the same technique as for mammals (Fig. 33).

    When Marek's disease affects the peripheral and/ or central nervous system it causes lameness, incoordination and paralysis. Avian encephalomyelitis (AE) (epidemic tremor) affects birds up to 3-4 weeks old from non-immune parents.

    ND may produce CNS disturbances in pigeons as well as respiratory, intestinal and reproductive lesions in chickens, pheasants, turkeys, and wild and cage birds of all ages. Bacterial infection (*Pasteurella, Salmonella, Staph.* etc.) and fungi (*Aspergillus*, etc.) also cause meningoencephalitis, occasionally in outbreak proportions.

    Vitamin E deficiency (avian encephalomalacia) causes lesions in the cerebellum (soft, dark areas) which may be visible grossly.

    Arsanilic acid and other feed additives and toxins may cause CNS disturbances, while others like the ionophores and coffee weed seeds (Cassia) cause muscle damage that mimics paralysis.

Fig. 1
Dislocation of the atlantooccipital joint.

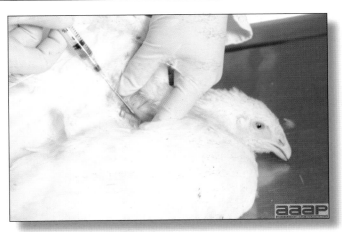

Fig. 2
Intravenous injection of barbiturate.

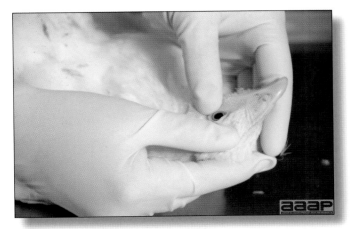

Fig. 3
Examination of the head.

Fig. 4
Cut across the upper beak .

Fig. 5
Sinus and turbinates exposed.

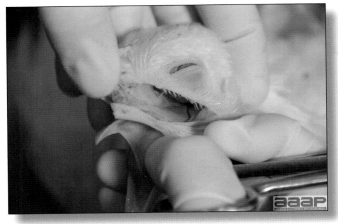

Fig. 6
Infraorbital sinus exposed.

Fig. 7
Scissors at commissure of the beak.

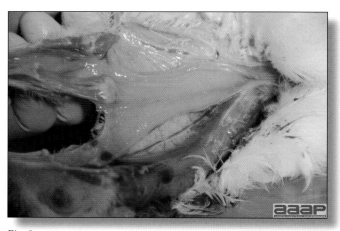

Fig. 8
Oesophagus being exposed.

Fig. 9
Trachea being cut opened.

Fig. 10
Check the feathers and skin under the wings and on the abdomen.

Fig. 11
Incision of the loose skin between the medial surface of each thigh and the abdomen.

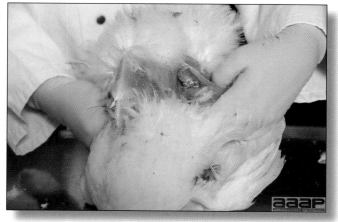

Fig. 12
Hip joint dislocation.

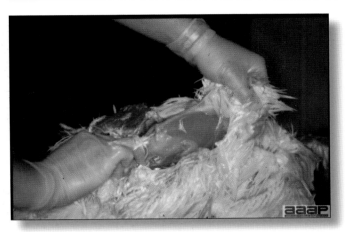

Fig. 13
Removal of the skin of the breast anteriorly.

Fig. 14
Breaking the metatarsal bone to verify for strength.

Fig. 15
Opening of the tibiotarsal joint and examination of the joint fluid.

Fig. 16
Longitudinal cut through the anterio-medial aspect of the tibial head to expose the growth plate.

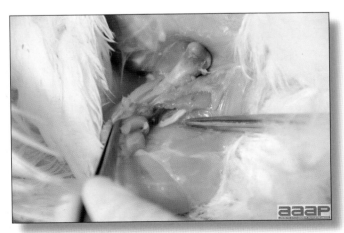

Fig. 17
Examination of the sciatic nerve .

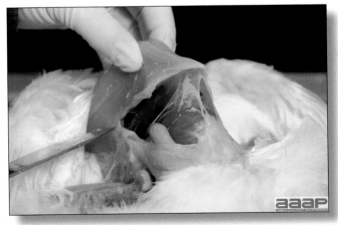

Fig. 18
Removal of the breast.

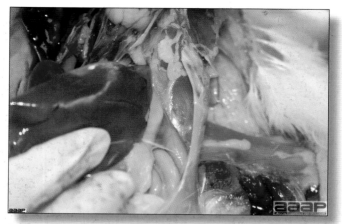

Fig. 19
Examination of the air sacs.

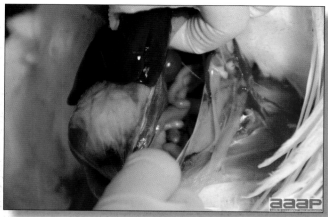

Fig. 20
Carefully pushing the liver and gizzard to the left to incise the peritoneum.

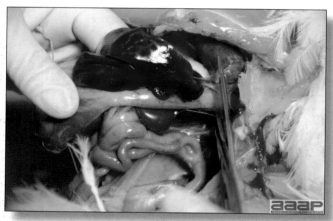

Fig. 21
Cutting the esophagus cranially to the proventriculus.

Fig. 22
Removal of the heart.

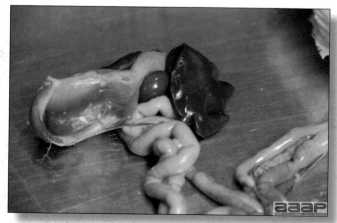

Fig. 23
Examination of the spleen located at the junction of the proventriculus and gizzard.

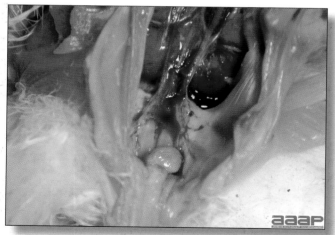

Fig. 24
Examination of the bursa located dorsally to the cloaca.

Fig. 25
Longitudinal incision through the proventriculus.

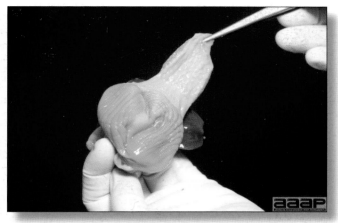

Fig. 26
Proventriculus and gizzard opened.

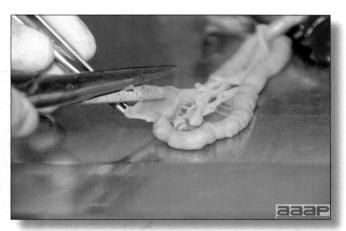

Fig. 27
Opening the duodenum.

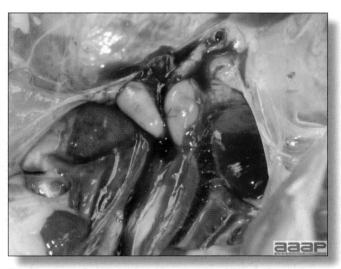

Fig. 28
Testicles.

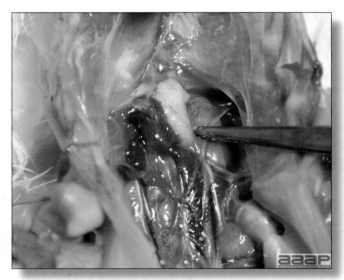

Fig. 29
Single left ovary in an immature female chicken.

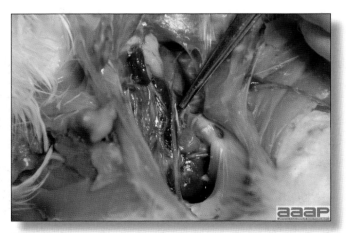

Fig. 30
Kidneys and ureters (scissors tip).

Fig. 31
Taking the lung out.

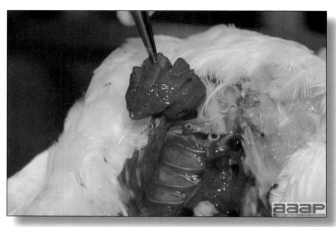

Fig. 32
Lung being exposed.

Fig. 33
Cerebrum and cerebellum being exposed.

Index

Hepatitis E virus (HEV, Hepatitis-Splenomegaly syndrome in chickens), 50-51, 240, 281

Hepatovirus, 18, 245

Herpesviridae, 59, 258

Herpesvirus, 30-31, 42, 59, 61, 231, 244, 246-249, 253, 256, 261

Heterakis gallinae, 158

Hexamitiasis, 70, 167-168, 249

Highlands J virus (HJV), 6-7, 243

Histomonas meleagridis, 158, 168, 178, 249

Histomoniasis (Blackhead, Enterohepatitis), 72, 158, 166-169, 175, 178, 235, 238, 242, 249, 251, 270-271, 287

Hypoglycemia-spiking mortality syndrome in broiler, 167, 180, 226, 229, 238, 246, 292

Hysteria, 219, 225, 240

I

Iatrogenic trauma, 209

Iltovirus, 59, 258, 261

Inclusion body hepatitis (IBH), 10-11, 16, 55-56, 238, 249, 251, 253, 279

Infectious bronchitis (IB, IBV), 52-54, 80, 87, 103, 106-107, 119, 171, 218, 222, 227, 239-240, 256, 258, 261-262, 268-269, 271, 281

Infectious bursal disease (IBD, IBDV, Gumboro disease), 11, 33, 39, 55-58, 101, 117, 134-135, 137, 156, 171, 201, 238-239, 251, 253, 259, 271, 281

Infectious coryza (Coryza), 80, 103-105, 183, 239-242, 257, 261-262, 268-269, 283

Infectious laryngotracheitis (ILT), 59-61, 103, 239-240, 268-269, 281

K

Keratoconjunctivitis (ammonia burn), 223, 238, 241, 268, 291

Knemidokoptes gallinae, 154

Knemidokoptes mutans, 154

L

Large intestinal roundworms, 158

Lesser mealworm, 156

Leukocytozoa, 159, 163

Lice, 153, 243, 259, 268-269, 286

Lymphoid leukosis (LL), 31-34, 37, 240, 252-253, 256, 270-271, 280

Lymphoproliferative disease of turkeys (LPD), 30, 33-34, 243, 280

M

Mareks' Disease (MD), 19, 30-36, 39, 83, 106, 156, 218, 238-240, 244, 246, 249, 252-253, 256, 269-272, 280, 293

Microsporum gallinae, 146

Mites, 132, 153-156, 160, 240, 243, 259, 268-270, 286

Mosquitos, 6, 17, 33, 46, 132, 156, 160, 234, 235, 237, 245, 259

mycobacterium avium, 77, 230, 235, 248, 251, 270

Mycoplasma gallisepticum (MG, Chronic Respiratory Disease, CRD, infectious Sinusitis of turkeys), 75, 80, 83, 103, 106-110, 112-113, 141, 169-170, 180, 186, 226, 236, 242, 258, 261, 263-268, 271, 284

Mycoplasma iowae, 106, 111

Mycoplasma meleagridis (MM), 72-73, 108-109, 114, 116, 153-154, 156, 172, 177-178, 183, 234, 236, 242, 247, 251-252, 284

Mycoplasma synoviae (MS, Infectious Synovitis, Tenovaginitis), 107-110, 115-116, 239, 242, 254, 270, 284

Mycoplasmosis, 22, 68, 106, 108, 239-240, 242-243, 256, 258, 284

Mycotoxicosis, 147-151, 247, 251, 269, 285-286

Myelocytomatosis, 31-32, 37, 218, 280

N

Necrotis enteritis (NE), 117-118, 164, 166-167, 200-201, 238, 240, 242, 249-250, 284

Nematodes, 158-159, 161-162, 177-178, 286

Newcastle disease (ND), 19, 22, 42-43, 45, 62-66, 75, 87, 95, 106, 154,

Acknowledgements

We gratefully acknowledge the contributions of the following resources:

American Association of Avian Pathologists (AAAP), *Avian Diseases*, 12627 San Jose Blvd, Suite 202, Jacksonville, Florida 32223-8638

American Association of Avian Pathologists (AAAP), Slide Study Sets 1 – 25, 12627 San Jose Blvd, Suite 202, Jacksonville, Florida 32223-8638

Saif, YM. (ed). 2008. *Diseases of Poultry*, 12th ed. Iowa State University Press, Ames, IA

Davis, JW, RC Anderson, L Karstad, and DO Trainer (ed). 1971. *Infectious and Parasitic Diseases of Wild Birds.* The Iowa State University Press, Ames, IA

Jordan, FTW. (ed). 2001. *Poultry Diseases*, 5th ed. W. B. Saunders, London; New York

Macwhirter, P. 1987. *Everybird. A Guide to Bird Health*, Inkata Press, Sydney, Australia.

Kahn, CM. 2005. *The Merck Veterinary Manual*, 9th ed. Merck and Co., Inc., Whitehouse Station, NJ.

Rosskopf, WJ and RW Woerpel (ed). 1996. *Diseases of Cage and Aviary Birds*, 3rd ed. Williams & Wilkins, Baltimore, MD.

Proceedings of the Western Poultry Disease Conference, 1973-2012. Cooperative Extension, University of California, Davis, CA

Dufour-Zavala, L, DE Swayne, JR Glisson, MW Jackwood, JE Pearson and WM Reed (eds). 2008. *A Laboratory Manual for the Isolation and Identification of Avian Pathogens*, 5th ed. AAAP, 12627 San Jose Blvd, Suite 202, Jacksonville, Florida 32223-8638

Randall, CJ. 1991. *Color Atlas of Diseases and Disorders of the Domestic Fowl and Turkey*, 2nd ed. Iowa State University Press, Ames, IA

Fletcher, OJ, T Abdul-Aziz. *Avian Histopathology.* 2008, 3rd ed. AAAP, 12627 San Jose Blvd, Suite 202, Jacksonville, Florida 32223-8638

Ritchie BW, GJ Harrison, and LR Harrison. 1994. *Avian Medicine: Principles and Application*, Wingers Publishing, Inc., Lake Worth, FL

Shane, SM. (ed.) 1995. *Biosecurity in the Poultry Industry*, AAAP, 12627 San Jose Blvd, Suite 202, Jacksonville, Florida 32223-8638

Steiner, CV and RB Davis. 1981. *Caged Bird Medicine*, Iowa State University Press, Ames, IA

Wobeser, GA. 1997. *Diseases of Wild Waterfowl*, Plenum Press, New York, NY

World Veterinary Poultry Association. *Avian Pathology*, Taylor & Francis Ltd., London

CONTRIBUTING AUTHORS

Martine Boulianne
Faculté de médecine vétérinaire
Université de Montréal
PO Box 5000
St Hyacinthe, QC Canada J2S 7C6

Marina L. Brash
Animal Health Laboratory
University of Guelph
PO Box 3612
Guelph, ON Canada N1H 6R8

Bruce R. Charlton
CAHFS Turlock Branch
University of California – Davis
3327 Chicharra Way
Coulterville, California 95311

Steve Fitz-Coy
Merck
PO Box 2074
Salisbury, MD 21802 United States

Richard M. Fulton
Diagnostic Center for Population
and Animal Health
Michigan State University
PO Box 30076
Lansing, Michigan, 48909

Mark W. Jackwood
Department of Population Health
College of Veterinary Medicine
The University of Gerogia
953 College Station road
Athen, Georgia 30602-4875

Richard J. Julian
Ontario Veterinary College
Dept of Pathobiology
Guelph, ON Canada N1G 2W1

Davor Ojkic
Animal Health Laboratory
University of Guelph
PO Box 3612
Guelph, ON Canada N1H 6R8

Linnea J. Newman
Merck Animal Health
17 Pine Street
North Creek, NY 12853

Jean E. Sander
Ohio State University-CVM
6828 Spruce Pine Drive
Columbus, OH 43235

H. L. Shivaprasad
CAHFS-Tulare Branch
999 E. Edgemont Drive
Fresno, CA 93720

Eva Wallner-Pendleton
Pennsylvania Animal Diagnostic Laboratory
System/PennState University
149 Centennial Hills Road
Port Matilda, PA 16870

Peter R. Woolcock
CAHFS - University of California – Davis
West Health Sciences Drive
Davis, California 95616